Markus Schmuck

Complex Heterogeneous Systems

## Also of Interest

*PEM Fuel Cells*
*Characterization and Modeling*
Jasna Jankovic, Jürgen Stumper (Eds.), 2023
ISBN 978-3-11-062262-1, e-ISBN (PDF) 978-3-11-062272-0,
e-ISBN (EPUB) 978-3-11-062360-4

*Chemical Energy Storage*
*2nd Edition*
Robert Schlögl (Ed.), 2022
ISBN 978-3-11-060843-4, e-ISBN (PDF) 978-3-11-060845-8,
e-ISBN (EPUB) 978 3 11 060859-5

*X-Ray on Electrochemical Systems*
*Synchrotron Methods for Energy Materials*
Artur Braun, 2017
ISBN 978-3-11-043750-8; e-ISBN 978-3-11-042788-2,
e-ISBN (EPUB) 978-3-11-079426-7

*Hydrogen Storage Alloys*
*With RE-Mg-Ni Based Negative Electrodes*
Shumin Han, Yuan Li, Baozhong Liu, 2017
ISBN: 978-3-11-050116-2; e-ISBN 978-3-11-050148-3,
e-ISBN (EPUB) 978-3-11-049838-7

Markus Schmuck

# Complex Heterogeneous Systems

Thermodynamics, Information Theory, Composites, Networks, and Electrochemistry

DE GRUYTER

**Author**
Dr. Markus Schmuck
Academic Visitor
Faculty of Engineering
Imperial College London
South Kensington Campus
London, SW7 2AZ
United Kingdom
m.schmuck@imperial.ac.uk

ISBN 978-3-11-057953-6
e-ISBN (PDF) 978-3-11-057954-3
e-ISBN (EPUB) 978-3-11-057952-9

**Library of Congress Control Number: 2023951289**

**Bibliographic information published by the Deutsche Nationalbibliothek**
The Deutsche Nationalbibliothek lists this publication in the Deutsche Nationalbibliografie; detailed bibliographic data are available on the Internet at http://dnb.dnb.de.


Cover image: urfinguss / iStock / Getty Images Plus
Typesetting: VTeX UAB, Lithuania
Printing and binding: CPI books GmbH, Leck

www.degruyter.com

This book is dedicated to my parents Annemarie and Hansjörg and to my wife Katrin and son Remo Alexander

# Acknowledgment

The scope and selection of topics addressed in the final form of this book were crucially inspired by mentors with their guidance and by giving me the freedom to find my own approach to answering questions and solving problems. Similarly, I would also like to thank colleagues, whose paths crossed at some point my academic journey, for supporting and motivating my line of thoughts and interests with invitations for extended research stays. In this context, I would like to thank Michael Struwe and Hans Christian Öttinger (both ETH Zürich), Andreas Prohl and Christian Lubich (both U. Tübingen), Martin Z. Bazant (MIT), Jan Haskovec (KAUST), Yongcheng Zhou (Colorado State U.), Peter Berg (U. of Alberta and Brock University), Daniel Peterseim (U. Augsburg), Ricardo Nochetto (U. of Maryland), Serafim Kalliadasis and Grigorios A. Pavliotis (both Imperial College, London), Bo Li (U. of California), Oliver Penrose (Heriot-Watt U.), Jacques Vanneste (U. of Edinburgh), Giles Richardson and Tiina Roose (both U. of Southampton), Grégoire Allaire (É. Polytechnique), Oliver Jensen (U. of Manchester), Andreas Herz (Ludwig-Maximilians-U. München), Peter Berg (U. of Alberta & Brock U.), Volker John and Jürgen Fuhrmann (both WIAS).

My research on the systematic and rigorous derivation of effective formulations in electrochemistry were only possible thanks to the grants awarded by the Swiss National Science Foundation (SNSF)[1] and by the Engineering and Physical Sciences Research Council (EPSRC), UK,[2] for supporting the build-up of my own research group. The understanding developed during these research projects represent a crucial part of this book, which never would have been possible without this kind of support.

I am also grateful for receiving a PECRE award (Grant PECRE-F13ERP01-0-0) funding six weeks of research at ETH Zürich and to the host Christoph Schwab. Thanks to discussions in this context on finite element tensor product approximations, I got the inspiration to develop the computational quantum algorithms for the numerical approximation of the linear systems of multiscale equations developed in this book.

The Heriot-Watt University and the Maxwell Institute for Mathematical Sciences allowed me to experience and mature on all possible aspects of everyday scientific research. Hence, I am grateful for the stimulating environment nurturing valuable impressions and experiences.

I would like to thank the members of the board of the Maxwell Institute Graduate School in Analysis and its Applications (MIGSAA)[3] and also the students who attended and actively participated in my homogenization courses during fall 2016 and spring 2017.

1 SNSF, Prospective Researcher Grant, PBSK-124596.

2 EPSRC, Grant EP/P011713/1.

3 Centre for Doctoral Training funded by the UK Engineering and Physical Sciences Research Council (grant EP/L016508/01), the Scottish Funding Council, Heriot-Watt University and the University of Edinburgh.

https://doi.org/10.1515/9783110579543-201

These courses have been extended and integrated into this book in the sections on modeling multiscale systems and their rigorous upscaling.

I want to express my gratitude to my current employer, the Swiss Government, for the highly interdisciplinary work on a real-time *Complex Heterogeneous System*, which represents a System with links to many related Sensor and Data Fusion Systems. The broad range of science and engineering topics involved in the daily work have extended my scientific horizon and hence also greatly influenced the direction and scope of this book over the last four years.

Finally, I am very grateful to Erich Carelli (Eastern Switzerland University of Applied Sciences) for the careful and time-consuming proofreading of this book, leading to many helpful comments valuably improving the readability and quality of this text.

# Preface

The original driver for this book is the author's research on electrochemical energy applications such as the development of reliable and systematic descriptions of batteries and fuel cells. Very quickly it became apparent that taking into account multiple scientific aspects using the language of physics and mathematics requires a broad but combined view of fundamental established scientific fields to systematically and reliably capture highly complex and interlinked physical, chemical, thermodynamic, mathematical, and practical engineering systems. Moreover, a clear understanding of involved processes such as chemical reactions is limited, since it is very difficult to obtain the preferred measurements/data from operational electrochemical devices, for instance. This difficulty of gaining operational data and measurements of real-time systems, without disturbing or modifying their inner workings, is a general challenge present in many applications such as chemistry (danger of modifying processes), biological multicell processes and signalling (e. g., broken communication), data-link networks (e. g., ghosts in sensor-fusion-control-actuator systems or degradation in quality of information such as the participants' locations) and social dynamics (e. g., invasive sensors preventing natural interaction). At the same time, many new technologies are increasingly a result of combining existing scientific developments and engineering products often by connecting them to smart networks. Examples are reaction networks (Catalysis), the Internet of Things (IoT), Command and Control (C2) Systems and their associated sensor and communication networks, and machine learning (ML) as well as artificial intelligence (AI) with their combination, extension, training, and development of novel neural networks and their corresponding training methods.

As a consequence, future technology represents itself often as interacting, network-like structures, which we systematically identify as CHeSs in this book. This broad and application-driven view is a result of various impressions, experiences, and efforts daily gained at international research places and different scientific and engineering departments, which had a crucial impact on the final scope and form of this book as motivated in the following subsequent paragraphs.

The scientific journey of this book starts at the Eidgenössische Technische Hochschule Zürich (ETH), the author's Alma Mater. The historically grown understanding and view, in part due to the Einstein's impact, that both Physics and Mathematics are not opposing fields, but rather synergetic scientific building blocks reaching their ultimate, optimal understanding by taking the different views from both fields into account. As a result, studying mathematics at ETH meant taking the same lectures in physics and mathematics as the physicists do and vice versa. At the turn of the millennium, the community of students and professors also induced and spread a general, ideal view on science, its fundaments, and its role in society. This certainly provided a motivating and healthy view to develop a scientific career. Moreover, the immediate transition from middle school classes (K-12) with at most 30 students to lectures at university with over hundreds of students can, depending on one's character, lead to a strong development

https://doi.org/10.1515/9783110579543-202

of independence in learning and understanding new and complex theories by working out the necessary concepts and solutions by oneself.

Originally, I started my dissertation under the supervision of Andreas Prohl at ETH on *Modeling, Analysis, and Numerics in Electrohydrodynamics*[4] late 2005. In 2007, A. Prohl kindly offered me the chance to continue my work in a different research environment. This means that I got the opportunity to complete and defend my dissertation on the rigorous development of reliable and convergent finite element schemes of the models, previously developed and analyzed at ETH, at the Universität Tübingen. My interdisciplinary PhD topic combines applications with rigorous mathematics supported by physical and chemical sciences and therefore immediately caught my interest and the necessary focus. Hence, it naturally builds a solid ground for my subsequent years of research leading to this book. In this context, I am grateful to A. Prohl for the initial shaping of my scientific thinking and for giving me the freedom to develop my independence by focusing on physically consistent and quantitative aspects of real-world problems. My arrival at the Numerical Analysis Group (NAG) in Tübingen was exactly at the time when Christian Lubich was finishing his book on numerical schemes for Schrödinger equations,[5] which served to me as an excellent and motivating entry into Quantum Mechanics. The computational view on Quantum Mechanics gained hereby together with the continuous advances in the development of Quantum Computers (QCs) over the last decade provided a motivating and solid basis for discussing the usefulness of QCs in the context of this book. In fact, the complexity of QC itself provides already a formidable example of a Complex Heterogeneous System (CHeS), deserving its own dedicated book.

Thanks to a prospective researcher grant awarded by the Swiss National Science Foundation (SNSF) and to my host, Martin Z. Bazant, who supported my planned research work, I could benefit from two intensive research years at the Massachusetts Institute of Technology (MIT), Cambridge. The research environment I had experienced so far was primarily driven by the goal of gaining a complete understanding by taking all possible aspects and details of a process of interest into account. Hence, it was enlightening to observe a primarily application-driven view for gaining useful physical and scientific insight by rewriting dimensional equations into their corresponding nondimensional form. This means that the naturally arising dimensionless parameters in such reformulations allow us to systematically discuss the physical and mathematical justification for neglecting certain terms (physical processes) depending on the scenario of interest.[6] This is what I have taken as the scientific basis of and the motivation

4 M. Schmuck, *Modeling, Analysis, and Numerics in Electrohydrodynamics*, Universität Tübingen, 2008.

5 M. Lubich, *From quantum to classical molecular dynamics: reduced models and numerical analysis*, European Math. Soc., 2008.

6 Related discussions I enjoyed with Roman Stocker (Professor at MIT at that time, now at ETH Zürich), whom I want to thank for his time and insights into his modeling approaches.

for the following general phrase building a fundamental cornerstone of MIT's research culture, i. e., "*Keep It Simple*".[7] Also during my time at MIT, I met Peter Berg (University of Alberta and Brock University) when he visited Cambridge, MA. Our discussions during that time extended my focus in electrochemistry to look into working principles of fuel cells. Thanks to Peter Berg's previous efforts in fuel cell research, we identified a fundamental prototype catalyst layer formulation allowing us to reliably and systematically derive effective catalyst layer equations for Proton-Exchange Membrane (PEM) fuel cells. Moreover, we rigorously derived how *microscopic, incompressible fluid systems* can become *compressible on the macroscale.*

Toward the end of my SNSF-funded research project at MIT in 2011, I came across a highly interesting research theme offered in a joint Research Associate position at Imperial College, London, by Serafim Kalliadasis (Chemical Engineering) and Grigorios A. Pavliotis (Mathematics). My research efforts led me to acquire and to extend a *Renormalization Group* technique and to combine it with a *Maximum Entropy Principle* toward a novel *Stochastic Mode Reduction* strategy in the context of the *Kuramoto–Sivashinksy equation.*[8] Finally, London's size with its Universities and Colleges as well as its history provide a truly international, scientific, and cultural melting pot. The stimulating and supporting environment in both research departments offered peaceful and inspiring working conditions. This finally allowed me to secure my first permanent academic position as Lecturer/Assistant Professor in Edinburgh in 2013.

Motivated and driven by historical developments such as the redesign of the steam engine by James Watt and its subsequent driver for the theory of Thermodynamics as well as the microscopic and probabilistic view on mechanics by James Clerk Maxwell and his contributions to Electrodynamics, I slowly acquired step by step fundamental aspects about the nature of heat in complex systems. In this context, I feel also fortunate for getting the opportunity to talk to Oliver Penrose[9] about the rigorous application of concepts from Thermodynamics and Statistical Mechanics. Hence, the research environment at the Maxwell Institute for Mathematical Sciences and the Heriot-Watt University provided an inspiring and calm enough environment for young aspiring researchers to pursue their ambitious scientific goals. Triggered by discussions on $Li_xFePO_4$-batteries in M. Z. Bazant's research group in 2009, I finally got the time to elaborate and develop a basic understanding of the important thermodynamic concept referred to as *Phase Transitions*. In fact, the highly heterogeneous character of $FePO_4$-electrodes represents

7 "Keep It Simple" means here that we should not work with unnecessarily complex model equations if the application of interest implies that certain physical forces/terms are negligible. Indeed, the requirement of simplicity is much older and often referred to as Occam's razor, in honor of the late 13th- to early 14th-century English philosopher William of Ockham (= Oak Hamlet).

8 M. Schmuck et al., *New Stochastic Mode Reduction Strategy for Dissipative Systems*, PRL 110:244101 (2013).

9 O. Penrose, *Foundations of Statistical Mechanics – A Deductive Treatment*, Pergamon Press, 1970.

a formidable prototype problem where systematically derived effective macroscopic descriptions by techniques such as upscaling and homogenization are of great practical interest. These efforts turned into the derivation of the effective macroscopic Cahn–Hilliard phase field formulation and its rigorous verification by error estimates. These error estimates additionally provide *convergence rates* quantifying how quickly the full microscopic description approaches the upscaled Cahn–Hilliard equation while increasing the heterogeneity (that is, $\epsilon \to 0$). Starting with investigations on modeling and analysis in electrohydrodynamics during my PhD, fundamental thermodynamic building blocks such as free energies appeared naturally as useful analytical quantities, e. g., an *entropy law* obtained in the process of deriving *a priori estimates* for establishing the solvability (i. e., the existence of solutions). The deeper study and understanding of free energies, in particular, their main usage in equilibrium thermodynamics, during my time in Edinburgh suddenly raised the following central and crucial question: *What are the systematic principles governing the consistent nonequilibrium thermodynamic (i. e., dynamic) description of systems relying on both reversible and irreversible processes?* At first, I believed the answer is to define the dynamics as a gradient descent of the system's free energy in the appropriate function space.[10] However, despite the fact that this still leaves us with a choice between different possible gradient descent schemes, the answer to the original question is still not convincing enough. After coming across the General Equations for Non-Equilibrium Reversible–Irreversible Coupling (GENERIC) developed by Hans Christian Öttinger and Miroslav Grmela, it is currently the most systematic framework available (to the best of my knowledge) for the physical derivation of reliable dynamic descriptions of time-dependent physical processes. On this occasion, I would like to express my gratitude to Hans Christian Öttinger (ETH Zürich) for his hospitality and the discussions on GENERIC for electrohydrodynamic and phase field formulations, which are elaborated and presented in this book, respectively.

In 2018, triggered by an E-Mail invitation from the ETH Alumni network about a rare opportunity to work on a uniquely designed radar environment taking the geographic features of the Swiss Alps into account, I took this chance to work as a governmental scientific analyst and researcher[11] in a highly interdisciplinary environment of a Command and Control (C2) system. The sensor-oriented nature of radar systems naturally depend on essential aspects such as Measurement Theory, Noise Filtering, and Uncertainty Quantification. These practical concepts strongly motivated me to account for the observer/measurement system in the formulation of Complex Heterogeneous Systems

10 The (continuous) gradient flow formulation is the vanishing limit of a time discretization parameter associated with the descent scheme. Indeed, gradient flows are frequently exploited in mathematical analysis, e. g., G. Perelman, *The entropy formula for the Ricci flow and its geometric applications*, arXiv:math/0211159v1, (2002), or R. S. Hamilton, *Three manifolds with positive Ricci curvature*, J. Diff. Geom. 17:255–306 (1982).

11 Official role reads Scientific Research & Development Engineer.

(CHeSs), which is the main driver and focus of this book. The highly interdisciplinary nature of the underlying C2 system, with different sensor and control applications requiring data and information fusion realized on dedicated computational resources such as Data Centers (DCs), represents a truly interesting and stimulating working environment, to which the author is grateful for being able to contribute.

Obviously, the highly interdisciplinary character of CHeSs and the goal of the book is to foster a general interest in establishing a concise theory and framework for capturing universal principles naturally emerging from the underlying structure of CHeSs. This motivates that interdisciplinary scientists, scientific engineers, and graduate and postgraduate students with an affinity in mathematical concepts are the most likely target audience for this book. The reader will benefit by having a basic knowledge in functional analysis and calculus of variations, both with a focus on partial differential equations, as well as in quantum mechanics. Certain sections of the book can well serve as introductory courses in “Homogenization”, “Smart Interacting Systems”, and “Quantum Computing”, for instance.

Finally, this review of essential impressions and experiences, leading to the present form of the book, does not intend to be complete, and hence, the author would like to thank any person not explicitly mentioned in this preface or the previous acknowledgement for the hospitality, inspirations, and stimulating discussions that might have influenced in some form the content of this book. Hopefully, this text stimulates future research on CHeSs and related topics and leads to novel promising and practically useful findings and publications. Naturally, the author appreciates readers giving reference to such stimuli.

Dübendorf, February 19, 2024

Markus Schmuck

# Contents

# Part I: Foundations and examples of complex heterogeneous systems

# 1 Introduction and motivation

The motivation for this book emerged from the study of electrochemical systems such as *batteries* and *fuel cells*. It quickly became apparent that applied electrochemistry represents a system of complex systems, which are interlinked by various subsystems and processes such as reactions in different phases, mass and charge transport in different media, and dynamics and gradients of participating quantities. Clearly, electrochemistry is a highly inter- and multidisciplinary research field. If we consider a battery as an abstract object, i. e., without going into the details of its single components, we are immediately concerned with the following crucial phenomena: *electric conduction* (electron transport), *ionic conduction* (in the form of mass and charge transport), and *chemical reactions* (generally on phase interfaces but not excluded from the bulk such as the *electrolyte* or the *electrodes*). From an application point of view, the multidisciplinarity grows further to include *optimal charging* and *discharging* in dependence of the battery's/cell's *state of charge*. As a consequence, we arrive at *battery management systems* relying on principles such as *optimal control and design* up to *machine learning*. Hence, battery research requires a broad scientific understanding not only ranging from thermodynamics, mechanics of classical and preferably also quantum mechanical systems to complex and coupled *reaction*, *convection*, and non/linear *diffusion* problems, but also taking into account *non/linear*, *random*, *dynamic*, and *network-based interactions*, just to mention a few.

In fact, already connecting systems such as different media/material phases creates interfaces that generally render the resulting *composite* a complex system and even more so under interactions with external systems/surroundings. Such interfaces represent transition regions that are very difficult to experimentally observe due to their very small scale compared to the size of the corresponding bulk phases. This challenge relies on the fact that we generally measure specific quantities in the bulk of the phase of interest. Hereby, we immediately enter other specific fields investigating the origin and propagation of *measurement errors*, i. e., the *theory and thermodynamics of measurement* (e. g., [134]) and *uncertainty quantification*. In physical and, more generally, scientific models, interfaces are usually described by appropriate boundary conditions accounting for processes such as ad- and desorption, material transport, and reactions. Such models provide a straightforward translation into *graph/network-oriented formulations*, where the bulk phases represent the nodes/vertices, and the interphases or boundaries become links/edges, see Section 6.2. For a truly *multiscale view*, the *complexity of interfacial processes* will make it necessary, that we carefully resolve the previously suggested link/edge representation of interfaces by multiple interlinked network descriptions, as required to resolve the different *time and length scales* of the underlying *interfacial interactions*. Indeed, many naturally appearing networks/graphs show a characteristic behavior associated with complex systems developing phase transitions, percolation events, or the emergence of new and often unexpected dynamics, see Section 3.2 on coarsening, for instance. Such a macroscopic graph-based representation, where the

https://doi.org/10.1515/9783110579543-001

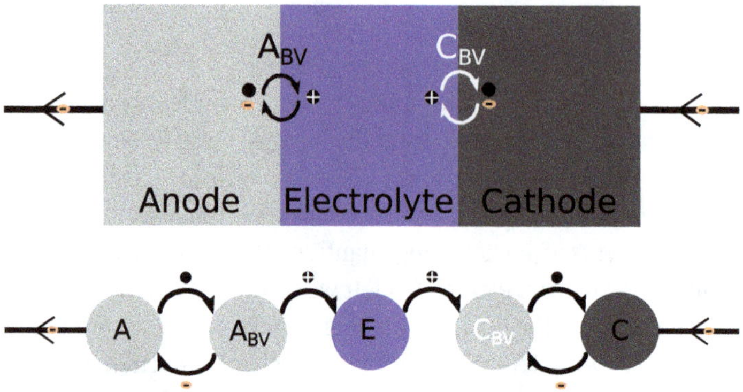


**Figure 1.1:** *Macroscopic view of a discharging battery. Top:* Single cell with negative electrode (anode), positive electrode (cathode), and electrolyte (separator). *Bottom:* Graph representing the above single cell with the main macroscopic elements, e. g., anode A, cathode C, and electrolyte E, and processes such as anodic and cathodic Butler–Volmer reactions $A_{BV}$ and $C_{BV}$, respectively. The graph clearly shows ($Li^+$-)ion conduction in E and electron conduction outside the cell and in A and C. Depending on the application, such single pouch cells will be composed in serial, parallel, or mixed form to achieve a specifically required performance. In the latter context, we arrive at a more complex graph with additional control elements for battery management.

importance of the electrochemical reactions is accounted for by assigning dedicated nodes/vertices to them instead of just links/edges, is depicted in the context of a single cell battery in Fig. 1.1.

Next to the multidisciplinarity of electrochemical systems mentioned above, the same catalytic processes involved are found in a wide range of other systems such as biological life forms on earth, see the growth of vegetation in Fig. 1.2, for instance. Indeed, multicell living matter qualitatively depends on the same kind of processes such as transport, reaction, and diffusion, as well as on their corresponding length and time scales. Moreover, the example of vegetation growth (Fig. 1.2) clearly shows highly heterogeneous structures such as various porous media in the subsurface as well as fractal growth of roots and branches. Hence, we can unexpectedly find physical, chemical, and various scientifically measurable aspects characteristic to one system (e. g., electrochemical cells) in entirely different contexts such as nature (i. e., growth of vegetation) or human engineered devices and networks, where the latter are generally inspired by observations in nature. This motivates that new technologies are increasingly the result of combining and linking existing scientific and engineering results into new and smart network-like structures such as the *Internet of Things (IoT)*, *reaction networks (Catalysis)*, car batteries representing smart packs of pouch cells, virtual (controlled and steered) social networks such as LinkedIn or Facebook, and artificial and naturally grown neural networks in *Artificial Intelligence (AI)* applications and living beings. These accelerating developments and the availability of universal elements across different fields motivate us to collect, in a single reference, basic terms and concepts

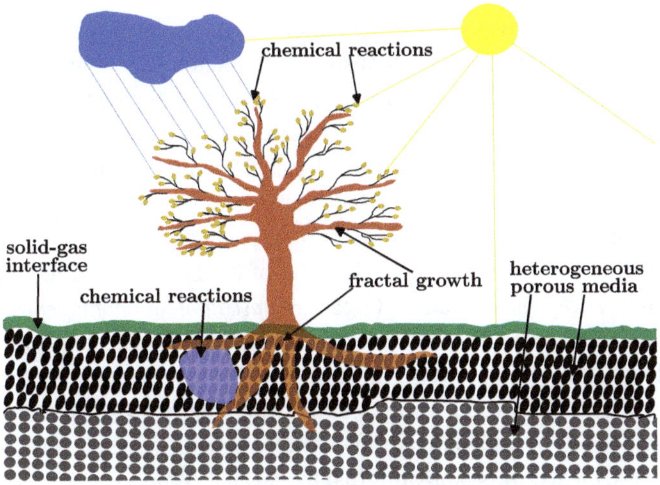


Figure 1.2: *Vegetation Growth as a CHeS:* Catalytic processes involved in the growth of plants and trees supported by fractal growth, interfacial reactions, diffusion, convection, imbibition, wetting, and heterogeneous porous media, just to mention a few.

as well as reliable and systematic methods for analyzing, understanding, and identifying possible universal behavior in high-dimensional, interacting, graph-like entities, which we refer to as *Complex Heterogeneous Systems (CHeSs)*, the main topic of this book.

The starting point of our investigation of CHeSs relies on the different scientific fields that coin and rely on terms such as *system* and *complexity*. It turns out that *thermodynamics* is a natural starting point and field where systematic terms and concepts have been introduced to study and analyze thermodynamic systems such as heat engines, see Section 2.1.2. The generality and ideas of these concepts go far beyond thermodynamics. For instance, it provides valuable insight by guiding a rigorous theory of economics as presented in [91, 125], see Fig. 1.3. Moreover, it seemingly introduces without specific attention the elementary concept of a single bit of information (Szilard's

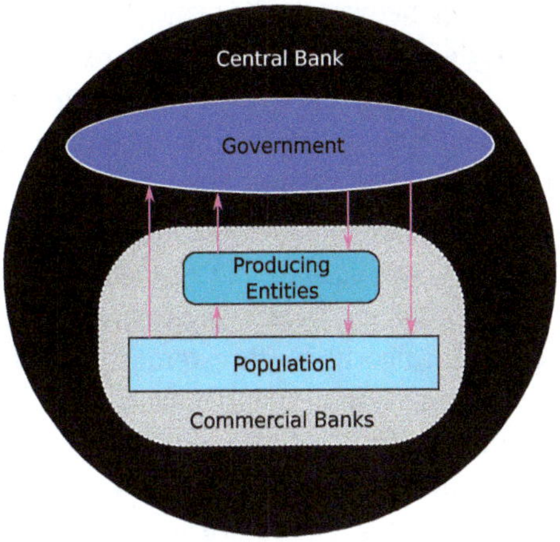


Figure 1.3: *Macroscopic CHeS of a state's economy:* Central bank as the central controlling unit of "value" evolution. Taxes, product, commodity and energy prices, investment credits, etc. are "value" flows between the key macroscopic entities such as Central Bank, Government, Production (Companies/Industry), and Consumption and Labor (Population). Clearly, we can immediately recognize a graph representation for these economic value flows, which become a highly intertwined network after its extension to foreign trade.

engine) and hereby paves the way toward the computational framework of the Turing machine. Hence, thermodynamics naturally established fundamental building blocks for defining the essential aspects of a system, see Section 2.1.1. In fact, the thermodynamic theory builds also the basis for the subsequent development of a microscopic view taken in *Statistical Mechanics*. This more particle-based description is the systematic deduction by Maxwell, Gibbs, Helmholtz, and Boltzmann, and it is consistent with the governing macroscopic observations of classical thermodynamics.

Powerful methodologies for taking multiscale features of CHeSs into account and for deriving reliable, effective representations of CHeSs are *upscaling* and *homogenization* (see Chapter 6). These allow us to rigorously derive simpler, effective descriptions, which can subsequently be exploited in lower-dimensional and simpler computational approximations. They are primarily applied to CHeSs showing highly heterogeneous, microscopic structures where effective/averaged, low-dimensional macroscopic formulations are of great interest. Clearly, this motivates that the level of complexity of a system depends on the scale at which it is observed. Hence, modeling and simulation with systematically and rigorously upscaled systems provide a way to qualitatively and quantitatively validate experimental measurements in highly heterogeneous media, where it is difficult to locate and assign bulk and interfacial regions.

Interestingly, also the term *complexity* seems to go back to thermodynamics, if again we recall Szilard's engine, see Section 2.5.1. This engine can be seen as a first simple example of an *information processing machine*, which naturally leads to the so-called *Algorithmic Information Content (AIC)*, a widely used (computational) measure for complexity, see Section 2.3.2. Comparing algorithms (in particular, across various possible encodings, programming languages, and even concepts of computation, e. g., classical or quantum) is a rather challenging task, and hence, there is a natural interest in a general and widely applicable definition going beyond a dependence on the computational concept applied (e. g., classical or quantum Turing machine).

Here, we advocate a *complexity measure* relying on graph-theoretic concepts such as the *cyclomatic complexity* (see Section 2.2.1) and its appropriate extensions/generalizations. In fact, our approach to *CHeSs* promotes graph-/network-based principles as a systematic and general framework to characterize, define, and describe *complex systems*. Nevertheless, we believe that algorithmic complexity classes are very important for the comparison of different algorithmic/computational approaches together with their realization on different computational platforms, e. g., on a *Personal Computer (PC)* (see Section 3.3.1 and Fig. 1.4) up to a *Data Center (DC)* (see Section 3.3.2 and Fig. 1.5), a *High Performance Computer (HPC)* (see Section 3.3.3 and Fig. 1.6), and even a *Quantum Computer (QC)* (see Section 3.3.4 and Fig. 1.7). Indeed, the recently growing computational performance of QCs and the availability of quantum algorithms, threatening presently considered secure encryption schemes, demonstrate the importance of computational complexity for preserving privacy and economic stability.

In particular, the globally and continuously increasing interest in quantum computing over the last decade motivated us to investigate the importance and the value of

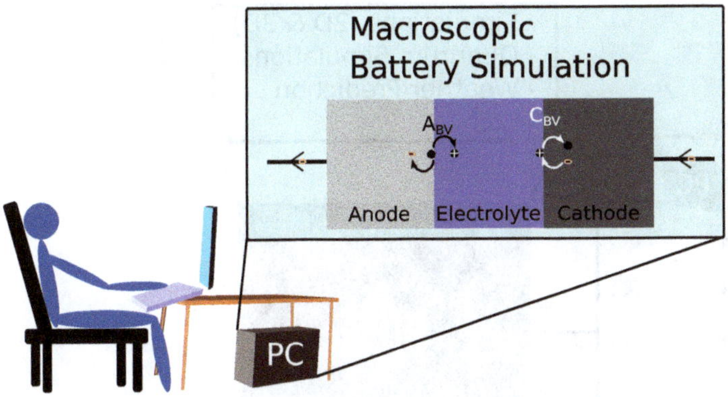


**Figure 1.4:** *Example PC application:* Macroscopic simulations such as coarse battery models not resolving the single particle level.

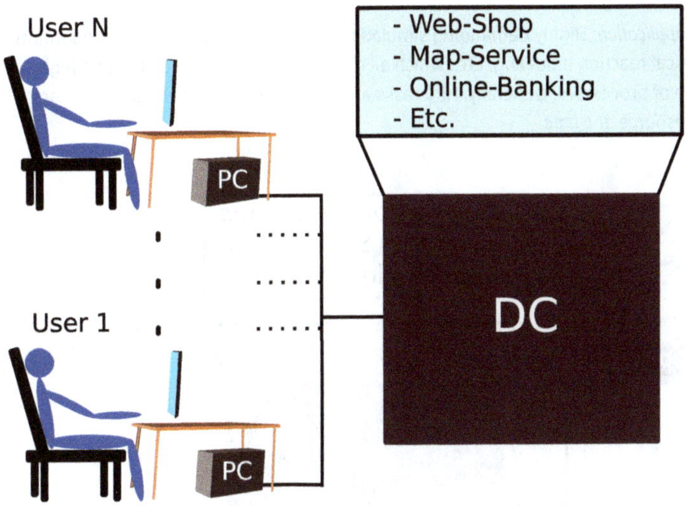


**Figure 1.5:** *Example DC application:* Multiple users access via devices such as PCs, mobile phones, and tablets cloud services such as search engines, web applications such as cloud-based MS Office all hosted by a DC. A DC's challenge obviously is *variable and uncertain parallel customer load/demand*.

the concepts and theories presented in this book beyond the time where QCs become available and useful for solving practical problems. Starting with basic concepts and elements allowing us to define quantum algorithms, we present the fundamental algorithms necessary for solving linear systems of equations in the quantum sense. The applied and rigorous *Upscaling* and *Homogenization* methods (see Sections 6.3, 8.3.6, 9, 10, and 11) discussed in this book allow for systematic and reliable derivations of lower-dimensional and compact formulations, which still lead to high-dimensional computational problems due to the high resolution required by the original problem. Hence both,

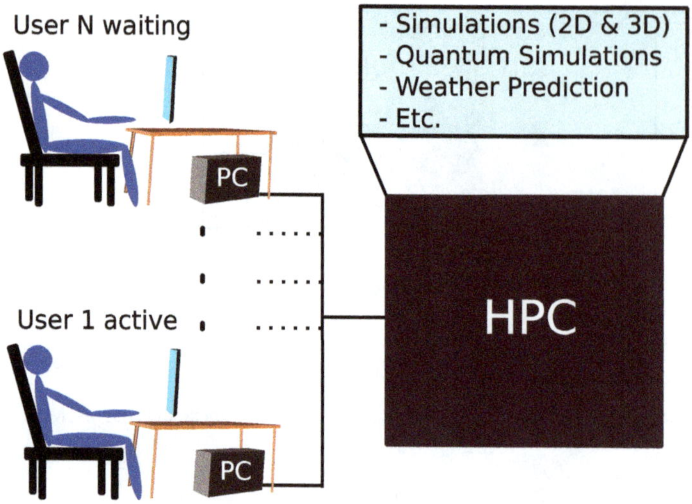


Figure 1.6: *Example HPC application:* Highly demanding simulations such as weather prediction, quantum-based material and chemical reaction modeling, etc., which all require unusual high memory and high number and performance of processors. Generally, HPC tasks are so demanding that *only one user* can access a dedicated HPC resource *at a time*.

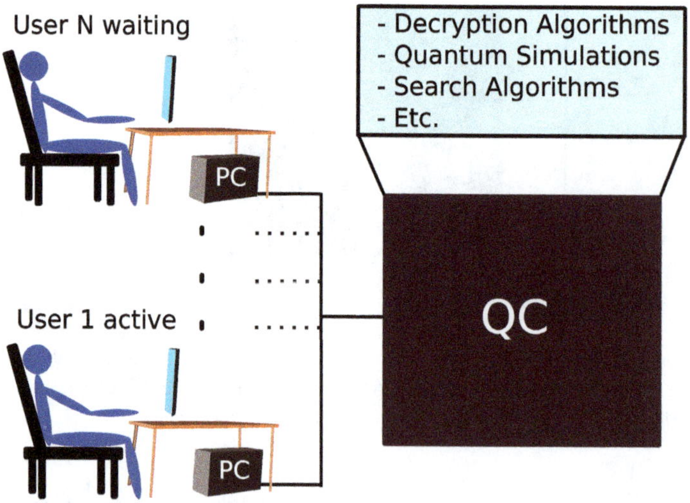


Figure 1.7: *Prospective QC application:* Going *beyond HPC* in terms of dimensional resolution and (parallel) processing speed, see Section 3.3.4 regarding algorithms and programming. Due to the high demand on resources, facilities, and skilled personnel, it can be expected that QC will generally or at least in its initial phase be provided as a cloud service. A graphic (Fig. 3.6) showing qualitative differences between PC, DC, HPC, and QC is presented in Section 3.3.

the still high-dimensional nature of upscaled/homogenized problems and their representation as linear systems of equations[1] make the use of quantum computers and the development of efficient algorithms with their parallelism a very promising direction for resolving high-dimensional CHeSs and their highly heterogeneous multiscale character.

Another topic of widely growing interest, despite the dominance of *artificial intelligence (AI)* and *machine learning (ML)* in the public and the media, is a systematic framework for understanding and describing *smart interactions* between simple systems up to CHeSs, e. g., human–robot interaction as depicted in Fig. 1.8. The term "systematic" means that we first aim to understand interactions between systems from a scientific point of view, and only in a subsequent step, i. e., after the underlying scientific principles are formulated, validated, and understood, we start to think about adding/replacing certain *data/experience-related subsets* with a dedicated neural network formulation.[2] Our main interest lies in developing a thorough understanding of the underlying *input–output mechanisms and processes* as well as the presence of and transition between characteristic states representing a CHeS, and not primarily to quickly find recipes for realizing decent (consumer) applications, without knowing how to improve them beyond an exploitation of more and more accurate data, being generally the result of a lack in system- and process-oriented understanding.

**Figure 1.8:** *Smart and autonomous interactions:* Human and robot playing paper-rock-scissors.

1 Note that solving nonlinear equations generally reduces to iteratively solving appropriate linear equations.

2 This is in contrast to a more industrial/engineering point of view that gears toward quickly developing new consumer products. This latter interest immediately explains the frequent news about new AI applications, which are a result of widely applied brute-force "plug problem-specific (training) data into various available and modified designs of neural networks and validate trained networks with test data"-approaches to exploit low-hanging fruits for the fast creation of applicable results.

As a consequence, our approach to *smart interactions* is motivated by *variational principles* present in nature such as *least action* and *maximum entropy*, see Section 2.8. This means that we believe that each CHeS can be characterized by an intrinsic *objective function* (also referred to as goal/cost function) such that a deviation from its preferred path incurs costs to it. Hence, the explicit definition of a system's cost function will define a *CHeS's specific character and behavior*. To account for the interaction of multiple systems, we accordingly need to identify a so-called *joint characteristic cost function*. This immediately allows us to impose actions to corresponding systems for them to minimize expenses/costs, which ultimately express their individual character and behavior. Note that *thermodynamics* is a leading classical macroscopic scientific field that systematically aims to rigorously assign material and, in general, system-specific cost functions referred to as *free energies* such as the *Gibbs* and *Helmholtz* free energies, see Section 2.1 for such fundamental aspects concerning CHeSs.

For simplicity, we present a basic framework, generalized to CHeSs, based on well-accepted terms and methods from game theory for determining so-called *reasonable interactions* deduced with the help of a specific *objective function* taking all aspects of possible interactions and its consequences into account, see Section 2.8. Hence, this approach gives participating/interacting systems a framework for *autonomous behavior*, see Definition 2.8.1, for instance. This either leads to *pure (inter)actions* in the case of a so-called *saddle point* or *mixed (inter)actions* otherwise. Both types of (inter)actions imply the existence of an underlying *interaction equilibrium*. Moreover, in the context of *fair interactions* (see Example 2.8.16, for instance) we also observe well-known physical and mathematical concepts such as *symmetry*, *phase transition*, and *interfaces*. A fair interaction represents an absolute symmetry between the Casino (roulette) and a betting party being a machine or a human, for instance. In fact, fairness forms the *interface* between the two possible *asymmetric interactions* such as the interaction favors the Casino (roulette) or its corresponding competitor.

Ultimately, the previously motivated *interaction concepts* turn into highly useful applications, if *numerical and computational schemes* are available, that allow us to compute *optimal/reasonable interactions* between systems up to CHeSs. Motivated by the *equivalence* between an interaction formulation relying on interaction (cost) matrices and a corresponding Linear Programming description, as exploited in the rigorous proof of the Fundamental Theorem of Game Theory (see Theorem 2.8.19), we present Dikin's method for computing *reasonable interactions*.[3] At the end, we show an application of this interacting systems framework for optimizing material design based on an available set of design choices (system $\mathcal{S}_1$) and accessible materials (system $\mathcal{S}_2$), see Section 8.2. In fact, this optimal material design approach relies on competing between material performance (conductivity versus price) and design (geometry versus price), which means

3 Note that a detailed investigation of pros and cons of various linear programming algorithms is beyond the scope of this book.

to look for an equilibrium between these possibly opposing aspects. Depending on the preferred performance and design options, we end up with *pure* or *mixed designs*.

A way to go beyond the widely accepted but practically often too idealized concept of *rationally/reasonably interacting parties* (see Definition 2.8.1 and Remark 2.8.2) is the framework of the *Fictitious Interaction Learning Method (FILM)* presented in Section 2.8.1. As a consequence, the theory of *Smart Interacting CHeSs* represents the *intelligence* for choosing the right action in a particular scenario (e. g., Dikin's method), and the ability to learn from previous experiences by FILM.

These aspects of CHeSs have motivated us to present a broad selection of fundamental methods and ideas discussed in this book together with a general and large set of tools as well as their application in a wide range of problems providing a promising source for many applicable innovative concepts with theoretical and practical tools in both applied and theoretical electrochemistry and various scientific and engineering fields striving toward a global understanding of CHeSs. Finally, let us give a brief overview about the structure and accessibility of the book with the help of the following graphical outline, see Fig. 1.9.

## 1.1 Notation, definitions, and dimensionless formulations

For the presentation of results and essential mathematical methodologies, we will apply the following color scheme:

**Mathematical, Physical, and Chemical Results**
Physical and Chemical Laws and Rigorous Theoretical Derivations such as Lemmata, Propositions, Theorems, and Corollaries will be presented in blue boxes.

**Axioms, Conventions, and Mathematical Definitions**
In such a box, we write concise Axioms and Definitions.

---

**General Methodologies and Tools**
Specific, new, or powerful methods are explained/described in such a black and white box. **!**

---

Let us start with the basic notation based on a prototype example of diffusive transport in two different phases $\mathcal{D}^1$ and $\mathcal{D}^2$ such that $\mathcal{D} := \mathcal{D}^1 \cup \mathcal{D}^2 \subset \mathbb{R}^d$, which means, under Einstein's summation convention,

$$\rho^\alpha c^\alpha \frac{\partial \theta^\alpha}{\partial t} - \frac{\partial}{\partial x_i}\left( d_{ij}^\alpha \frac{\partial \theta^\alpha}{\partial x_j} \right) = 0 \quad \text{for } \alpha = 1, 2\,, \tag{1.1}$$

where $c^\alpha$ and $\rho^\alpha$ are specific heat and density of material $\alpha = 1$ and $\alpha = 2$, respectively, $\theta^\alpha$ stands for the temperature in phases $\alpha = 1$ and $\alpha = 2$, and the dimension of space is

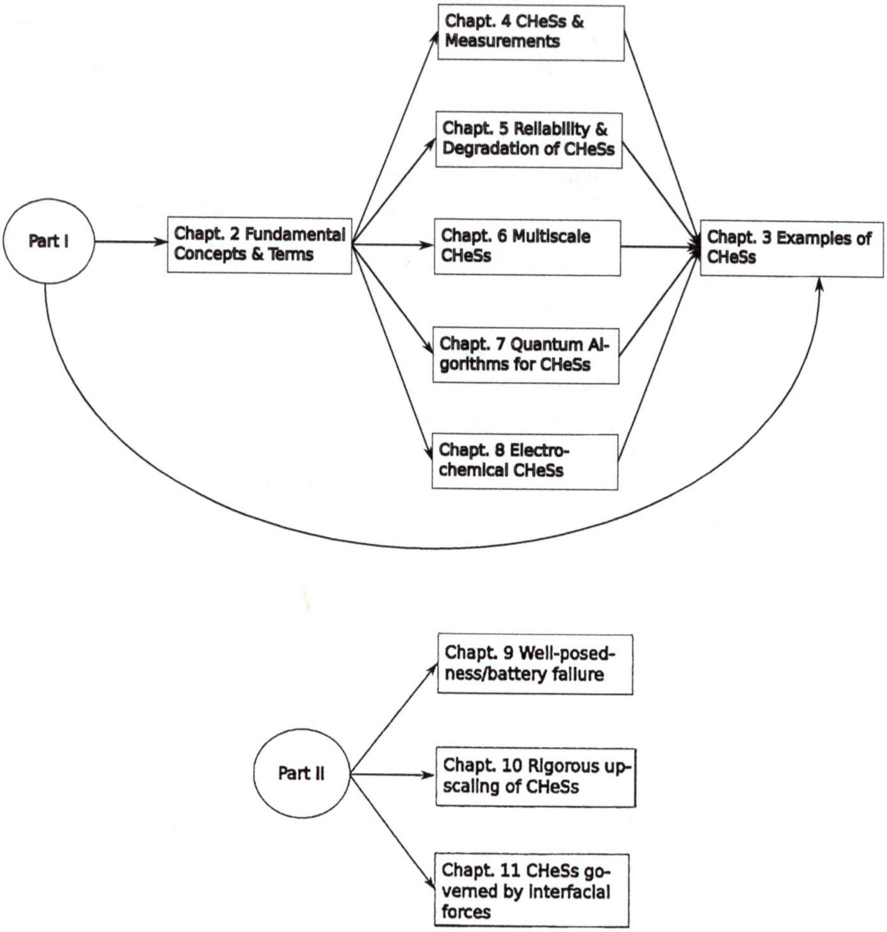


Figure 1.9: The two parts, Part I and Part II of this book, can be read independently. Nevertheless, Part II will be more interesting to read with the knowledge from Part I. Part I is already structured in a deductive manner so that following one chapter after the other will give the best experience. As depicted, the examples discussed in Chapter 3 can also be consulted earlier for inspiration and motivation, but they will be more intuitive after having digested Chapter 2.

$1 \leq d \leq 3$. For simplicity, we subsequently work in 2D, but the results can be established along the same lines for $1 \leq d \leq 3$. $\hat{\mathrm{D}}^{\alpha}$ denotes the material specific (heat) conductivity tensor in three dimensions,

$$\hat{\mathrm{D}}^{\alpha} := \begin{bmatrix} d_{11}^{\alpha} & d_{12}^{\alpha} & d_{13}^{\alpha} \\ d_{21}^{\alpha} & d_{22}^{\alpha} & d_{23}^{\alpha} \\ d_{31}^{\alpha} & d_{32}^{\alpha} & d_{33}^{\alpha} \end{bmatrix} . \tag{1.2}$$

The tensor $\hat{\mathrm{D}}^{\alpha}$ is anisotropic and nonhomogeneous with possibly spatially depending components, i. e., $\hat{\mathrm{D}}^{\alpha} = \hat{\mathrm{D}}^{\alpha}(\mathbf{x})$. We note that the well-posedness of equation (1.1) together with suitable boundary and initial conditions is generally established for symmetric positive definite tensors $\hat{\mathrm{D}}^{\alpha}$ based on the Faedo–Galerkin method, see [54], for instance. In this context, the notion of strongly/uniformly elliptic tensors is also often used.

**Strongly elliptic tensors**

**Definition 1.1.1.** We call a tensor $\hat{A}(\mathbf{x}, \mathbf{y}) := \{a_{ij}(\mathbf{x}, \mathbf{y})\}_{i,j=1}^{d}$ strongly elliptic, if there exist positive constants $0 < \alpha_1 \le \alpha_2$ such that,

$$\alpha_1 |\mathbf{r}|^2 \le \sum_{i,j=1}^{d} a_{ij}(\mathbf{x}, \mathbf{y}) r_i r_j \le \alpha_2 |\mathbf{r}|^2 , \quad \text{for all } \mathbf{r} \in \mathbb{R}^d . \tag{1.3}$$

Note that (1.3) implies $\hat{A} \in [L^{\infty}(\mathcal{D} \times Y)]^{d \times d}$.

We will frequently make use of *characteristic functions* $\chi_A(\mathbf{x})$, which are one if $\mathbf{x} \in A$ and zero elsewhere. This allows us to skip the index $\alpha$ in (1.1). This means that if we consider the stationary form of (1.1), then we have,

$$-\operatorname{div}(\hat{\mathrm{D}}(\mathbf{x}) \nabla \theta) = -\frac{\partial}{\partial x_i}\left( d_{ij}(\mathbf{x}) \frac{\partial \theta}{\partial x_j} \right) = 0 , \quad \text{in } \mathcal{D} = \mathcal{D}^1 \cup \mathcal{D}^2 , \tag{1.4}$$

where $d_{ij} = d_{ij}^1 \chi_{\mathcal{D}^1}(\mathbf{x}) + d_{ij}^2 \chi_{\mathcal{D}^2}(\mathbf{x})$. Elliptic equations of the form (1.4) will serve as the starting point and motivation for the wide range of problems for which homogenization theory is a reliable and systematic method to pertain physically relevant characteristics in a low-dimensional effective macroscopic formulation, see Section 6.3.

*Vectors* such as the spatial coordinate $\mathbf{x}$ are written in boldface. *Tensors and matrices* will be indicated by the $\hat{\cdot}$-symbol. A wide spread ingredient in homogenization theory is *Y-periodicity*, where $Y \subset \mathbb{R}^d$ stands for a characteristic reference cell defining elementary microscopic properties of materials such as pore geometry in porous media or composition of elastic or conducting materials.

***Y*-periodic function**

**Definition 1.1.2.** For a reference cell $Y := (0, l_1) \times (0, l_2) \times \cdots \times (0, l_d)$, we say that a function $u(\mathbf{x})$ defined a. e. on $\mathbb{R}^d$ is *Y-periodic*, if and only if,

$$u(\mathbf{x} + k l_i \mathbf{e}_i) = u(\mathbf{x}) , \quad \text{for a. e. } \mathbf{x} \in \mathbb{R}^d ,\ \forall k \in \mathbb{Z} , \text{ and } d = 1, 2, 3 . \tag{1.5}$$

Let $I_{12} := \partial\mathcal{D}^1 \cup \partial\mathcal{D}^2$ denote the interface between phases $\alpha = 1$ and $\alpha = 2$. If $I_{12}$ represents an ideal interface, then we have the continuity of the temperature $\theta^{\alpha}$ and flux $\mathbf{J}^{\alpha} = [J_1^{\alpha}, J_2^{\alpha}, \ldots, J_d^{\alpha}]'$, i. e., $J_i^{\alpha} := d_{ij}^{\alpha} \frac{\partial \theta^{\alpha}}{\partial x_j}$, $i = 1, 2, \ldots, d$, across the interface $I_{12}$, i. e.,

$$\begin{cases} \theta^1 = \theta^2 & \text{on } I_{12} , \\ \mathbf{n}_{12} \cdot \mathbf{J}^1 = \mathbf{n}_{21} \cdot \mathbf{J}^2 & \text{on } I_{12} , \end{cases} \tag{1.6}$$

where the unit normal $\mathbf{n}_{12}$ on $I_{12}$ points from phase $\alpha = 1$ into phase $\alpha = 2$, and $\mathbf{n}_{21} = -\mathbf{n}_{12}$.

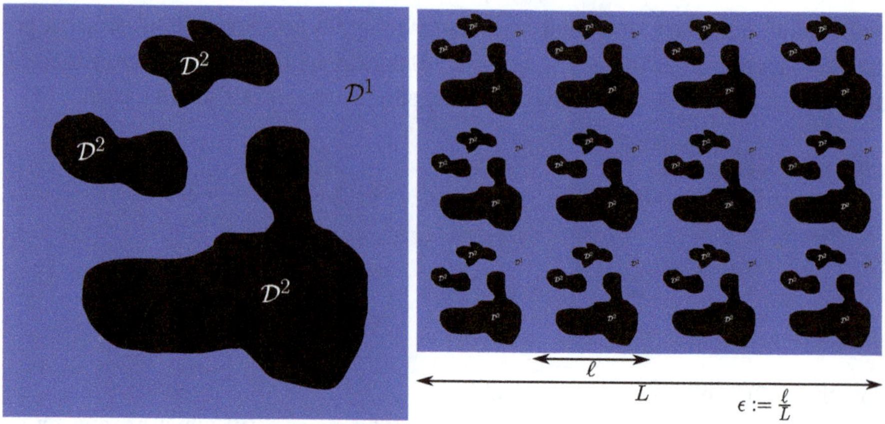


**Figure 1.10:** *Left:* A simple composite $\mathcal{D} := \mathcal{D}^1 \cup \mathcal{D}^2$. *Right:* A strongly heterogeneous periodic composite $\mathcal{D}^\epsilon$ with the characteristic heterogeneity $\epsilon := \frac{\ell}{L}$, where $\ell$ is the microscopic length of the material specific volume element, and $L$ denotes the macroscopic length of the medium.

We can rewrite (1.1) in a dimensionless form by identifying characteristic values of specific physical quantities such as the reference specific heat $c$, reference conductivity $d$, reference density $\rho$, reference temperature $\theta$, reference length scale $\ell_r$, and a reference time scale $\tau$. For a periodic composite material (Fig. 1.10, right), we can identify a microscopic length scale $l$ associated with a reference cell or representative volume element $Y \subset \mathbb{R}^d$, $1 \leq d \leq 3$, as well as a macroscopic length scale $L$ associated with the whole composite. An example of a reference cell is depicted in Fig. 1.11. This allows us to identify the so-called heterogeneity $\epsilon$ of the composite by $\epsilon := \frac{l}{L} \ll 1$. Herewith, we immediately can define two different diffusion times, i. e., $\tau_l := \frac{l^2 \rho c}{d}$ and $\tau_L := \frac{L^2 \rho c}{d}$.

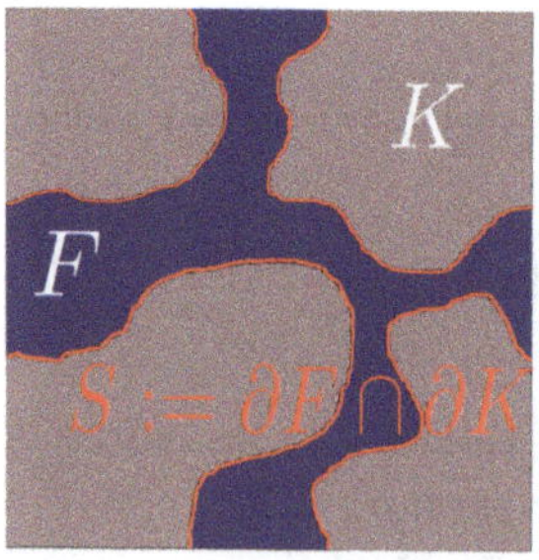


**Figure 1.11:** Periodic reference cell $Y$: $K$ stands for a solid catalyst and $F$ for a fluid phase.

If we choose the *macroscale L as a characteristic/reference length scale* for the diffusion time and the coordinates, i. e.,

$$\tilde{\mathbf{x}} := \frac{\mathbf{x}}{L}, \quad \tilde{t} := \frac{t}{\tau_L}, \quad \tilde{\theta}^\alpha := \frac{\theta^\alpha}{\theta}, \quad \tilde{\rho}^\alpha := \frac{\rho^\alpha}{\rho}, \tag{1.7}$$

$$\tilde{c}^a := \frac{c^a}{c}, \quad \tilde{d}^a_{ij} := \frac{d^a_{ij}}{d}, \tag{1.8}$$

then we end up with the following dimensionless heat conduction,

$$\tilde{\rho}^a \tilde{c}^a \frac{1}{\tau_L} \frac{\partial \tilde{\theta}^a}{\partial \tilde{t}} - \frac{1}{\tau_L} \frac{\partial}{\partial \tilde{x}_i} \left( \tilde{d}^a_{ij} \frac{\partial \tilde{\theta}^a}{\partial \tilde{x}_j} \right) = 0, \quad \text{for } a = 1, 2. \tag{1.9}$$

On the other hand, if we are interested in the *macroscopic time scale* $\tau_L$ but on the level of the microscopic length scale $l$ of the reference cell $Y$, that is, $\tilde{\mathbf{x}} = \frac{\mathbf{x}}{l}$, then equation (1.1) becomes,

$$\frac{\tau_l}{\tau_L} \tilde{\rho}^a \tilde{c}^a \frac{\partial \tilde{\theta}^a}{\partial \tilde{t}} - \frac{\partial}{\partial \tilde{x}_i} \left( \tilde{d}^a_{ij} \frac{\partial \tilde{\theta}^a}{\partial \tilde{x}_j} \right) = 0, \quad \text{for } a = 1, 2. \tag{1.10}$$

Since $\epsilon^2 = \frac{\tau_l}{\tau_L}$, the time derivative term in (1.10) is scaled by $\epsilon^2$. For notational convenience, we will generally skip the $\tilde{\cdot}$ from dimensionless variables.

# 2 Fundamental building blocks for CHeSs

The investigation, understanding, prediction, and control of *Complex Heterogeneous Systems (CHeSs)* relies on a highly interdisciplinary mix of reliably developed and systematically validated fundamental theories and methods from science such as Thermodynamics, Mathematics, Physics, Chemistry, Computer Science, and Engineering. To handle such a large amount of concepts, frameworks, and tools, we subsequently focus on the essential and basic building blocks that naturally allow for a broad view, but in a streamlined framework for defining, analyzing, understanding, and extending our knowledge of CHeSs, and ultimately for controlling CHeSs.

## 2.1 Thermodynamic principles as a systematic pillar

The development of the theory of thermodynamics has been driven by fundamental questions such as *What is heat?*, *Can heat be used to perform mechanical work?*, or *How does heat influence (chemical) reactions?*. Early developments in thermodynamics were strongly inspired by James Watt's modification of Newcomen's steam engine in enterprises [43] back in 1776. Related revolutionary thoughts, generally considered as the origins of economic theory, are contained in Adam Smith's book *The Wealth of Nations* [159] in 1776. These novel ideas and the technological progress induced by the steam engines are generally considered as the origins of the industrial revolution. In fact, economics itself is a field that aims to shed light on various kinds of CHeSs such as local, national, and world-wide economies. In this context, a physical and thermodynamic view provides a deeper insight and understanding on economic concepts and elementary principles, see the books *The Second Law Of Economics* by Kümmel [91] or *Econodynamics* by Pokrovskii [125], for instance.

### 2.1.1 Key concepts: system, heat, work, energy, and entropy

Thermodynamics can be considered to be a *macroscopic* theory, since it emerged from various systematic attempts to explain observations of a system made on the length scale of centimeters to meters and a time-scale of seconds to hours. We note that "system" generally is an abbreviation for *thermodynamic system*, which relies on the following formal definition.

**Thermodynamic system**

**Definition 2.1.1.** Let $\mathcal{M}$ be a set containing everything that exists, e. g., the $d$-dimensional set $\mathbb{R}^d$.[a] We call an open bounded subset $\mathcal{S} \subset \mathcal{M}$ to be a *thermodynamic system*, if we are interested in interactions (involving heat and work) with its open complement $\mathcal{N} := \overline{\mathcal{S}}^c = \mathcal{M} \setminus \overline{\mathcal{S}}$,[b] called the *surroundings*, across the joint interface $\mathcal{I}_{\mathcal{S},\mathcal{N}} := \overline{\mathcal{S}} \cap \overline{\mathcal{N}} = \partial\mathcal{S} \cap \partial\mathcal{N}$. These interactions generally represent energy or material

https://doi.org/10.1515/9783110579543-002

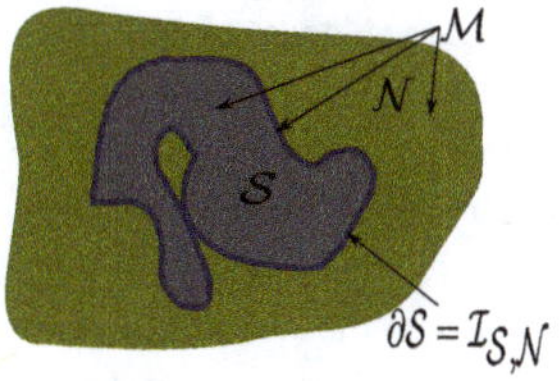


Figure 2.1: Graphical definition of *thermodynamic system* $\mathcal{S}$, interface $\partial\mathcal{S}$, surroundings $\mathcal{N}$, and universe $\mathcal{M} = \mathcal{S} \cup \mathcal{N}$.

exchange (e. g., interfacial reactions) over the boundary of the system, i. e., $\partial\mathcal{S} := \overline{\mathcal{S}} \backslash \mathring{\mathcal{S}}$, generally referred to as the interface $\mathcal{I}_{\mathcal{S},\mathcal{N}}$, see Fig. 2.1.[c]

*a* $\mathcal{M}$ is frequently called the universe.

*b* $\overline{\mathcal{S}}$ denotes the closure of a set $\mathcal{S}$.

*c* $\mathring{\mathcal{S}}$ denotes the interior of a set $\mathcal{S}$.

**Remark 2.1.2** (Real observable systems). Systems $\mathcal{S}$ characterized by Definition 2.1.1 represent so-called *ideal systems*. However, we have to work with *measurable/observed systems* $\tilde{\mathcal{S}}$ in our real world, i. e., the universe $\mathcal{M}$, that is, systems of which we only have limited and generally erroneous information available, e. g., due to the lack of sensor/measurement technologies and measurement errors such as thermal noise. ◇

This leads to the following refined notion of an *observable system*, see also Figure 2.2.

**Observable system**

**Definition 2.1.3.** Consider a triplet $\mathcal{R}_{\mathcal{S}} := (\tilde{\mathcal{S}}, \mathbf{I}, O)$ that consists of a system $\tilde{\mathcal{S}} \subset \mathcal{M}$, where $\mathcal{M}$ denotes the universe, which comprises everything possible (including the unknown/unobservable) but at least $\tilde{\mathcal{S}}$ and its complement $\tilde{\mathcal{N}} := \tilde{\mathcal{S}}^c$, called the surroundings, an observer $O$, and information flow $\mathbf{I}$.[a]

We call $\mathcal{R}_{\mathcal{S}}$ a *real/observable system*, if and only if there is an observer $O$ receiving information $\mathbf{I} = \dot{\tilde{\mathbf{z}}}$ through some kind of measurement process $\mathbf{h}$, i. e.,

$$\dot{\tilde{\mathbf{z}}} = h(\tilde{\mathbf{s}}) + \tilde{\mathbf{w}}\,, \tag{2.1}$$

performed on $\tilde{\mathcal{S}}$ and its surroundings $\tilde{\mathcal{N}}$, where $\tilde{\mathbf{w}} \in \mathbb{R}^{d_{\mathbf{I}}}$ accounts for $d_{\mathbf{I}}$-dimensional white noise error sources, and $\tilde{\mathbf{z}} \in \mathbb{R}^{d_{\mathbf{I}}}$ denotes the $d_{\mathbf{I}}$-dimensional vector of measurable quantities. Herewith, the observer gains knowledge of the $d_{\tilde{\mathcal{S}}}$-dimensional *state* $\tilde{\mathbf{s}} \in \mathbb{R}^{d_{\tilde{\mathcal{S}}}}$ of $\tilde{\mathcal{S}}$ and of its surroundings $\tilde{\mathcal{N}}$, allowing to associate a so-called *state transition/evolution* $\mathbf{f} \in \mathbb{R}^{d_{\mathcal{S}}}$, that is,

$$\dot{\tilde{\mathbf{s}}} = f(\tilde{\mathbf{s}}) + \tilde{\mathbf{v}}\,, \tag{2.2}$$

where $\tilde{\mathbf{v}} \in \mathbb{R}^{d_{\tilde{\mathcal{S}}}}$ accounts for intrinsic error sources by white noise.

*a* The term *information flow* means that information (e. g., encoded in bits and bytes) is moving in a particular direction, see Fig. 2.2.

Note that Definition 2.1.3 also accounts for the information exchange between $\tilde{\mathcal{S}}$ and $\tilde{\mathcal{N}}$ represented, for instance, by mass, energy, and information transport or a mixture of these.

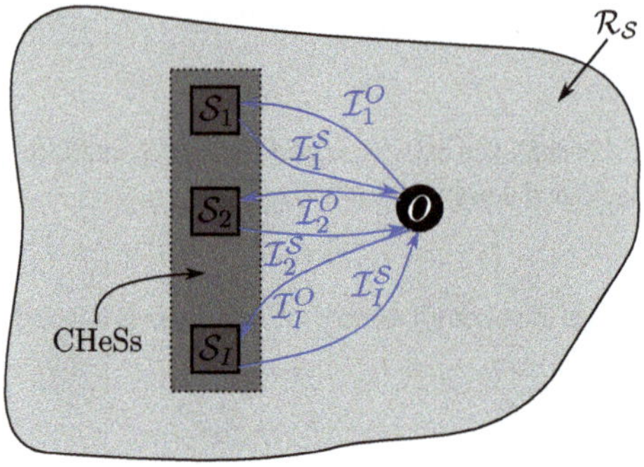


Figure 2.2: A real/observable system $\mathcal{R}_S$ consists of a system of interest (e. g., CHeSs consisting of subsystems $\mathcal{S}_1, \tilde{\mathcal{S}}_2, \dots, \tilde{\mathcal{S}}_I$ taking into account information flows $\mathcal{I}_i^\eta$ for $i \in \{1, 2, \dots, I\}$ and $\eta \in \{O, \mathcal{S}\}$). Hence, a possible information/material flow back to the involved subsystem is considered as well, depending on the observation method applied. This means that certain forms of observation can alter the systems observed. Finally, we note that for simplicity, possible dependencies between the subsystems $\mathcal{S}_i, i \in \{1, 2, \dots, I\}$, which represent a CHeSs of interest, are not explicitly shown.

A simple system that will play a fundamental role for the demonstration of basic concepts relies on a volume filled with an ideal gas, which we define as follows.

**Ideal gas, $\mathcal{S} = \mathcal{S}_{ig}$**

**Definition 2.1.4.** An *ideal gas* is a system $\mathcal{S}_{ig}$ representing a volume $V$ filled with identical particles satisfying the following assumptions:

**(IG1)** $\mathcal{S}_{ig}$ consists of a large number of the same particles/species, atoms, or molecules,

**(IG2)** the particles/species are point masses (no volume extension),

**(IG3)** the particles/species move randomly,

**(IG4)** there are no interparticle forces except for elastic collisions between each other and the boundary of the system.

We will work with the notion of *thermodynamic systems* if not explicitly stated otherwise, since the above term of *observable real systems* depends on the additional identification of a measurement/observation process, which hence requires the additional notion of information/knowledge.

Important classifications of systems are the following three basic definitions.

**Isolated, closed, and open system**

**Definition 2.1.5.** Let $\mathcal{S} \subset \mathcal{M}$ be an open bounded subset (referred to as a thermodynamic system)[a] of a universe $\mathcal{M}$ according to Definition 2.1.1.

We call $\mathcal{S}$ to be *isolated*, if it does not exchange mass and energy with $\mathcal{N}$.

For $\mathcal{S}$ to be *closed*, it only exchanges energy but not mass.

Finally, we call $\mathcal{S}$ to be *open*, if it exchanges both energy and mass.

---

*a* Note that the notions *open* and *closed* have the usual topological meaning in the context of mathematical sets. However, here the system-related meaning applies, i. e., as soon as a set fulfills Definition 2.1.1 of a thermodynamic system, *open* and *closed* adopt the system-specific notions defined here.

The most important thermodynamic state of a system $\mathcal{S}$ is the following.

**Equilibrium state**

**Definition 2.1.6.** A thermodynamic system $\mathcal{S}$ is said to admit an *equilibrium state*,[a] if all the energy available for doing useful work (mechanical/$pV$-work) is expended.

*a* The subsequently introduced concept of *local thermodynamic equilibrium* will play a crucial role in the upscaling/homogenization of nonlinear, complex heterogeneous systems discussed later on in this book.

**Remark 2.1.7** (Equilibrium). A system $\mathcal{S}$ at equilibrium is stationary, that is, it only changes its state by forces/energies from the surroundings $\mathcal{N}$ acting on it. As soon as the Second Law of Thermodynamics has been introduced (see Section 2.1.1.2), we can define the state of *thermodynamic equilibrium* as the state of *maximum entropy*, i. e., the state in which the entropy $S$ does not change anymore. ◇

Note that for a general system $\mathcal{S}$, it usually takes an infinite amount of time to assume thermodynamic equilibrium. As a consequence, experiments in thermodynamics require a lot of time. Nevertheless, since a system in (thermodynamic) equilibrium can be more reliably measured compared to a system constantly changing its state,[1] much of *classical thermodynamics* is devoted to the study of equilibrium systems. To describe the *state of a thermodynamic system*, the following concepts play a fundamental role.

**State variables, functions, and equations**

**Definition 2.1.8.** The minimum number of variables assuming fixed values while approaching a thermodynamic equilibrium of a system $\mathcal{S}$ are called *state variables*.

The state variables are generally not independent, whereas their dependence becomes apparent by the following associated *state equation*,

$$f(\mathbf{s}_d) = 0\,, \tag{2.3}$$

where $f : \mathbb{R}^{n_d} \to \mathbb{R}$, and $\mathbf{s}_d := [s_1, s_2, \ldots, s_{n_d}]'$ is the vector of $n_d \in \mathbb{N}$ dependent state variables describing $\mathcal{S}$.

Rewriting the state equation (2.3) just for the independent state variables $\mathbf{s} := [s_1, s_2, \ldots, s_n]'$, where $n \in \mathbb{N}$ with $n < n_d$, leads to a so-called *state function* $F(\mathbf{s})$, which has the important property that it does not change (its value) in an isolated system $\mathcal{S}$, that is,

$$F(\mathbf{s}) = \text{const.} \tag{2.4}$$

in a system $\mathcal{S}$ not exchanging (energy and mass) with the surroundings $\mathcal{N}$.

**Example 2.1.9** (Ideal gas). Consider an ideal gas $\mathcal{S}_{ig}$. In experiments with real gases, helium well approximates the assumptions of an ideal gas $\mathcal{S}_{ig}$. The state variables $\Theta$, $P$, $V$ can be identified to satisfy the following state equation,

$$f_{ig}(\Theta, P, V) := PV - nR\Theta = 0\,, \tag{2.5}$$

1 A well-known non-equilibrium system is a material/body experiencing heat conduction driven by temperature gradients, for instance.

where $n := N/N_A$ is the number of moles for a given number of gas particles $N$, $R := 8.314$ [$\text{JK}^{-1}\text{mol}^{-1}$] is the universal gas constant, $V$ is the volume of $\mathcal{S}_{ig}$, $P$ is the pressure, and $\Theta$ is the temperature.

A fundamental state function of an arbitrary system $\mathcal{S}$ is the internal energy $U$, see the subsequent Section 2.1.1.1. For the ideal gas, the internal energy can be identified in three different ways due to the number of dependent (i. e., 3) and independent (i. e., 2) state variables. We have the following three expressions for the internal energy $U = U_{ig}$, that is,

$$U(\Theta, V) := nc_V\Theta\,, \tag{2.6}$$

$$U(\Theta, P) := \frac{c_V}{R}\sqrt{nR\Theta}\sqrt{PV}\,, \tag{2.7}$$

$$U(P, V) := \frac{c_V}{R}PV\,, \tag{2.8}$$

where each admits the same value for the same state $\mathbf{s} \in \{[\Theta, V]', [\Theta, P]', [P, V]'\}$, and where $c_V$ denotes the *heat capacity at constant volume.*[2] ◇

To introduce a central property for the analysis of many applied and practically relevant systems, we need the following concept of the *Helmholtz free energy* $A$. We have that,

$$A(\Theta, V, N) := U(\Theta, V, N) - S(U, V, N)\Theta\,, \tag{2.9}$$

where $S(U, V, N)$ denotes the macroscopic/thermodynamic entropy consistent with Boltzmann's microscopic/configurational (and hence more intuitive) entropy introduced further in Definition 2.1.20.

**Remark 2.1.10** (Original macroscopic entropy $S$). Due to the more application-driven view in this Part I establishing the foundations of CHeSs, we limit the amount of technical arguments. Hence, we refer the reader interested in the mathematical motivation of the macroscopic entropy $S(U, V, N)$ based on Carathéodory's principle to Guggenheim [67, Section 1.19, p. 16], for instance.

Similarly, the essential *thermodynamic potentials,* such as the *enthalpy* and the *Gibbs and Helmholtz free energies,* can be rigorously motivated with the help of Legendre transformations. The former two generally play an important role in investigations about thermodynamic equilibrium, see [36, Section 5.3, p. 121], for instance. ◇

Frequently, one associates with (2.9) a so-called *free energy density* $a(\Theta, V, \rho)$, where $\rho := \frac{N}{N_{\text{tot}}}$ represents the particle density associated with the total number of particles $N_{\text{tot}}$. This immediately leads to another important thermodynamic quantity,

2 Note that the derivation of (2.8) is an immediate consequence of the internal energy consideration discussed in the subsequent Example 2.1.15 and the state equation (2.5).

$$\mu := \frac{\partial a}{\partial \rho}, \tag{2.10}$$

which is referred to as a *chemical potential.* Herewith, we can state an important thermodynamic principle applicable to a large class of frequently used material/mass transport formulations.

**Local thermodynamic equilibrium**

**Definition 2.1.11.** Let $\mu(\mathbf{s}_d)$ be the *chemical potential* of a system $\mathcal{S}$, where $\mathbf{s}_d$ denotes the system's dependent state vector. Moreover, a so-called reference element/representative volume element $Y$ characterizes the microscopic, characteristic, and periodic features of the corresponding macroscale $\mathcal{S}$, see Fig. 1.10 or Fig. 1.11, for instance. In this context, we employ the variables $\mathbf{y} \in Y$ on the microscale and $\mathbf{x} \in \mathcal{S}$ on the macroscale. The microscale $Y$ belonging to the macroscale $\mathcal{S}$ is then said to be in *Local Thermodynamic Equilibrium (LTE)* if the chemical potential $\mu(\mathbf{s}_d)$ satisfies the following property,

$$\frac{\partial \mu(\mathbf{s}_d)}{\partial x_k} = \begin{cases} 0\,, & \text{on the reference cell } Y\,, \\ \frac{\partial \mu(\mathbf{s}_d)}{\partial x_k}\,, & \text{on the macroscale } \Omega\,. \end{cases} \tag{2.11}$$

**Remark 2.1.12** (Local thermodynamic equilibrium). The physical meaning of LTE as defined in (2.11) is that processes vary on the microscale so slowly (e. g., inside a microscopic representative volume element) that they are not visible on the macroscale (e. g., on the level of the medium or CHeS of interest). In the context of *nonlinear CHeSs*, the concept of LTE plays a crucial physical and mathematical role in the *rigorous upscaling/homogenization*, see [139–141, 143], for instance. ◇

A manifestation for why thermodynamics serves as a fundamental theory not only for describing a single complex system, but rather as a general framework to capture a *system of systems*, is the fact that it represents a self-contained axiomatic theory relying on the three laws of thermodynamics. We refer the interested reader to the literature starting with Carathéodory [32], Arens [4], Boyling [25], and Lieb and Yngvason [96]. The following less frequently appearing zeroth law, which fundamentally represents logic causality, is the fundamental basis of the concept of *temperature.*

**Transitivity of equilibria**

**Zeroth law of thermodynamics.** Consider three different thermodynamic systems. They are all in the same equilibrium state, if two of them are in equilibrium with the third one.

The notion of temperature, introduced and systematically defined in thermodynamics, naturally relies on the *equilibrium state* of a system $\mathcal{S}$, see Definition 2.1.6. This state allows us to make reliable measurements, albeit over long times, i. e., until the (macroscopic) system state does not change anymore. In fact, recall that the *state variables* are defined at equilibrium, see Definition 2.1.8.

#### 2.1.1.1 Internal energy $U$: the 1st law

The role of energy from a macroscopic system and application point of view can be distinguished by the following three main categories: transport, production/source, and consumption. In this context, the systematic physical and well-established concept of *internal energy U* plays a fundamental role.

**Internal energy $U$**

**Definition 2.1.13.** The internal energy $U$ of a system $\mathcal{S}$ represents the total energy of $\mathcal{S}$, that is, its total potential ($E_{\text{pot}}$) and kinetic ($E_{\text{kin}}$) energy,[a]

$$U(\mathbf{x}) = E_{\text{kin}}(\mathbf{x}) + E_{\text{pot}}(\mathbf{x}), \quad \mathbf{x} \in \mathcal{S}, \tag{2.12}$$

where $\mathbf{x}$ denotes any specific point constituting $\mathcal{S}$.

*a* Note that translations ($E_{\text{kin}}$) generally can be motivated with classical mechanics, but oscillations and rotations (both part of $E_{\text{pot}}$) generally rely on quantum mechanical considerations.

**Remark 2.1.14** (Measuring $U$). It is not possible to measure an absolute value of the internal energy $U$, but with the help of a defined reference energy, we can relate measurements to it. ◇

**Example 2.1.15** ($U$ for ideal and multiatom gases). The internal energy $U$ of the ideal gas has already been introduced in Example 2.1.9 as an example of a state function. We note that $U(\Theta, V)$ in (2.8) can be written in the more general form,

$$U(\Theta, V) = \frac{n}{2}(f_{\text{tra}} + f_{\text{rot}} + f_{\text{osc}})R\Theta, \tag{2.13}$$

so that we can identify the specific *heat capacity for constant volume* $c_V$ by,

$$c_V := \frac{f}{2}R, \tag{2.14}$$

where $f := (f_{\text{tra}} + f_{\text{rot}} + f_{\text{osc}})$ is the sum of the translational, rotational, and oscillating *degrees of freedom*, respectively. For multiatom gases, we can employ very simple models approximating the real situation, see Figure 2.3, for instance. ◇

With the eyes of a chemist, we naturally extend the focus of classical thermodynamics to account for less visible, microscopic participants in our world, i. e., atoms and molecules. In fact, Definition 2.1.13 of internal energy naturally accounts for these microscopic actors in a chemical system/process via *kinetic energy (movement/transport)* and *potential energy (atomic/molecular configuration)*. This immediately leads us to the following fundamental law in thermodynamics.

**Conservation of energy**

**First law of thermodynamics.** In an isolated system, the total (internal) energy is constant.

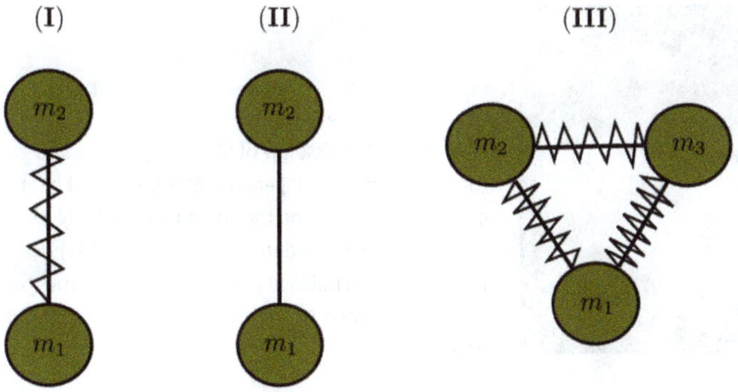


Figure 2.3: *(I) Spring model:* Two atoms connected via a spring allowing for oscillations along the link as a contribution to $E_{pot}$ as well as translations and rotations around the center of mass as contributions to $E_{kin}$. *(II) Dumbbell model:* Two atoms being a fixed distance apart only allowing for $E_{kin}$, i. e., only translations and rotations are possible with $f = f_{tra} + f_{rot} + f_{ocs} = 5$ degrees of freedom, where $f_{tra} = 3$ denotes the movement of the center of mass, $f_{rot} = 2$ represents the two angles defining the orientation of the molecule axis, and $f_{osc} = 0$ due to fixed distance between $m_1$ and $m_2$. Therefore $U = nf/2R\Theta$. *(III) Spring model:* Three atoms linked via springs able to oscillate and rotate. Naturally, oscillation patterns are more complex in this case.

In real applications, it is necessary to understand that changes in internal energy $\Delta U$, according to the 1st law, can only appear due to exchanges of energy, of which work is part of, with the surroundings $\mathcal{N}$. Therefore, the first law can be rewritten for practical purposes as follows,

$$\Delta U = Q^{\swarrow} + W^{\nearrow} = Q - W \,, \tag{2.15}$$

where $Q^{\swarrow} = Q > 0$ is the energy/heat absorbed by the system $\mathcal{S}$ from its surroundings $\mathcal{N}$, and $W^{\nearrow} = -W < 0$ represents the work done by $\mathcal{S}$ on $\mathcal{N}$, see Fig. 2.4. This is consistent with the following:

**Sign convention on $w$**

The work $w$ will follow the following sign conventions:

$$\begin{aligned} w^{\swarrow} &= w > 0: \quad \mathcal{N} \textit{ does work on } \mathcal{S}\,, \textit{ and} \\ w^{\nearrow} &= -w < 0: \quad \mathcal{S} \textit{ does work on } \mathcal{N}\,, \end{aligned} \tag{2.16}$$

where $w = W/m$ denotes the work per unit mass.[a]

*a* We silently apply the convention of using small letters for quantities being "per unit mass" in thermodynamic contexts.

The analysis of thermodynamic systems historically relies on the most important state, where a so-called *equilibrium* is reached or admitted, and which therefore is called *thermodynamic equilibrium*, see Definition 2.1.6.

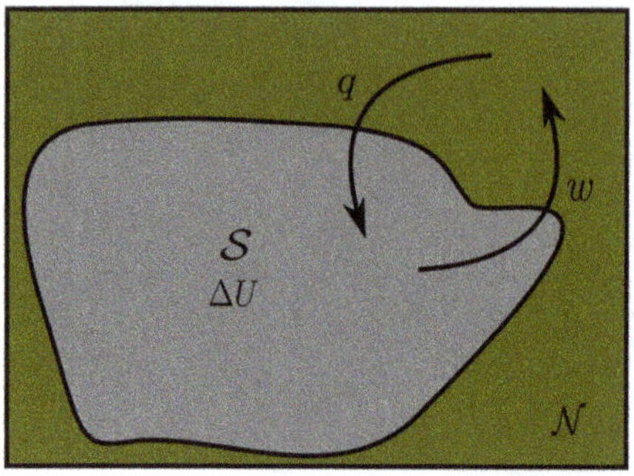


Figure 2.4: The first law of thermodynamics and the change in internal energy $\Delta U$ for influx of heat $q$ and work $w$ done on the surroundings $\mathcal{N}$. Note the sign convention defined in equation (2.16): $W^{\nearrow} = -W < 0$ holds, if the system $\mathcal{S}$ does work on the surroundings $\mathcal{N}$, which represents the scenario depicted here.

Finally, we recall often used terms of fundamental thermodynamic processes:

**Adiabatic process**

**Definition 2.1.16.** If a process happens under some form of *thermal isolation*, i. e., $Q = 0$, such that heat can neither escape nor enter the system $\mathcal{S}$, then we call this process *adiabatic*.

**Isothermal process**

**Definition 2.1.17.** If a process translates the energy/heat influx ($Q^{\swarrow} = Q > 0$) or out-flux ($-Q^{\nearrow} = -Q < 0$) into work done to the surroundings $\mathcal{N}$, i. e., $-W^{\nearrow} = -W < 0$, or on the system $\mathcal{S}$, i. e., $W^{\swarrow} = W > 0$, respectively, without changing the internal energy, that is, the *internal energy remains constant*, i. e., $\Delta U = 0$, then we refer to this as *an isothermal process*.

**Reversible process**

**Definition 2.1.18.** A thermodynamic process associated with a system $\mathcal{S}$ is said to be *reversible* if the system $\mathcal{S}$ *remains at equilibrium* during this process.

A more application-oriented motivation of a reversible process is that it does not show heat losses (energy dissipation) and hence the loss of the ability to do useful work on $\mathcal{N}$. In the context of the first law of thermodynamics, this amounts to $Q^{\nearrow} = Q^{\nearrow}_{\mathrm{rev}} = Q^{\nearrow}_{\mathrm{min}}$ and $W^{\nearrow} = W^{\nearrow}_{\mathrm{max}}$. This means that since the system is isolated (property of the first law), it has all the available energy (heat $Q$) expended to do useful work $W$ and hence reaches a state of minimal energy/heat $Q_{\mathrm{min}}$. Hence, the system has performed the maximum possible amount of work $W_{\mathrm{max}}$.

**Remark 2.1.19** (Reversible process). A careful consideration of Definition 2.1.18 immediately raises a concern about a possible contradiction, that is, how can there be a process for a system at equilibrium? This apparent contradiction highlights the fact that it is very unlikely (in fact, impossible with present means and understanding) to realize a reversible process in practice.

However, in theory, we can realize reversible processes by an infinitesimal deviation of the system $\mathcal{S}$ from equilibrium by an infinitesimal force (invisibly small distance away from equilibrium). Hence such processes are infinitesimally slow, see also Definition 2.1.11 of *local thermodynamic equilibrium.* ◇

#### 2.1.1.2 Entropy $S$: second law of thermodynamics

The consideration of and experimentation with thermodynamic systems inevitably leads to the general understanding that any attempt of entirely converting a certain amount of heat into (mechanical) work (see Section 2.1.2) will fail due to losses/dissipation of heat. Apart from steam and, more generally, heat engines, as the main driver of thermodynamics, there are also other scientific branches introducing entropy from slightly different angles, e. g., the molecular point of view by Boltzmann [21, 22] and a related gas theoretic point of view by Maxwell [106], or an information-theoretical view by Shannon [153]. Hence, let us state a formal definition of Boltzmann's and Maxwell's entropy concept.

**(Molecular/)Boltzmann Entropy**

**Definition 2.1.20.** Consider a system $\mathcal{S}$ that is in a specific state $\mathbf{s}$ consisting of independent state variables, and let $C(\mathbf{s})$ denote the number of ways the same macroscopic state $\mathbf{s}$ can be achieved, e. g., under different microscopic configurations. The so-called *Boltzmann entropy* $S = S_B$ then reads,

$$S := k_B \ln C(\mathbf{s}) , \tag{2.17}$$

where $k_B$ denotes the Boltzmann constant.

For a deductive approach to entropy and, in general, statistical mechanics, we refer to [123, 152]. Let us apply this definition to a very simple scenario.

**Example 2.1.21** (Boltzmann entropy of a puzzle). Consider a puzzle consisting of $N$ pieces. The correctly composed puzzle taking all $N$ necessary pieces into account is only realized in one unique way such that the *Boltzmann entropy* becomes,

$$S = k_B \ln C(\mathbf{s}) = k_B \ln 1 = 0 . \tag{2.18}$$

On the other hand, the Boltzmann entropy of the same puzzle but with an arbitrary single piece missing is as follows,

$$S = k_B \ln N , \tag{2.19}$$

since such a puzzle can be realized in $N$ different ways. ◇

In physics, the notion of entropy generally enters in the form of the *second law of thermodynamics*, where it plays the role of an arrow of time. That means, the second law states in simple words that in real systems the entropy increases while time passes.

**Entropy increases**

**Second law of thermodynamics.** Consider an isolated system $\mathcal{S}$ with possible irreversible processes. The system is characterized by a state function, which is always increasing due to the irreversibility.[ab] Such a state function is called entropy.

*a* This is equivalent to the following **Kelvin–Planck formulation**: *No heat engine(/not any device) can convert all the heat it receives(/operating on a cycle receiving heat from a single reservoir) to useful work(/net amount of work)*, see [34] and Definition 2.1.22.
In fact, this also implies that *no heat engine can have a thermal efficiency of 100 %.*
Note that the wording "in a cycle" refers to the fact that the engine remains unaltered by the process.

*b* There is a second equivalent formulation for refrigerators, the so-called **Clausius statement**, see [34].

**Entropy vanishes at 0 Kelvin**

**Third law of thermodynamics.** Consider arbitrary two large equilibrium systems while reducing the temperature to absolute zero ($\Theta = 0$ K). Under these circumstances, each of these systems will approach the same entropy value ($S = 0$).

### 2.1.2 Efficiency: heat engines and generalizations

Let us first introduce the fundamental concept of a *heat engine*, which not only played a fundamental role in the industrial revolution (steam engine) but also in establishing thermodynamics as a powerful macroscopic and scientific framework.

**Heat engine**

**Definition 2.1.22.** A (thermodynamic) system $\mathcal{S}_E$, that receives heat $Q_H^{\swarrow}$ from a hot reservoir/surroundings $\mathcal{N}_H$, performs work $W^{\nearrow}$ to the surroundings $\mathcal{N}_w$, and releases heat $Q_C^{\nearrow}$ to a cold reservoir/surroundings $\mathcal{N}_C$, is called a *heat engine*, see Fig. 2.5.

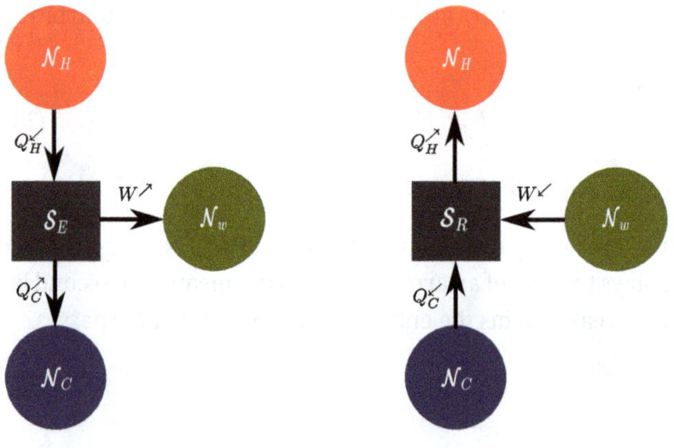


Figure 2.5: *Left:* Heat engine $\mathcal{S}_E$. *Right:* Refrigerator $\mathcal{S}_R$. Hence, refrigerators are just heat engines running in inverse mode.

We note that heat engines generally operate in cycles. However, the same term is also frequently used for car engines, which are internal combustion engines in difference to steam engines etc. representing external combustion engines. An ideal and important class of heat engines is characterized as follows.

**Carnot engine**

**Definition 2.1.23.** A reversible heat engine that operates only between two heat reservoirs is called a *Carnot engine*.

A major interest in heat engines is to elucidate what are the limitations on the efficiency. To this end, we can immediately state the following canonical definition.

**Efficiency of heat engines**

**Definition 2.1.24.** For heat engines according to Definition 2.1.22, we can canonically define their (thermodynamic) efficiency by,

$$\eta := \frac{W^{\nearrow}}{Q_H^{\swarrow}}, \tag{2.20}$$

where $W^{\nearrow}$ is the work done to the surroundings $\mathcal{N}_w$, and $Q_H^{\swarrow}$ is the heat received from the hot reservoir.

All kinds of heat engines, independently whether they operate in an ir- or reversible manner, do admit the following efficiency relation,

$$\eta = 1 - \frac{Q_L}{Q_H}, \tag{2.21}$$

where $Q_L$ and $Q_H$ denote the heat energies transferred to a lower-temperature reservoir and from a higher-temperature reservoir, respectively, see Fig. 2.5 with corresponding $Q_C$ (for cooler) and $Q_H$ (for hotter) reservoirs. For the class of Carnot engines, we can prove the following statement [33].

**Carnot's Theorem I. (Optimality)**

**Theorem 2.1.25.** *A Carnot heat engine (see Definition 2.1.23) shows an optimal efficiency $\eta_{\text{rev}}$, that is,*

$$\eta_{\text{irr}} \leq \eta_{\text{rev}} = 1 - \frac{T_L}{T_H}, \tag{2.22}$$

*where $T_L$ and $T_H$ are the temperatures of released, unused, lower-temperature heat $Q_L$ and the higher-temperature heat $Q_H$ consumed from the external heat reservoir, i. e., $T_L < T_H$.*

*Proof.* Our strategy to establish inequality (2.22) is by contradiction, which will imply a violation of Kelvin's statement of the second law of thermodynamics. Therefore, we assume that,

$$\frac{W_{\text{rev}}}{Q_H} = \eta_{\text{rev}} < \eta_{\text{irr}} = \frac{W_{\text{irr}}}{Q_H}, \tag{2.23}$$

so that $W_{\mathrm{irr}} > W_{\mathrm{rev}}$. According to this constellation, we have two heat engines, a reversible one and an irreversible one, which are denoted by $E_{\mathrm{rev}}$ and $E_{\mathrm{irr}}$, respectively. Now we reverse the mode of operation of the reversible heat engine $E_{\mathrm{rev}}$, which performs the work $W_{\mathrm{rev}}$ to the surroundings while consuming the heat $Q_H$ and releasing the lower amount of heat $Q_{\mathrm{L,rev}}$ to the surroundings. Hence, it becomes a refrigerator $R$ that consumes work $W_{\mathrm{rev}}$ while taking up the lower-temperature heat $Q_{\mathrm{L,rev}}$ and releasing the higher-temperature heat $Q_{\mathrm{H}}$ to the environment. Adding this refrigerator $R$ now to our irreversible heat engine $E_{\mathrm{irr}}$, it performs the work $W_{\mathrm{irr}}(> W_{\mathrm{rev}})$ to the environment while consuming the heat $Q_H$ from the heat reservoir at temperature $T_H$ and releasing the lower amount of heat $Q_{\mathrm{L,irr}}$ to the lower-temperature heat reservoir at temperature $T_L$. This leads to a heat engine $E$ performing the net work $W_{\mathrm{irr}} - W_{\mathrm{rev}} > 0$ to the surroundings while consuming the heat $Q_L$ from the single heat bath at lower temperature $T_L$. However, this implies that we have obtained a heat engine performing positive work $W_{\mathrm{irr}} - W_{\mathrm{rev}} > 0$ while only being connected to a single heat reservoir. This violates the second law of thermodynamics, Kelvin's statement. □

The readers interested in rigorous investigations of thermodynamic efficiency and heat engines (e. g., Carnot cycle) are referred to [166], for instance.

**Carnot's Theorem II. (Uniqueness)**
**Theorem 2.1.26.** *Heat engines that are reversible and operate between the same two heat reservoirs show the same thermodynamic efficiency.*

The subsequent proof relies on the same strategies as for establishing Carnot's Theorem I, Theorem 2.1.25.

*Proof.* By contradiction, we assume without loss of generality that the efficiency of one of the two reversible heat engines is smaller, i. e.,

$$\frac{W_1}{Q_H} = \eta_1 < \eta_2 = \frac{W_2}{Q_H}. \tag{2.24}$$

Therefore, we have $0 < W_1 < W_2$. Note that $Q_H$ is the heat influx into the two engines $E_1$ and $E_2$ from the reservoir at the higher temperature $T_H$. At the same time, heat engine $E_1$ releases the energy $Q_{1,L}$ to the joint reservoir at the lower temperature $T_L$. Now, we run the reversible heat engine $E_1$ in inverse mode as a refrigerator $R$ that consumes work $W_1$ and heat $Q_{1,L}$ from the heat reservoir at temperature $T_L$. Combining now the resulting refrigerator $R$ with the heat engine $E_2$ leads to a heat engine $E$, which produces the net amount of work $W_2 - W_1$, while consuming the heat $Q_{1,L} - Q_{2,L} > 0$ from a single reservoir and loosing a connection to a second reservoir. This implies the wanted contradiction to Kelvin's statement of the second law of thermodynamics. □

#### 2.1.2.1 Thermodynamic temperature scale and Clausius inequality

To obtain statements independent of the working medium used in heat engines, the following result on temperature scales plays a crucial role.

**Thermodynamic Temperature Scale**

**Proposition 2.1.27.** *Consider a reversible heat engine (Carnot heat engine) working in a cycle between two heat reservoirs of which one is at a high temperature $T_H$ and the other at a lower temperature $T_L$. A* thermodynamic temperature scale, *independent of the associated substance, can be established in the following way,*

$$Q_T = \phi(T)\,, \tag{2.25}$$

*where $Q_T$ stands for an amount of heat available from a heat reservoir at temperature $T$. The function $\phi$ is an arbitrary continuous function independent of the substance/medium of the heat reservoir at temperature $T$.*

*Proof.* The proof relies on the uniqueness of the efficiency of Carnot's heat engines as stated in Theorem 2.1.26. To this end, we consider two heat reservoirs, one being at a high temperature $T_H$ and the other having a lower temperature $T_L$. Between these two reservoirs, we have two reversible heat engines $E_1$ and $E_2$. Engine $E_1$ receives the heat $Q_H$ from the high-temperature reservoir and releases the heat $Q_A$, so that we arrive at an intermediate temperature $T_A$ such that $T_L < T_A < T_H$. Moreover, there is a second reversible heat engine $E_2$, which receives the heat $Q_A$ and releases the heat $Q_L$ to the reservoir at the lower temperature $T_L$.

According to the Carnot's uniqueness theorem (Theorem 2.1.26), these two heat engines can be combined to a single reversible heat engine $E$ having the same efficiency $\eta_E = 1 - \frac{Q_L}{Q_H}$ as $E_1$ and $E_2$ combined, since it is connected to the same two heat reservoirs at temperatures $T_H$ and $T_L$. That is, engine $E$ receives the heat $Q_H$ and releases also the heat $Q_L$ to the lower-temperature reservoir.

Moreover, the uniqueness of efficiencies of heat engines connected to the same two heat baths allows us also to deduce that there exists a function $f(T_L, T_H)$ such that,

$$\frac{Q_L}{Q_H} = f(T_L, T_H)\,. \tag{2.26}$$

Moreover, that fact that we can combine the two reversible heat engines $E_1$ and $E_2$, which show the efficiencies $\eta_1 = 1 - \frac{Q_A}{Q_H}$ and $\eta_2 = 1 - \frac{Q_L}{Q_A}$ described above, allow us to write with the help of the function $f$ motivated in (2.26) the following relation,

$$f(T_L, T_H) = \frac{Q_L}{Q_H} = \frac{Q_A}{Q_H}\frac{Q_L}{Q_A} = f(T_A, T_H)f(T_L, T_A)\,. \tag{2.27}$$

This allows us to conclude that the functional form of $f(T_L, T_H)$ satisfies,

$$f(T_L, T_H) = \frac{\phi(T_L)}{\phi(T_H)} \tag{2.28}$$

for an (at least) continuous function $\phi$. □

**Clausius inequality.** Our previous discussion on heat engines allows us to establish the Clausius inequality with the help of a particular design involving a cyclic heat engine $\mathcal{E}_C$ for which the lower-temperature heat bath is replaced with a purely mechanical work-producing system $\mathcal{S}_W$.

In the next section, we will motivate how the concept of heat engines allows us to establish well-known thermodynamic characterizations and properties such as the Clausius inequality.

### 2.1.3 Fluctuations

Like an arbitrary system, CHeSs can also be investigated at different length and time scales. Generally, we speak of three essential length scales, that is, the micro-, meso-, and the macroscale, see the beginning of Section 2.4. In this section, we discuss the validity of the so-called *thermodynamic limit* with the help of *fluctuations*. In fact, we will derive here the key property of fluctuations required to establish this *thermodynamic limit*, which allows us to accept a physical/thermodynamic property established on the microscale to be also valid on the macroscale.

#### 2.1.3.1 Canonical ensemble

Instead of first defining the *microcanonical ensemble* despite its theoretical basis for other ensembles such as *canonical* and *grand canonical*, we immediately introduce the practically relevant *canonical ensemble*, since temperature can be more easily controlled in experiments than internal energy being part of the macroscopic state $(U, V, N)$ of the microcanonical ensemble. Therefore, we recall that the *canonical ensemble* of a macroscopic state $(\Theta, V, N)$ of a system of interest consists of a large number of (hypothetical) identical thermodynamic systems showing the same macroscopic state $(\Theta, V, N)$, see [36]. This allows us to characterize the canonical ensemble with the help of the probability of a microstate being visited in a system at constant temperature $\Theta$, that is,

$$p_m := \frac{\exp(-\beta E_m)}{Q(\Theta, V, N)}, \tag{2.29}$$

where $\beta := \frac{1}{k_B \Theta}$ is the usual inverse thermal energy parameter with the Boltzmann constant $k_B$. The index $m$ in (2.29) refers to the microstate, and $Q$ is the *canonical partition function*,

$$Q(\Theta, V, N) := \sum_{\text{all } n \text{ at } V,N} \exp(-\beta E_n), \tag{2.30}$$

which is related to the *Helmholtz free energy* $A$ as follows,

$$A = -k_B \ln Q. \tag{2.31}$$

#### 2.1.3.2 Averages/expectation of a quantity

Let us consider a quantity $\chi$ such as total, kinetic, or potential energy, or the distance between a specific pair of atoms. The *canonical ensemble average* of a certain quantity $\chi$ at a specific temperature is given by,

$$\mathbb{E}[\chi] := \sum_{\text{all } n \text{ at } V,N} \chi_n p_n\,, \tag{2.32}$$

where $\chi_n$ denotes the value of the quantity $\chi$ in the microstate $n$. Generally, the average $\mathbb{E}[\chi]$ is denoted by $\langle\chi\rangle$ in the related literature on statistical mechanics.

#### 2.1.3.3 Distribution of a quantity $\chi$

The probabilities $p_m$ associated with microcanonical states $m$, as defined in Section 2.1.3.1, allow us to immediately give an expression of the distribution of a quantity $\chi$,

$$p[\chi] := \sum_n p_n \delta_{\chi_n,\chi}\,. \tag{2.33}$$

Recall that the delta distribution $\delta_{x,y} := \delta(x-y)$ is defined such that for a continuous integrable function $f(x)$ we have that,

$$\int_{-\infty}^{\infty} f(x)\delta(x-y)\,dx = f(y)\,. \tag{2.34}$$

Hence, it provides a way to go from a discrete probability distribution over to a continuous probability distribution. Moreover, $\delta$ is generally referred to as the Dirac $\delta$-function.

#### 2.1.3.4 Density of states and entropy

A quantity that accounts for degeneracies of microstates having the same energy is the so-called *density of states*,

$$\Omega(E,V,N) = \sum_{\text{all } n \text{ at } V,N} \delta_{E_n,E}\,. \tag{2.35}$$

Note that (2.35) represents a counting measure by counting the number of microstates with energy $E$. Quantity (2.35) immediately gives the Boltzmann entropy,[3]

$$S(E,V,N) := k_B \ln \Omega(E,V,N)\,. \tag{2.36}$$

3 This elegant entropy definition goes back to Planck [124] but has been previously suggested by Boltzmann in a less compact form.

#### 2.1.3.5 Distribution of energy levels

Using the expressions for probability (2.29) and partition function (2.30) of a *canonical ensemble* allows us to write the distribution of energies with (2.35) as,

$$p[E] = \frac{\Omega(E, V, N)\exp(-\beta E)}{Q(\Theta, V, N)}, \tag{2.37}$$

which is a consequence of setting $\chi = E$ in Section 2.1.3.3,

$$\begin{aligned} p[E] &= \sum_n p_n \delta_{E_n, E} \\ &= \sum_n \frac{\exp(-\beta E_n)}{Q(\Theta, V, N)} \delta_{E_n, E} \\ &= \frac{\exp(-\beta E_n)}{Q(\Theta, V, N)} \sum_n \delta_{E_n, E} \\ &= \frac{\Omega(\Theta, V, N)\exp(-\beta E)}{Q(\Theta, V, N)}, \end{aligned} \tag{2.38}$$

where $E_n$ is the energy associated with a microstate $n$. Finally, substituting $\Omega$ into (2.37) with the entropy relation (2.36) gives,

$$p[E] = \frac{\exp(S(E, V, N)/k_B - \beta E)}{Q(\Theta, V, N)}. \tag{2.39}$$

#### 2.1.3.6 Thermodynamic limit: vanishing energy fluctuations

Our goal is to justify the validity of the subsequently introduced thermodynamic limit (see Definition 2.4.3), which amounts to passing the particle number $N$ to infinity. We will see in Definition 2.4.3 that this limit is accepted under the assumption that energy fluctuations $\sigma_E^2 = \mathbb{E}[(E - \mathbb{E}[E])^2]$ relative to $\mathbb{E}[E]$ are negligible, i. e., they vanish with growing particle number $N$ as follows,

$$\frac{\sqrt{\sigma_E^2}}{\mathbb{E}[E]} = \frac{\sqrt{c_V k_B \Theta^2}}{\mathbb{E}[E]} \sim \frac{\sqrt{N}}{N} \sim \frac{1}{\sqrt{N}}, \tag{2.40}$$

where we used the definition of the constant-volume heat capacity (per particle $c_V$), that is, $C_V = Nc_V = (d\mathbb{E}[E]/d\Theta)_V$.[4]

To justify (2.40), let us first give meaning to,

$$\begin{aligned} \mathbb{E}[E] &= \sum_{\text{all } n \text{ at } V,N} E_n p_n = \sum_{\text{all } n \text{ at } V,N} E_n \frac{\exp(-\beta E_n)}{Q} \\ &= -\frac{\partial}{\partial \beta} \sum_{\text{all } n \text{ at } V,N} \frac{\exp(-\beta E_n)}{Q} = -\frac{1}{Q}\frac{\partial Q}{\partial \beta} = -\frac{\partial \ln Q}{\partial \beta}. \end{aligned} \tag{2.41}$$

4 Recall that $E, V, N, S$, etc. are extensive variables, and hence $c_V$ is also an extensive variable.

Similarly, looking at the constant-volume heat capacity per particle allows us to relate it to $\sigma_E^2 = \mathbb{E}[(E - \mathbb{E}[E])^2]$, i. e.,

$$
\begin{aligned}
Nc_V &= \left(\frac{d\mathbb{E}[E]}{d\Theta}\right)_V = \left(\frac{d\mathbb{E}[E]}{d\beta}\frac{d\beta}{d\Theta}\right)_V = -\frac{1}{k_B\Theta^2}\frac{d\mathbb{E}[E]}{d\beta} \\
&= -\frac{1}{k_B\Theta^2}\frac{d}{d\beta}\left(\sum_{\text{all } n \text{ at } V,N} E_n \frac{\exp(-\beta E_n)}{Q}\right) \\
&= -\left(\frac{\sum_{\text{all } n \text{ at } V,N} E_n^2 \exp(-\beta E_n) Q + \frac{dQ}{d\beta}\sum_{\text{all } n \text{ at } V,N} E_n \exp(-\beta E_n)}{k_B\Theta^2 Q^2}\right) \\
&= \frac{1}{k_B\Theta^2}\left(\sum_{\text{all } n \text{ at } V,N} E_n^2 \frac{\exp(-\beta E_n)}{Q} - \mathbb{E}[E]\sum_{\text{all } n \text{ at } V,N} E_n \frac{\exp(-\beta E_n)}{Q}\right) \\
&= \frac{1}{k_B\Theta^2}\left(\mathbb{E}[E^2] - [E]^2\right) = \frac{\sigma_E^2}{k_B\Theta^2}\,.
\end{aligned}
\tag{2.42}
$$

Finally, we note that based on the latter equation and the definition of the heat capacity $c_V$, we immediately see that $\mathbb{E}[E] \sim N$, so that (2.40) follows.

## 2.2 Complexity

A general observation about the usage of "complex" or "complexity" is that it is often only loosely connected to adjectives such as *difficult, highly interconnected and dependent*, and *going beyond intuition*. For example, the term *complexity* has been defined in [53] with the help of the broadly accessible and descriptive guidance provided by the Engineering and Physical Sciences Research Council (EPSRC) of the United Kingdom for when to label research as being complex. We quote it here again for the reader's convenience.

**Complexity (see [53, p. 5])**
**Emergent properties:** Properties at system level consist of interaction-induced cooperative emergence. Interacting components lead to hierarchical structures with different causations at different levels.

**Adaptation:** A multiple-component system evolves and adapts as a consequence of internal and external dynamic interactions. The system keeps becoming a different system, and the demarcation between the system and its surroundings evolves.

**Many levels:** Complexity science bridges the gap between the individual and the collective, for example, from psychology to sociology, from organism to ecosystems, from genes to protein networks, from atoms to materials, from the PC to the World Wide Web, and from citizens to society.

**Feedback to manipulation:** When a multiple-component system is manipulated, it reacts. The manipulator and the complex system inevitably become entangled, for example, as a farmer harvests what he sows and cultivates. Complexity research attempts to understand the sum of the multiple causes.

To investigate the complexity of a system of interest, we first need to understand the main processes that cause a certain appearance or dynamics of the system. Therefore, a natural starting point and the main interest are the different components building the fundamental structure of a system as well as their interactions and states.

As a consequence, our main goal is to motivate the importance of the search for a systematic definition of *complexity* showing enough generality to be applicable to arbitrary systems and allowing us to compare the complexity of different systems with the help of an associated meaningful measure, that is, a quantitative concept. Hence, the ultimate aim is to develop a *complexity concept together with a "universal" measure* being simple enough for applications. Based on the current literature, we consider the subsequent and intuitive graph-based formulation to be most promising toward a general definition that is able to classify different levels of complexity in a meaningful and rigorous way. Finally, we note that complexity is rather a relative concept in that it allows us at most to assess whether one system $\mathcal{S}_1$ is more complex than another system $\mathcal{S}_2$. However, we cannot hope to find a generally valid threshold $\kappa_\zeta > 0$ with respect to a *universal complexity measure* $\zeta$, above which any system $\mathcal{S}$ is referred to as being complex, i. e., any system $\mathcal{S}$ satisfying the inequality,

$$\zeta(\mathcal{S}) > \kappa\,. \tag{2.43}$$

Clearly, already these initial thoughts motivate an intuitive, graph-based description to capture the essential structural principles governing the behavior of systems. Along this line of thought goes a concept proposed by McCabe [107] in 1976, the *cyclomatic complexity*. Note that the book [53], cited at the beginning of this section, focuses on *networks* and motivates the fact that general networks/graphs are indeed complex. Before we can define cyclomatic complexity, we need to introduce fundamental terms such as the notions of "graph" and "component of a graph".

**Graph** *G*

**Definition 2.2.1.** For a set of nodes/vertices $V$ together with a set of edges/links $E$ connecting these nodes, we refer to the combination of these two sets as the graph $G = G(V, E)$. If edges/links $e_{ij} \in E$, that connect vertex/node $i$ with node $j$, do not show a particular direction, then $G$ is called an *undirected graph*. If the link set $E$ additionally contains information such as a direction or weight, then the resulting graph is called *directed* or *weighted*, respectively.

This allows us to immediately establish the second term.

**Component of a graph** *G*

**Definition 2.2.2.** Let $G$ be an undirected graph. A subgraph $C \subset G$, of which any two nodes/vertices are only connected to nodes/vertices of $C$ by paths/edges but to no nodes of $G \setminus C$, is called a *component of G*.

**Example 2.2.3** (Component of a graph). Figure (2.44) represents a graph with two components: the component consisting of the nodes/vertices from 1 to 6 and the second component showing the nodes from 7 to 8, i. e.,

$$\begin{array}{lcr} 1-2-3 & \quad & 7 \\ \quad / & & \backslash \quad \\ 4-5-6 & & 8 \end{array} \tag{2.44}$$

◇

### 2.2.1 McCabe's complexity: cyclomatic complexity

McCabe introduced in 1976 a graph-based concept in the context of software engineering. We believe that this provides a systematic basis for the development of a general complexity measure. Hence, we rely on McCabe's measure as follows.

**Cyclomatic Number**

**Definition 2.2.4.** The *cyclomatic number* of a *strongly connected graph* $G^a$ representing the essential components, i. e., vertices/nodes and interactions, that is, edges/links, describing a system $\mathcal{S}$ of interest, reads as follows,

$$c_{\text{cyc}}(G) := e - n + p\,, \tag{2.45}$$

where $e$, $n$, and $p$ denote the numbers of edges/links, nodes, and components of the graph of $\mathcal{S}$, respectively.

*a* Strongly connected means that the graph's starting point/entry is connected to the graph's exit point.

As already noted, Definition 2.2.4 has been originally introduced for quantitatively measuring the complexity of software. To this end, we identify basic/elementary code blocks of a program by nodes and branches, e. g., "IF-statements" become edges. In fact, equation (2.45) is valid for programs where the entry node is connected with the exit node (i. e., strongly connected). In case that a program shows a *directed graph*, the underlying code is a structured program that can be represented by a so-called *control-flow graph*. In this case, the cyclomatic complexity (2.45) needs to be refined to,

$$c_{\text{cyc}}(G) := e - n + 2p\,. \tag{2.46}$$

Obviously, the graph-based definition of complexity in Definition 2.2.4 allows us to apply it not only to software/algorithms, but also to general systems. To motivate this generalization, let us look at some examples.

**Example 2.2.5** (Cyclomatic number of scissors). Consider generic scissors as depicted in Fig. 2.6. Their graph $G$ associated to their main functional/processing, macroscopic components consists of a lower blade (1), an upper blade (2), and the joint (3) keeping the

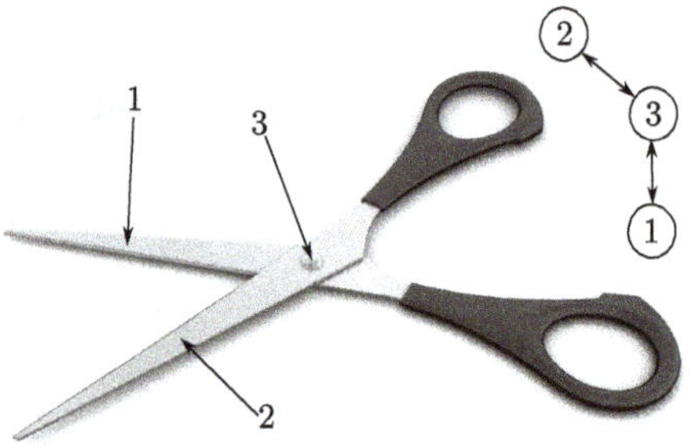


Figure 2.6: *Scissors. Left:* Image showing the key components/nodes. *Right:* representative graph with the number of edges/interactions being $e = 4$ and the number of nodes is $n = 3$. Finally, the graph consists of a single component, i. e., $p = 1$. Therefore, we have that $c_{cyc}(G) = 2$.

blades together and allowing them to move/rotate. The edges/links represent interactions between these components, and these are considered to be two links/edges accounting for the influence of each component on its connected/adjoint component. ◇

**Example 2.2.6** (Cyclomatic number of a kick-scooter). Consider a kick-scooter as depicted and analyzed in Figure 2.7. By restricting the scope to functional macroscopic components such as main frame, steering, breaks, and wheels, we end up with a cyclomatic complexity of $c_{cyc} = 13$. ◇

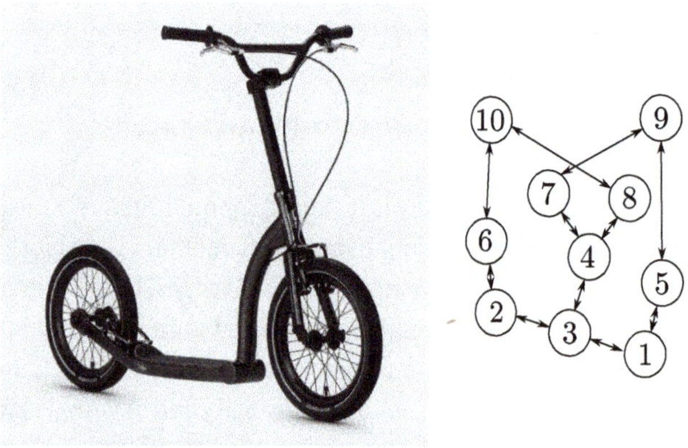


Figure 2.7: *Kick-Scooter. Left:* Image showing ten basic macroscopic elements/nodes: front wheel *(1)*, back wheel *(2)*, main frame *(3)*, steering part *(4)*, front brakes *(5)*, back brakes *(6)*, front brake handle *(7)*, back brake handle *(8)*, front brake cable *(9)*, and back brake cable *(10)*. *Right:* Graph $G$ associated with these main elements with resulting cyclomatic complexity $c_{cyc}(G) = e - n + p = 22 - 10 + 1 = 13$.

**Remark 2.2.7** (Complexity is a matter of scale). We note that the sensitivity for detail (i. e., the scale/level of resolution) plays a crucial role in the systematic, quantitative, and qualitative identification of the complexity of an observed CHeS. In fact, the previous Example 2.2.6 allows us either to even further increase the cyclomatic complexity by also taking the tire and the rim into account or to decrease it by neglecting the different parts of the brakes and replace them with a single node for the front and the rear. Therefore, *complexity* is naturally tight to the employed *level of observation*. ◇

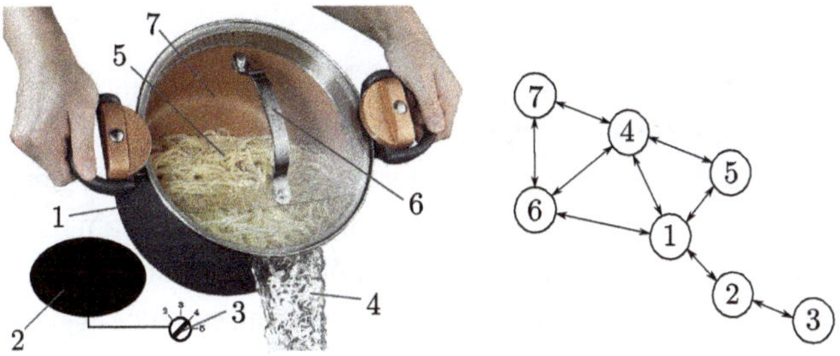


Figure 2.8: *Noodle Cooking. Left:* The main macroscopic elements such as pot *(1)*, stove *(2)*, temperature control *(3)*, water *(4)*, noodles *(5)*, pot lid *(6)*, and steam *(7)*. *Right:* Graph $G$ based on these main elements implies the complexity $c_{cyc}(G) = e - n + p = 18 - 7 + 1 = 12$.

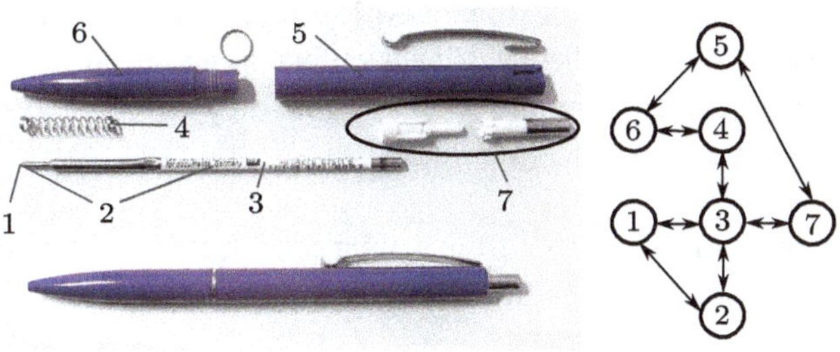


Figure 2.9: *Ballpoint Pen. Left:* The main macroscopic elements such as ballpoint *(1)*, ink *(2)*, ink storage *(3)*, spring *(4)*, upper casing *(5)*, lower casing *(6)*, and thrust button *(7)*. *Right:* Graph $G$ based on these main elements implies the complexity $c_{cyc}(G) = e - n + p = 15 - 7 + 1 = 9$.

**Example 2.2.8** (Cyclomatic number of noodle cooking). We consider the noodle cooking with all the elements as depicted in Figure 2.8. Interestingly, the noodle cooking shows with $c_{cyc}(G) = 12$ almost the same complexity as Example 2.2.6 of a kick-scooter. ◇

**Example 2.2.9** (Cyclomatic number of a ballpoint pen). Finally, we look at a ballpoint pen in Figure 2.9. Also, here it seems not immediately obvious that noodle cooking (cyclomatic complexity of 12) is more complex than a ballpoint pen (cyclomatic complexity of 9). ◇

A promising aspect of Definition 2.2.4 is that it provides the basis for systematic extensions such as assigning so-called interaction complexities to links/edges, e. g., based on the leading order term of the underlying interaction law and the number of interaction processes taken into account. This demonstrates again that complexity necessarily is tight to the level of detail applied in observations, see Remark 2.2.7. In fact, nodes in the graphs can consist themselves of multiple subsystems, for which each can show a different complexity. Such kind of multiscale aspects immediately imply the need of a systematic and reliable coarse-graining framework for accurately averaging *microscopic complexities* into an *effective macroscopic complexity* of the large-scale/macroscopic/coarse-grained system.

Interestingly, a major part of research in complexity science relies on the so-called concept of *Algorithmic Information Content (AIC)*. Therefore, we devote our next and last section to this topic.

## 2.3 Algorithmic Information Content (AIC)

In difference to the previous approach, this concept is based on the representation of a system $\mathcal{S}$ by a program running on a *Turing machine*. Clearly, this imposes certain constraints regarding the description of the system $\mathcal{S}$ of interest; e. g., we generally need to choose an encoding of the components comprising $\mathcal{S}$. Moreover, this also amounts to decide for a relevant observation scale and a model describing interactions between the components of $\mathcal{S}$, for instance. Hereby, the generality (and as a consequence also the applicability) is restricted, since "systems" depend on some form of the chosen approximation realized by a (computer) program running on a Turing machine.[5] Note that the concept of the Turing machine has also been extended to quantum computers. As a consequence, a complexity concept based on algorithmic principles naturally depends on the underlying computational approach applied and hence suffers from a potential loss of uniqueness. For instance, is it possible to uniquely and systematically assess the complexity of problems that currently show promise for quantum supremacy over classical computers?

Despite these concerns about AIC in the context of system complexity, we motivate that it is part of the important research on computational complexity, which allows us to classify the computational demand of algorithms executed on classical computers. In fact, the potential arrival of a quantum computer being able to solve real-world problems with quantum supremacy imposes some threats on classical encryption standards. In this context, computational complexity allows us to assess and estimate the level of risk imposed by this threat.

The definition of AIC relies on the *Turing machine*, and therefore we will describe it first.

### 2.3.1 The Turing machine

The basic form of a Turing machine consists of a *long enough tape*, which contains cells that either are *empty/blank*, i. e., showing the symbol "B", or contain an input sequence of bits, that is, the characters "0" or "1". There is a *head*, which can read and write to the cells and move to the left or right according to the instructions given by a set of rules. These rules represent a *program* for the Turing machine. Herewith, we immediately arrive at the following definition.

---

5 Note that it is exactly the subject of *Applied and Numerical Analysis* to rigorously, qualitatively and quantitatively analyze mathematical models of systems and their (computational) numerical approximations. In this context, uniqueness plays a central role.

**Basic operations**

**Definition 2.3.1.** A Turing machine is equipped with the following *basic operations*:

**(O1)** *Reading/writing from/to* a cell a symbol $a \in A$, where $A := \{0, 1, B\}$ is referred to as the alphabet/set. Reading the symbol $a = B$ finishes the program/computation.

**(O2)** *Moving the head to the left (L) or to the right (R)*, depending on the movement instruction $m \in M$ given, where the set $M := \{L, R\}$ contains all the possible movements.

These basic operations of Definition 2.3.1 are executed according to a program containing specific instructions for an action $(w, m) =: \alpha \in \mathcal{A} := A \times M$ depending on the read input $r \in A$ and the state $s \in S$, in which the system currently is. Let us collect these aspects of a rule in the following definition.

**Program of a Turing machine: rules/instructions**

**Definition 2.3.2.** A *program* of a Turing machine represents a map,

$$\begin{aligned} \mathcal{R} : S \times A &\to A \times M \times S\,, \\ (s_n, r_n) &\mapsto (w_{n+1}, m_{n+1}, s_{n+1})\,, \end{aligned} \tag{2.47}$$

that is, a Turing machine in processing step $n \in \mathbb{N}_{\geq 0}$ occupies the state $s_n \in S$ and reads the input $r_n \in A$ such that the pair $\zeta_n := (s_n, r_n) \in S \times A$ represents the input argument of the map $\mathcal{R}$ whose image $\mathcal{R}(s_n, r_n) = \rho_{n+1} := (w_{n+1}, m_{n+1}, s_{n+1}) \in A \times M \times S$ is the *rule/instruction* resulting from this input pair to occupy the subsequent system state $s_{n+1}$ at the processing step $n + 1$.

Note that such a program $\mathcal{R}$ of a Turing machine can be represented by a matrix/table with rows representing the current state $s_n \in S$ and columns representing the input $r_n \in A$ read by the head. Hence, the instructions $\rho_{n+1} = \mathcal{R}(\zeta_n)$ are the components of the matrix associated with the specific pairs $\zeta_n = (s_n, r_n)$. Let us visualize this in the following example.

**Example 2.3.3** (Program of a Turing machine: rules/instructions). Let us consider a Turing machine loaded with the program $\mathcal{R}$ as defined in Table 2.1. The operation of this program is depicted in Figure 2.10. ◇

| $s_n \backslash r_n$ | 0 | 1 | B |
|---|---|---|---|
| 0 | 1R1 | 0L2 | END |
| 1 | 1L0 | 1R2 | END |
| 2 | 1L0 | 1L1 | END |

Table 2.1: A program $\mathcal{R}$ that can be used with a Turing machine.

### 2.3.2 Definition of AIC

As noted in the introduction, the AIC relies on the fundamental concept of computation as provided by the Turing machine introduced in the previous Section 2.3.1. Herewith, we are equipped to define the well-established AIC based on a string of symbols $s$, e. g.,

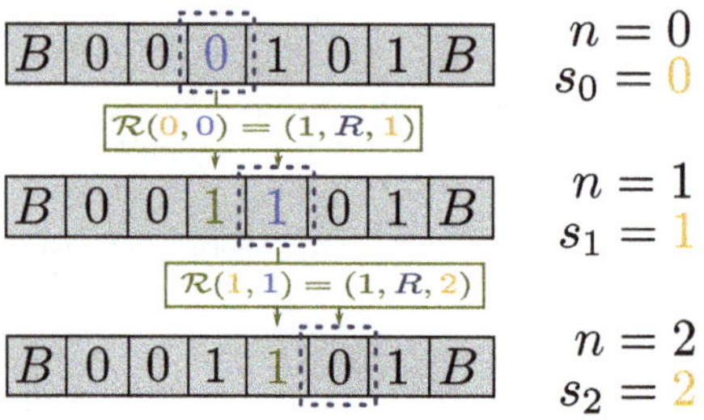


Figure 2.10: Operation of a Turing machine based on the program given in Table 2.1.

containing a sequence of characters available in the form of an alphabet $A$. That is, $s = a_1a_2a_3 \dots a_\ell$ is a string of length $\ell$ with symbols/characters $a_i \in A$.

**Algorithmic Information Content (AIC)**

**Definition 2.3.4.** The *Algorithmic Information Content (AIC)* of a symbolic string $s$ is defined by,

$$K(s) := \min\big\{|\mathcal{R}| \,\big|\, s = C_T(\mathcal{R})\big\}, \tag{2.48}$$

where $\mathcal{R}$ denotes a program of a Turing machine (see Definition 2.3.2), and $|\mathcal{R}|$ is its length. Finally, $C_T(\mathcal{R})$ represents the result obtained after running the program on a Turing machine.

Since the AIC has been frequently recommended as a quantitative measure for identifying the complexity of a system, we collect different aspects about AIC in the following remark.

**Remark 2.3.5** (AIC as a complexity measure). **i.** We note that the AIC defined in (2.48) is only exact for strings of infinite length.

**ii.** We can see that the AIC admits the maximal value $K_{\text{max}} = |s|$ if the string $s$ is free from correlations, i. e., $s$ is entirely random.

**iii.** A major drawback of AIC to describe complexity is based on the fact that it increases monotonically from a perfectly regular sequence to a complete random sequence, as also motivated in [62]. In fact, Gell-Mann and Lloyd [62] also connect complexity to Maxwell's demon and its extension towards Szilard's engine (which also relates to the scale/detail of observation, see Remark 2.2.7) and propose the so-called *Effective Complexity (EC)*, which should not be confused with the multiscale-based one proposed at the end of Section 2.2.1. Indeed, EC also shares the same deficiencies as AIC by depending on a chosen/identified encoding. ◇

We believe that this short introduction into *complexity* demonstrates the dividing views on the subject and the fact that we are still far from having a well-accepted and general understanding of how to rigorously establish the complexity of systems.

## 2.4 Networks and graphs: from the theory of systems to CHeSs

From a more historical perspective, it is a natural approach to consider CHeSs under a broad thermodynamic scope. For instance, thermodynamics very much relies on the concept of a system (e. g., isolated, closed, open), its observation (in the sense of measurements), and its control (e. g., by maintaining a certain system state). The systematic and formal thermodynamic theory developed during more than a century makes it a natural and fundamental pillar for the definition and description of general systems. The fact that elementary interactions with a system, e. g., measurement and control, imply an exchange and processing of information, whose concept seems to go back to Maxwell's demon and Szilard's engine (see Section 2.5), further motivates a thermodynamic point of view on systems.

However, in a first attempt to consistently define *CHeSs*, we need a well-developed theory for describing complexity of general systems. Unfortunately, a careful search for such a theory currently leads to the surprising discovery that despite the wide spread use of the term *complexity*, a unique and generally applicable definition seems still to be out of reach at this time. In fact, as we have seen in Section 2.2, the core of research seems to rely on a system's algorithmic/computational realization for the quantification of its complexity. However, already the fact the we are able to define not just one but multiple Turing machines, e. g., a classical and a quantum Turing machine (see [167, 41]), motivates that such an approach to complexity lacks uniqueness. This means that any new type or class of information processing machine will make it necessary to reassess and adapt any notion of computational complexity.

On this occasion, we note that Deutsch also indicates deficiencies in the definition of *complexity* motivated by his development of the *quantum Turing machine* [41]. A deeper study of the Church–Turing hypothesis[6] along these lines demonstrates that the *uniqueness* of a *system's simulation* is not explicitly stated but necessary for a unique quantification by a specific *complexity measure*. As a consequence, any argument about complexity needs to rigorously establish the uniqueness. Additionally to proving the uniqueness, we also note that any notion about complexity of a system needs to explicitly discuss the level of detail, that is, the level of scale (e. g., in time, space, etc.) for which the system is of interest.[7] Otherwise, for highly heterogeneous CHeS, we can reduce complexity by approaches such as *Upscaling* or *Coarse Graining*, see Sections 6 and 10. Moreover, the computational/algorithmic complexity takes a computational modeling perspective, which, by its general and crucial dependence on physical and numerical approximations of a system of interest, makes it difficult to assess differences between

6 Every finitely realizable physical system can be perfectly simulated by a universal model computing machine operating by finite means.

7 This fact about the matter of scale is also directly related to the choice of the system's encoding and the model/description of its interacting components necessary for its "simulation" on a Turing machine.

models and their numerical schemes applied next to the observation level/scale applied, e. g., micro-, meso-, or macroscale.

In Section 2.2, we therefore investigated important developments and achievements for identifying complexity. Here, we will define CHeSs with a general structure (i. e., network/graph) allowing us to compare various CHeSs by a systematic, intuitive, and qualitative *complexity measure*, that is, the cyclomatic complexity introduced in Definition 2.2.4. In fact, once a system such as CHeS has been defined, the applications and interest with respect to this system might not be restricted to thermodynamic and information related aspects. For instance, the geometry of CHeSs, which is defined by its defining subsystems and their arrangement, can be investigated from a purely structural point of view without any direct links to thermodynamics and information theory.

Using Definition 2.1.1 of a single system, we are able to look at the next level, that is, the composition of various single systems into a highly complex combination.

**Complex Heterogeneous System (CHeS)**

**Definition 2.4.1.** Let $\mathcal{S}_i \subset \mathcal{M}$, $1 \leq i \leq I$, be $I \in \mathbb{N}$ (thermodynamic) systems of a universe $\mathcal{M}$ according to Definition 2.1.1. We can compose these $I$ systems by linking them via a so-called *adjacency matrix*,

$$A := (a_{ij})_{i,j}^{I,I} , \tag{2.49}$$

where

$$a_{ij} := \begin{cases} 0, & \text{if there is no link/edge from system } i \text{ to } j , \\ 1, & \text{if there is a link/edge from system } i \text{ to } j . \end{cases} \tag{2.50}$$

The resulting graph $G = (V, E)$, which consists of the vertex set $V$, that is, the set of systems $\mathcal{S}_i$, $1 \leq i \leq I$, and of the set of links/edges $E$ whose elements read $(i, j)$ for $1 \leq i, j \leq I$, is referred to as a *Complex Heterogeneous System (CHeS)*. Note that if the adjacency matrix is symmetric, then the graph representing the CHeS is *undirected*.

**Remark 2.4.2** (Components $\mathcal{S}_i$ of CHeSs). In general, the single system components $\mathcal{S}_i$, $1 \leq i \leq I$, of CHeSs can represent very primitive elements/objects with their only ability of admitting different states such as variations in density or a discrete and finite number of integer values representing a thermodynamic phase (e. g., solid, liquid, or gas) for instance. In fact, the $\mathcal{S}_i$ can refer to single atoms for which the state of interest could be the corresponding spins, number of electrons (in a specific shell), etc.

In specific examples such as electrochemical systems, it immediately becomes apparent that the identification of these elementary systems $\mathcal{S}_i$ is a matter of scale, see also Remark 2.2.7. This means that a CHeS depends on whether we look for a detailed microscopic description or for a rather averaged macroscopic formulation of the underlying microscopic electrochemical processes. For instance, we can consider a single electrode such as the cathode as an elementary system $\mathcal{S}_i$ or we can zoom into the electrode and refer to the different phases/chemical compounds appearing as basic systems $\mathcal{S}_i$. ◇

This generalized system view of CHeSs formulated in Definition 2.4.1 takes multiple interacting systems into account and accounts for the increasing complexity and the growing connectivity of arbitrary types of systems such as in various scientific, engineering, and industrial applications, e. g., Internet of Things (IoT). Its purpose is to elaborate, collect, and aggregate crucial and key scientific tools and methods for investigating and establishing a reliable and systematic understanding of CHeSs, in particular, due to their continuously increasing scope and range of applications. An example of a CHeS, which was the original driver for the much more general point of view in this book, is *electrochemical systems*, e. g., *primary* (nonrechargeable) and *secondary* (rechargeable) batteries.

Before we consider specific applications that represent CHeSs, let us first come back to the fundamental pillars such as thermodynamics providing a systematic approach to describe and understand the elementary unit of CHeSs, i. e., a single system. A motivating aspect of thermodynamics and its laws is that it can be described by the (fundamental)[8] theory of *Statistical Mechanics*. Since a proper understanding of microscopic interactions/chemical processes (the basis of Statistical Mechanics) in electrochemical systems is still out of reach, the current research crucially depends on the experimental search on various length and time scales for finding materials/chemicals with a low price and allowing for best performances in their optimal composition. Hence, thermodynamics provides us with the necessary macroscopic/measurable guidance to optimize battery design and composition by describing complex systems by establishing balance and transfer relations between them and the surroundings in the form of mathematical equations.

Generally, a many-body/multiparticle system is considered to be macroscopic, if we are observing a volume element $Y \subset \mathbb{R}^3$ of the size of $1\,\mathrm{cm}^3$ containing $1\,\mathtt{mole} = 6.022 \cdot 10^{23}$ particles, which represents the Avogadro constant $N_A = 6.022 \cdot 10^{23}\,\mathtt{mole}^{-1}$. The continuously increasing ability to observe processes/interactions on the *nanoscale*, motivates the search for possible novel, scale-dependent physical principles, which ultimately might allow for novel industrial applications. Thermodynamic systems at the nanoscale are frequently referred to as *mesoscopic systems*. Therefore, we have the following *classification of thermodynamic systems*:

1. **Microscopic systems** show very small numbers of atoms/particles leading to a system where temperatures are independent of particle numbers.
2. **Mesoscopic systems:** the thermodynamic properties(/internal structure) do not depend on particle/atom number and their structure/geometrical arrangement, despite the higher particle/atom number compared to microscopic systems.

8 In fact, Statistical Mechanics allows us to explain Thermodynamics.

3. **Macroscopic systems** have a very large atom/particle number such that their thermodynamic properties are independent of both particle number and particle locations/geometrical arrangements.[9]

According to [152], for real/practical systems, the applicability of the term *macroscopic* needs to be justified by establishing the *smallness* of *fluctuations* $\Delta\chi$ in the macroscopic quantity of interest. Let $\chi$ be such a quantity. Then, this requirement means that,

$$\frac{\langle\Delta\chi\rangle}{\langle\chi\rangle} = \frac{\sqrt{\sigma_\chi^2}}{\mathbb{E}[\chi]} \sim \frac{1}{\sqrt{N}} \approx 10^{-12}, \tag{2.51}$$

where $\langle\cdot,\cdot\rangle$ denotes some form of average, e. g., the average over a *canonical ensemble* (see Section 2.1.3.1 for more detail), and $\mathbb{E}$ is the expectation with respect to the probability associated with the canonical ensemble, see Section 2.1.3 and in particular Section 2.1.3.6 on *energy fluctuations*. A rigorous microscopic justification of thermodynamics based on statistical mechanics relies on the following definition.

**Thermodynamic limit**
**Definition 2.4.3.** The process of passing to the limit of infinite atom/particle number $N$, i. e., $N \to \infty$, is generally referred to as the *thermodynamic limit*. The validity of this limiting property is considered to be justified under the asymptotic behavior $\frac{1}{\sqrt{N}}$ of relative fluctuations such as energy, i. e., for $\chi = E$ in (2.51).

## 2.5 From thermodynamics to information

There are at least two historical concepts that motivate us to elaborate a link between "thermodynamics" and "information theory", that is,

(i) *Maxwell's demon* and subsequent developments such as Szilard's engine followed by Brioullin's measurement argument, Landaur's "energy for erasure"-principle, and Bennet's "no work for measurement but work for erasure"-argument, and
(ii) *thermodynamic entropy* versus *information entropy*.

Motivated by the physical investigation of mechanical systems, we introduce a so-called phase space $\mathcal{Z}_S$, which represents all the possible states $z \in \mathcal{Z}_S$ that can be admitted/realized by a particular system $S$. The likelihood of observing a specific state $z_k \in \mathcal{Z}_S$, where $k \in \mathbb{N}$ with $1 \leq k \leq |\mathcal{Z}_S|$, and $|\mathcal{Z}_S|$ denotes the number of possible/realizable states, is described by the following probabilities,

9 There are no clear limits allowing to clearly separate these three categories except for the concept of *thermodynamic limit* (increasing the system size by keeping the density constant). Hence, thermodynamic properties do still change while going from a *mesoscopic* to a *macroscopic system*.

$$\mathbb{P}[z_k] = \begin{cases} p(z_k)\ \textit{discrete probability}, & \text{if } \mathcal{S} \text{ is a discrete system},\\ f(z_k)\ \textit{continuous probability density}, & \text{if } \mathcal{S} \text{ is a continuous system}. \end{cases} \tag{2.52}$$

This allows us to define the *entropy* as follows.

**Information content and entropy**

**Definition 2.5.1.** Consider a system $\mathcal{S}$ with phase space $\mathcal{Z}_{\mathcal{S}}$ to which belongs the associated probability space $(\Omega, \mathcal{F}, \mathbb{P})$. The information content of a system $\mathcal{S}$ of a specific state realization $z_k \in \mathcal{Z}_{\mathcal{S}}$ reads as follows,

$$I[z_k] := \begin{cases} -\ln \frac{1}{p(z_k)}, & \text{if } \mathcal{S} \text{ is a discrete system},\\ -\ln \frac{1}{f(z_k)}, & \text{if } \mathcal{S} \text{ is a continuous system}, \end{cases} \tag{2.53}$$

where $p(z_k)$ and $f(z_k)$ are discrete and continuous probability densities, respectively. The so-called (Shannon) *information entropy* $S_I$ is defined as the *expected information content of a system* $\mathcal{S}$, that is,

$$S_I[\mathcal{S}] := \mathbb{E}\big[I[z]\big], \tag{2.54}$$

for $z \in \mathcal{Z}_{\mathcal{S}}$. ◇

We can also assert an information entropy $S_I[\boldsymbol{\mathcal{S}}_n]$ to a combination/collection of $n$ systems, that is, to $\boldsymbol{\mathcal{S}}_n := \mathcal{S}_1 \times \mathcal{S}_2 \times \cdots \times \mathcal{S}_n$. To this end, the likelihood for the combined $n$ systems $\boldsymbol{\mathcal{S}}_n$ to be in a particular state $\mathbf{z}_k \in \boldsymbol{\mathcal{S}}_n$, $1 \le k \le |\mathcal{S}_n|$, is described by a joint probability distribution,

$$\mathbb{P}[\mathbf{z}_k] = \mathbb{P}[z_1^k, z_2^k, \ldots, z_n^k], \tag{2.55}$$

which induces $n$ marginal distributions,

$$\mathbb{P}_i[z_i] := \begin{cases} \sum_{\mathring{\mathbf{z}} \in \boldsymbol{\mathcal{S}} \setminus \mathcal{S}_i} p(z_i, \mathring{\mathbf{z}}), & \text{if } \boldsymbol{\mathcal{S}}_n \text{ is discrete},\\ \int_{\boldsymbol{\mathcal{S}} \setminus \mathcal{S}_i} f(z_i, \mathring{\mathbf{z}})\, d\mathring{\mathbf{z}}, & \text{if } \boldsymbol{\mathcal{S}}_n \text{ is continuous}, \end{cases} \tag{2.56}$$

for $z_i \in \mathcal{S}_i$, where $p$ is a discrete probability, and $f$ is a continuous probability distribution. As a consequence, the definition of (Shannon's) information entropy, see Definition 2.5.1, can be immediately extended to a combination of $n$ systems with the help of probabilities (2.55)–(2.56).

Moreover, with the help of the marginal probability distributions (2.56), it can be immediately verified that the (Shannon) information entropy is subadditive, i. e.,

$$S_I[\boldsymbol{\mathcal{S}}_n] \le \sum_{i=1}^{n} S_I[\mathcal{S}_i]. \tag{2.57}$$

### 2.5.1 Maxwell's demon

Consider a classical ideal gas contained in a box without energy or mass exchanges.[10] Moreover, the gas is in thermodynamic equilibrium at constant temperature $\theta > 0$ (Kelvin scale). The so-called demon appears in the form of a feedback control mechanism that relies on the following operations:

**(M1)** *keeping particles separated by a barrier dividing the box into two equally sized parts after thermal equilibrium has been reached in the original box;*

**(M2)** *measuring the speed of gas particles; the demon places particles from the right box with kinetic energy higher than the equilibrium thermal energy into the left part and takes the ones in the left part with lower than the equilibrium energy into the right part.*

Clearly, the demon induces a temperature gradient between the left and right compartments with operations **(M1)** and **(M2)**, see Fig. 2.11. As a consequence, the entropy of the gas is decreased despite the lack of mass and energy exchanges with the surroundings. This clearly implies a contradiction to the *second law of thermodynamics* without a suitable consideration of the demon's role. This contradiction is generally referred to as the *paradox of Maxwell's demon.*

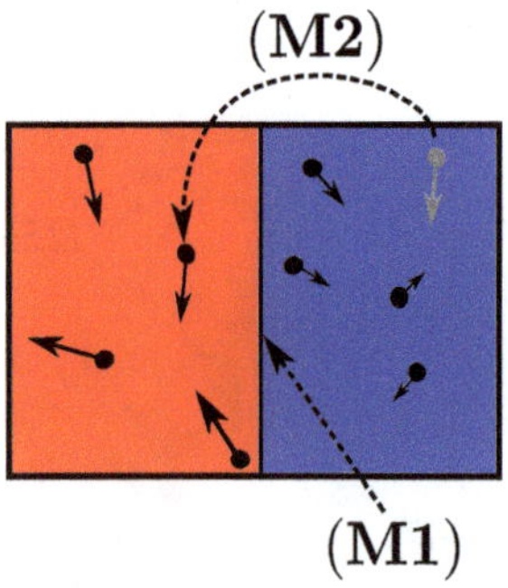


Figure 2.11: *Activities of Maxwell's demon:* (**M1**) introduces a barrier dividing a box of ideal gas molecules into two equal compartments; (**M2**) measures the speed of molecules in the right compartment and puts molecules with kinetic energy above the thermal equilibrium energy to the left compartment, from which he takes particles with kinetic energy lower than equilibrium energy to the right compartment. *Result:* The left compartment becomes warmer and the right compartment cooler due to control on the level of fluctuations.

As we will subsequently see, the scientific basis explaining systematically Maxwell's *Gedankenexperiment* is far from trivial and led to different views explaining why there is no violation of the *second law of thermodynamics.*

#### 2.5.1.1 Szilard's engine: linking information and thermodynamics

The complexity of Szilard's engine (SE) is slightly larger than Maxwell's demon with the benefit of obtaining a quantitative measure for energy/information extraction. The im-

10 Recall that a system showing neither energy nor mass exchanges between itself and the surroundings is called an *adiabatic system.*

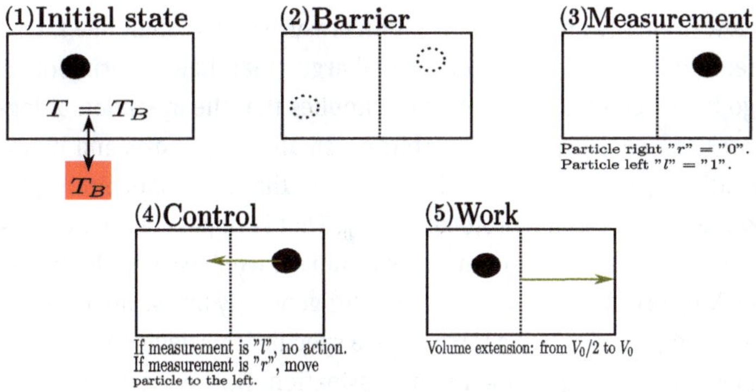
(1)Initial state
$T = T_B$
$T_B$
(2)Barrier
(3)Measurement
Particle right "r" = "0".
Particle left "l" = "1".
(4)Control
If measurement is "l", no action.
If measurement is "r", move particle to the left.
(5)Work
Volume extension: from $V_0/2$ to $V_0$

Figure 2.12: *Szilard's engine:* Explained in five steps. Note that after performing the 4th step, the system is in the same state as in the context of Maxwell's demon, where the entropy is decreased due to the increased order in the system (particle on the left-hand side). Szilard then concluded that the entropy has to increase somewhere else at least about the same amount in order for the 2nd law of thermodynamics to hold.

portant contribution of this "Gedankenexperiment" is that it provides a simple link between information and thermodynamics, i. e., in the form of energy/work.

The cycle of SE consists of the following five steps (see [133] and Fig. 2.12): **(SE1) Initial state:** single gas particle in a box in thermal equilibrium. **(SE2) Placement of a barrier** in the center of the box (ideally/quasi-statically performed without work): location of the particle unknown. **(SE3) Measurement:** information content $I[p] = -\ln\frac{1}{p} = \ln 2$ for the probability $p = p(\text{“}l\text{”}) = p(\text{“}r\text{”}) = 0.5$ of finding the particle in the left part "$l$" or the right part "$r$". **(SE4) Control:** if the measurement result is "$l$", then the demon does nothing; if the measurement gives "$r$", then the demon moves the particle quasi-statically (i. e., without work) to the left part. **(SE5) Work:** By moving the barrier quasi-statically to the left (i. e., an isothermal and quasi-static expansion) the system is able to perform work in the context of an ideal gas ($pV = nk_BT$). This means that SE executes the following amount of work,

$$W_{SE} = \int_{V_0/2}^{V_0} p\, dV = \int_{V_0/2}^{V_0} \frac{k_B T}{V}\, dV = k_B T \ln 2\,. \tag{2.58}$$

As noted in Sagawa [133], it became apparent that there are measurement processes that consume energy [28, 29]. Moreover, Landauer [93] proposes that the erasure of information dissipates energy and identifies $k_BT \ln 2$ as the smallest amount of energy dissipation required to irreversibly erase one bit of information. Note that this lower limit appears in *logical irreversible computation* relying on "OR", "AND", and "NAND" gates, for instance, see also Chapter 7. To remove Landauer's lower energy limit for erasure, Bennett [13] introduces a form of computation based on *reversible logic*. Hence Bennett's work [13] divided the scientific community into a group supporting the existence of Landauer's lower energy limit on information erasure and another one favoring Bennett's argument. In fact, López-Suárez et al. [99] argue that their experiment, in which they clearly go below Landauer's lower limit, implies that the operation of logically irreversible gates can be achieved with arbitrary small energy costs and hence seems to favor Bennett's theory. At the same time, we note that the argument in [99] relies on control parameters such as the protocol time $\tau_p$. That is, to pass far below Landauer's $k_B T$-limit, a long enough protocol time $\tau_p$ is required, which seems closely related to arguments making use of "quasi-static" steps. This generally brings up the question about the time and length scales being intrinsic to a specific logic device and consequently their influence on energy requirements and dissipation. This raises the question whether Landauer's limit is rather related to classical scales/mechanics whereas logic devices entirely operating on quantum mechanical scales would need to be related to a so-called quantum-corrected Landauer limit, for instance, with the help of Planck's constant $\hbar$?

Note that these thoughts about SE can be extended under a system of systems point of view as motivated by CHeSs. In fact, this amounts to introducing an observer that watches the controlling demon as well as the box with the single gas particle and the movable barrier. We leave this exercise to the interested reader.

## 2.6 Percolation: criticality and phase transitions

The term *percolation* stands for the investigation of the existence of a system/medium penetrating phase. For instance, consider the creation of a *random composite* by placing a conducting phase with probability $p$ and hence the isolating phase (that is, no link) with probability $1 - p$, see Figure 2.13. In this example, the crucial question is what is the *critical probability* $p_c$, according to which a conducting link/bond is placed in a square grid, for the resulting composite medium to become conducting? This question immediately reveals what we mean by *phase transition* in the context of percolation, that is, the change of a characteristic property (i. e., a conducting or isolating medium).

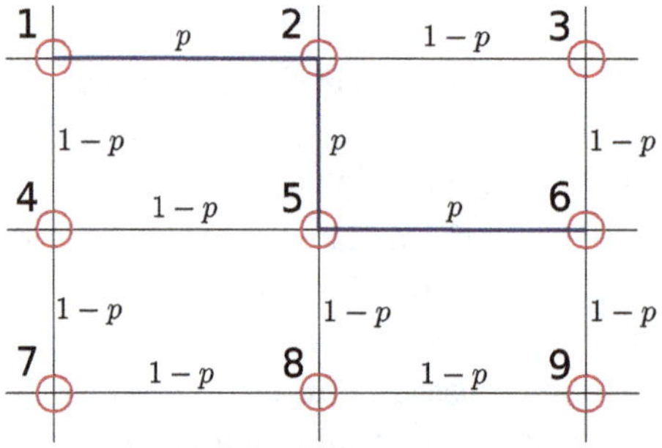


Figure 2.13: *Bond percolation a on rectangular grid with 9 nodes:* A sample realization of a conducting phase obtained by placing a conductor with probability $p$ and an isolator/no link with $1 - p$. For simplicity, we neglect diagonal links here.

In this section, we focus on the special case of bond percolation based on a rectangular grid. Our goal is to present a constructive approach to determine a lower bound of the critical probability $p_c$ for bond percolation to happen. The general methodology reads as follows.

---

**Bond percolation: computation of lower bounds**

Consider the problem of bond percolation in a graph consisting of $N \in \mathbb{N}$ nodes representing the medium of interest. To compute a lower bound of the critical probability $p_c$ for percolation to happen, follow the following steps:

1. **Identify the transition matrix $\hat{C}$:** each component $c_{ij}$ of the transition matrix $\hat{C} := \{c_{ij}\}_{i,j}$ is identified by $c_{ij} = p$ if there exists a link between the nodes $i$ and $j$ to indicate its probability of being conducting. If there does not exist any kind of link between bond $i$ and $j$ at all, then we set $c_{ij} = 0$. This latter case happens with probability $1 - p$.
2. **Apply the transition matrix to a suitable initial vector:** multiply an initial vector **u**, which represents one end of the medium (e. g., the left-hand side), i. e., the numbers of the nodes forming this (left-hand) side are the components of **u** containing the number 1. By applying the transition matrix $\hat{C}$ from Step 1 $m$ times, we arrive at the vector,

$$\mathbf{v} := \hat{C}^m \mathbf{u}, \tag{2.59}$$

where $m$ is defined subsequently. The definition of the transition matrix $\hat{C}$ implies that the components of the vector **v** in (2.59) represent the probabilities of reaching the nodes of the opposite end (the right-hand side), which have the same integer label as the corresponding components of **v**. Since we focus on a lower bound here, the integer $m \in \mathbb{N}$ represents the minimal number of bonds to reach from a specific node component of **u**, e. g., $u_i$, from the left-hand side one on the right-hand side, e. g., $v_i$, for instance.
3. **Identifying the lower bound of the percolation threshold $p_c$:** by adding up the values found in the components of **v** representing the opposite end (i. e., the right-hand side, $\mathcal{RHS}$) and solving the equation,

$$\sum_{i \in \mathcal{RHS}} v_i(p) = 1, \tag{2.60}$$

with respect to $p$, we obtain the lower bound $\underline{p}_c$ of $p_c$.

---

Let us apply this approach to a specific example.

**Example 2.6.1** (Critical percolation in a $(3 \times 3)$-square grid). Consider the square grid depicted in Figure 2.13. We follow the three steps introduced above for computing a lower bound $\underline{p}_c$ on the percolation threshold $p_c$:

**Step 1.** Looking at the bond network in Figure 2.13, we can immediately define the following transition matrix:

$$\hat{C} := \begin{bmatrix} 0 & p & 0 & p & 0 & 0 & 0 & 0 & 0 \\ p & 0 & p & 0 & p & 0 & 0 & 0 & 0 \\ 0 & p & 0 & 0 & 0 & p & 0 & 0 & 0 \\ p & 0 & 0 & 0 & p & 0 & p & 0 & 0 \\ 0 & p & 0 & p & 0 & p & 0 & p & 0 \\ 0 & 0 & p & 0 & p & 0 & 0 & 0 & p \\ 0 & 0 & 0 & p & 0 & 0 & 0 & p & 0 \\ 0 & 0 & 0 & 0 & p & 0 & p & 0 & p \\ 0 & 0 & 0 & 0 & 0 & p & 0 & p & 0 \end{bmatrix} . \tag{2.61}$$

**Step 2.** Next, we compute the vector $\mathbf{v} = \hat{C}^m \mathbf{u}$ for the minimal number of transitions $m = 2$ to reach from a node on the left-hand side a node on the right-hand side. The vector $\mathbf{u}$ reads as follows:

$$\mathbf{u} := [1, 0, 0, 1, 0, 0, 1, 0, 0]' , \tag{2.62}$$

where the components with 1 represent the nodes of the left-hand side, that is, the nodes 1, 4, and 7. Hence we obtain

$$\mathbf{v} := \hat{C}^2 \mathbf{u} := \hat{C}\mathbf{v}' = [p^2 + 2p^2, 2p^2, p^2, 3p^2, 4p^2, p^2, 3p^2, 2p^2, p^2] , \tag{2.63}$$

where $\mathbf{v}' = [p, p, 0, 2p, p, 0, p, p, 0]$.

**Step 3.** Finally, after adding up the components 3, 6, and 9, that is, $v_3 = p^2$, $v_6 = p^2$, and $v_9 = p^2$, we solve the resulting equation

$$\sum_{i \in \{3,6,9\}} v_i(p) = 3p^2 \stackrel{!}{=} 1 . \tag{2.64}$$

As a result, we obtain $\underline{p}_c = \frac{1}{\sqrt{3}} = 0.5774$.

Our analytically computed lower bound on the critical probability $p_c$ for percolation to happen is very close to the primarily computationally (e. g., by Monte Carlo methods) evaluated value of $p_c = 0.59274\ldots$. Finally, we have plotted the growth of the lower bound $\underline{p}_c$ with respect to the grid size $m$, see Figure 2.14. ◇

Since computer systems such as PCs, DCs, and HPCs (see Section 3.3) represent CHeSs to which humanity is exposed almost daily in western civilizations, the interest in *cybersecurity* continuously increases. Hence, we motivate that the analysis methodology provided by the lower percolation threshold $\underline{p}_c$ provides a systematic way of obtaining a lower bound on the likelihood that a certain (information) system (CHeS) has been penetrated by a hacker. In fact, this information can be systematically exploited as a guiding principle for the minimum amount of protection that should be applied to prevent an enemy to penetrate (i. e., to percolate) a cyberphysical system of interest.

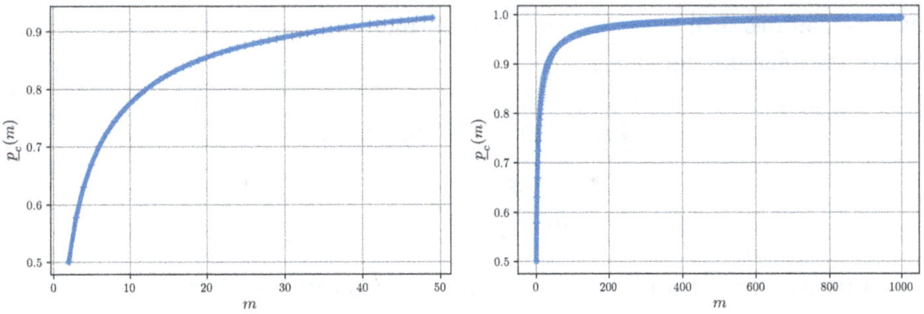


Figure 2.14: *Percolation threshold $p_c$ of a $(m \times m)$-square grid:* Growth of the lower bound $\underline{p}_c(m)$ with respect to the size $m$ of the $m \times m$-square grid.

## 2.7 Nonequilibrium CHeSs: GENERIC and Variational Principles

The description of thermodynamic systems very much relies on its key quantities of interest such as energy (first law of thermodynamics) and entropy (second law of thermodynamics) as introduced in Section 2.1. In fact, the definition of so-called *thermodynamic free energies* can be motivated as a systematic function (state function) that naturally incorporates the laws of thermodynamics under a calculus of variations point of view. That is, a classical variation of the state function $F(\mathbf{s})$ with respect to the state vector $\mathbf{s}$ describing the system $\mathcal{S}$ generally leads to a stationary/time independent solution representing the state $\mathbf{s}$ of the system $\mathcal{S}$.

To arrive at a dynamic description of the system $\mathcal{S}$, a straightforward idea taking the laws of thermodynamics into account (due to the definition of the state function) is to iteratively minimize the free energy $F(\mathbf{s})$ by identifying the resulting iterates $\mathbf{s}^k$ as the system's state at a specific time $t_k \in \mathbb{R}$. Obviously, the crucial question is now what is the correct iteration process/strategy such that the hereby obtained iterates $\mathbf{s}^k$ accurately represent the actual system state at the corresponding time steps $t_k$. In fact, an intuitive approach is to apply a steepest descent strategy leading to a *gradient flow*. A so-called energy variational analysis [50] aims to extend classical calculus of variations toward a variational framework for dynamic problems by explicitly making use of an Eulerian frame, see (2.67) in Section 2.7.1.

In this context, Jordan, Kinderlehrer, and Otto [80] rigorously demonstrate the convergence toward the continuous Fokker–Planck equation of a possible steepest descent approach by exploiting concepts from optimal transport. However, despite the convincing presentation of the application of optimal transport as a step toward a systematic description of nonequilibrium systems, such optimal transport-based convergence proofs are limited to a specific system and do not represent a required, general nonequilibrium thermodynamic framework valid for a broad class of physically known systems.

For such kind of aspirations, Grmela and Öttinger [65, 119, 121] have proposed a very general framework, called GENERIC, which suitably connects the essential thermodynamic quantities such as the state vector $\mathbf{z}$, the total energy $E(\mathbf{z})$, and the total entropy $S(\mathbf{z})$ by the following equations,

$$\begin{cases} \mathbf{z}_t = \mathcal{L}(\mathbf{z})\frac{\delta E(\mathbf{z})}{\delta \mathbf{z}} + \mathcal{M}(\mathbf{z})\frac{\delta S(\mathbf{z})}{\delta \mathbf{z}}\,, \\ \mathcal{L}(\mathbf{z})\frac{\delta S(\mathbf{z})}{\delta \mathbf{z}} = 0\,, \\ \mathcal{M}(\mathbf{z})\frac{\delta E(\mathbf{z})}{\delta \mathbf{z}} = 0\,, \end{cases} \tag{2.65}$$

where the *antisymmetric matrix* $\mathcal{L}$ and the *symmetric positive semidefinite matrix* $\mathcal{M}$ are the so-called *Poisson* and *friction* matrices, respectively. Moreover, $\mathcal{L}$ fulfills the Jacobi identity.

Recall that the energy $E$ and the Poisson matrix $\mathcal{L}$ take reversible contributions into account and similarly the entropy $S$ and the friction matrix $M$ represent the irreversible elements. Hence, the degeneracy requirements $(2.65)_2$–$(2.65)_3$ are required for cases where reversible and irreversible quantities are mixed.

### 2.7.1 Attempts inspired by GENERIC: dilute solutions from minimal energy and maximum dissipation

Over about a decade ago, the authors of [75] promoted an alternative variational methodology to GENERIC by keeping for the reversible part of the dynamics an energy-based/Hamiltonian least action principle and by additionally introducing the so-called *Maximum Dissipation Principle (MDP)*. This shows that a broadly accepted nonequilibrium thermodynamic formulation is still lacking. In fact, ultimately, this variational approach should be consistent with the more algebraic structure advocated by GENERIC, and hence a key step toward establishing various approaches into well-accepted theories is to rigorously achieve their equivalence in a systematically identified physical/practical context.

Dilute solutions are governed by the following total (Helmholtz, i. e., $V = |\mathcal{U}| =$ const. and $T =$ const. in contrast to Gibbs with $p =$ const. and $T =$ const.) action functional,

$$F(\rho, \{\rho_i\}_{i=1}^N, \mathbf{v}) := \int_0^T \int_{\mathcal{U}_t} e_k(\rho, \mathbf{v}) + e_I(\{\rho_i\}_{i=1}^N) + s(\{\rho_i\}_{i=1}^N, u)\, d\mathbf{x}\, dt\,, \tag{2.66}$$

which consists of the (barycentric) kinetic energy density $e_k(\rho, \mathbf{v}) := \frac{1}{2}\sum_{i=1}^N \frac{\mathbf{M}^2}{\rho}$, the interaction energy density $e_I(\{\rho_i\}_{i=1}^N) := \frac{q^F(\mathbf{x},t)}{2\varepsilon}\int_{\mathcal{U}} G(\mathbf{x}, \mathbf{x}')q^F(\mathbf{x}', t)\, d\mathbf{x}'$, and the energy density $s(\{\rho_i\}_{i=1}^N, u) := -R\theta \sum_{i=1}^N \frac{\rho_i}{\rho_{\mathrm{ref}}}(\ln \frac{\rho_i}{\rho_{\mathrm{ref}}} - 1)$ for a reference density $\rho_{\mathrm{ref}}$.

The free energy function in (2.66) uses the Eulerian material coordinates $\mathbf{x}(\mathbf{X}, t) \in \mathcal{U}_t$, where $\mathcal{U}_t$ is the evolution of the initial domain $\mathcal{U} = \mathcal{U}_0 \subset \mathbb{R}^d$ under the flow map,

$$\begin{cases} \dot{\mathbf{x}}(\mathbf{X}, t) = \mathbf{v}(\mathbf{X}, t)\,, \\ \mathbf{x}(\mathbf{X}, 0) = \mathbf{X}\,, \end{cases} \tag{2.67}$$

for the fixed initial (Lagrangian) material coordinate $\mathbf{X} \in \mathcal{U} = \mathcal{U}_0 \subset \mathbb{R}^d$. We note that the Eulerian flow defines a deformation tensor by $\mathrm{F}(\mathbf{x}(\mathbf{X}, t), t) = \frac{\partial \mathbf{x}(\mathbf{X},t)}{\partial \mathbf{X}}$. Accordingly, the free energy functional (2.66) with respect to the reference/initial domain $\mathcal{U} = \mathcal{U}_0$ then reads.

$$\begin{aligned} F(\rho, \{\rho_i\}_{i=1}^N, \mathbf{v}) := \int_0^T \int_{\mathcal{U}_0} \Bigg( &\frac{1}{2}\rho(\mathbf{x}(\mathbf{X}, t), t)\dot{\mathbf{x}}^2(\mathbf{X}, t) \\ &+ \frac{q^F(\mathbf{x}(\mathbf{X}, t), t)}{2\varepsilon} \int_{\mathcal{U}_0} G(\mathbf{x}(\mathbf{X}, t), \mathbf{x}(\mathbf{X}', t))q^F(\mathbf{x}(\mathbf{X}', t), t)\, d\mathbf{X}' \\ &- R\theta \sum_{i=1}^N n_i(\mathbf{x}(\mathbf{X}, t), t)(\ln n_i(\mathbf{x}(\mathbf{X}, t), t) - 1) \Bigg) |\det \hat{\mathrm{F}}|\, d\mathbf{X}\, dt\,. \end{aligned} \tag{2.68}$$

With respect to the latter energy function, we can compute the reversible/Hamiltonian part of the evolution by a classical least action principle [75],

$$\frac{\delta E(\rho, \{\rho_i\}_{i=1}^N, \mathbf{v})}{\delta \mathbf{x}} = 0\,, \tag{2.69}$$

where $E$ represents the Hamiltonian part of (2.68) by removing the entropic part,

$$S(\rho, \{\rho_i\}_{i=1}^N, \mathbf{v}) := -R\theta \sum_{i=1}^N \int_0^T \int_{\mathcal{U}_0} s(\{\rho_i\}_{i=1}^N, u)|\det\hat{\mathrm{F}}|d\mathbf{X}\, dt\,. \tag{2.70}$$

In a last step, we determine the irreversible part with the help of a so-called Maximum Dissipation Principle (MDP) [75], that is, identifying

$$\frac{\delta S}{\delta n_i} = 0\,, \tag{2.71}$$

for $1 \leq i \leq N$.

## 2.8 Smart Interacting CHeSs: autonomy, decision, strategy, and learning

Let us first discuss what we mean by *interacting CHeSs* and what entails *autonomy*. Before looking at the details, we motivate the general systems' view that does not restrict itself to human or social interactions, but allows us to consider various interacting entities such as machine–machine, human–machine, government–government, government–alliance, and hence arbitrary system–system interactions.

**Autonomously and Reasonably Interacting CHeSs**

**Definition 2.8.1.** We call $I \in \mathbb{N}$ systems $\mathcal{S}_i$, $1 \le i \le I$, to be *interacting*, if each system $\mathcal{S}_i$ chooses an action $a_i \in A_i$ from a set of possible actions $A_i$ available to $\mathcal{S}_i$ by minimizing a characteristic *objective/goal function* $G_i(a_i; \{a_1, a_2, \dots, a_I\} \backslash \{a_i\})$, $1 \le i \le I$, associated with the possible interactions, i. e., it chooses a $G_i$-optimal action $a_i \in A_i$ by assuming the other systems to be reasonable, which means operating under the corresponding variational principle (of minimizing costs/maximizing gain). This defines a natural degree of *autonomy* under certain circumstances, see Definition 2.8.11. ◇

**Remark 2.8.2** (Uncertainty in objective/cost). The assumption about reasonable, i. e., rational systems, should also account for the uncertainty, i. e., randomness, of various involved systems which might assess or model their objective function (that is, the costs) slightly differently than anticipated. Naturally, a reliable description of uncertainty by randomness requires to carefully analyze each application of interest, since the complexity and demand grow due to the associated modeling and computational aspects. Finally, note that being rational under one utility/objective function appears to be nonrational under another. Hence, the label *rational* is directly linked to a particular cost/utility function. ◇

The definition of an appropriate objective/utility function for a particular scenario of interest is highly complex and can depend in various degrees on interlinked physical, social, and psychological aspects, see [15] for Bernoulli's *Utility Concept* and its correction by Kahnemann and Tversky in [81] with the so-called *Prospect Theory*.

**Example 2.8.3** (Scissors, rock, and paper choices). Consider two interacting systems $\mathcal{S}_1$ and $\mathcal{S}_2$ of which each can choose one of the following three elements: scissors $a_1$, rock $a_2$, and paper $a_3$. Both systems, $\mathcal{S}_1$ and $\mathcal{S}_2$, have the same objective to choose the element that is stronger than the one picked by the other system, see [170, 175], for instance.

This interacting system, which is more commonly referred to as a *two-player game*, shows the following natural *gain* and *loss* relations,

$$
\begin{aligned}
G_1(a_1; a_3) &> G_2(a_3; a_1) = -G_1(a_1; a_3), && \text{scissors } a_1 \text{ cuts paper } a_3, \\
G_1(a_2; a_1) &> G_2(a_1; a_2) = -G_1(a_2; a_1), && \text{rock } a_2 \text{ destroys scissors } a_1, \\
G_1(a_3; a_2) &> G_2(a_2; a_3) = -G_1(a_3; a_2), && \text{paper } a_3 \text{ wraps rock } a_2.
\end{aligned}
\tag{2.72}
$$

The >-relation in $(2.72)_1$ denotes that the player/interacting system $\mathcal{S}_1$, choosing the first argument $a_1$ of $G_1$ on the left-hand side of the first relation, wins +1, whereas the player/interacting system $\mathcal{S}_2$, choosing the second argument of $G_1$, i. e., $a_3$, on the right-hand side, has to pay to the other player/system +1 such that it incurs the loss $G_2 = -1$. It remains to describe the situation where both systems/players choose the same element $a_i \in A := \{a_1, a_2, a_3\}$. The natural assignment of gain/loss to this situation is the neutral value $G_1(a_i; a_i) = G_2(a_i; a_i) = 0$, $i \in \{1, 2, 3\}$, which expresses a draw for this interaction cycle. The common way to represent the effect of systems/players $\mathcal{S}_1$ and $\mathcal{S}_2$ with associated indices $i$ and $j$, respectively, to decide at every new interaction cycle for a specific action $a_i$ and $a_j$ from the set $A$, is defined by an *interaction matrix/game matrix*, see Table 2.15. ◇

| $i \backslash j$ | 1 | 2 | 3 |
|---|---|---|---|
| 1 | 0 | −1 | 1 |
| 2 | 1 | 0 | −1 |
| 3 | −1 | 1 | 0 |

Figure 2.15: *Symmetric Interaction/Game Matrix:* The values in this matrix are the gain for system/player 1 whose choice is represented by the index $i$. As a consequence, the positive values in this table denote a loss, and negative ones denote a gain for system 2/opponent, whose decision has the index $j$. Recall that $i = j = 1$, $i = j = 2$, and $i = j = 3$ denote scissors, rock, and paper, respectively. Hence, on the diagonal the game cycle will end in a draw. Finally, note the *skew symmetry* in gain for $\mathcal{S}_1$ (i. e., positive values) and for $\mathcal{S}_2$ (i. e., negative values) shows that this interaction is fair for the participants, who hence might choose the same reasonable action.

These first thoughts surely remind the reader familiar with *game theory* about the basic concepts therein, see [163, 170, 175], for instance. The motivation of the approach taken in this book relies on the fact that the classical term *game theory* seems to generally imply limits to its applicability, i. e., its application to games, and hence is sometimes considered to be something children play with.[11] As a consequence, we believe that our system-driven approach to *game theory* honors the broader scope of this very applied and powerful scientific field. Moreover, we believe that a physical motivation, i. e., a view of the subject motivated by general principles observed in nature, encourages the reader for the presented concepts and principles and makes them easily accessible and recallable.

In Example 2.8.3, we have collected all the necessary information that entirely characterizes the scope of possible interactions between two systems. But there, we did not address how we can exploit the *concept of autonomy* in such interactions. According to Definition 2.8.1, each system's/player's autonomy amounts to maximizing their corresponding profit/gain and the degree of autonomy is governed by the concept of Definition 2.8.11. To state the framework that allows us to assign a so-called ideal/reasonable action to each system/player, we depend on basic quantities such as *interaction/payoff/cost matrices* and their properties.

**Interaction/Cost/Payoff Matrix**

**Definition 2.8.4. 1.** Interactions between two systems $\mathcal{S}_1$ and $\mathcal{S}_2$ can be characterized by the gain for system $\mathcal{S}_1$ whose *actions/choices/strategies* $i$ refer to different rows $1 \le i \le I$ in an associated, charac-

11 Note that the long history of wargaming serves as a motivating counter example [116] and explains the growing use of the term "*serious game*".

teristic *interaction matrix* $\hat{G}$. In the most basic and classical form of so-called *zero-sum interactions*, the gains/payoffs $g_{ij} > 0$ to system $\mathcal{S}_1$ imply loss/costs $-g_{ij}$ to system $\mathcal{S}_2$, whose *actions/choices/strategies* $b_j$ are represented on different columns of $\hat{G}$, that is,

$$\hat{G} := \begin{bmatrix} g_{11} & g_{12} & \cdots & g_{1J} \\ g_{21} & g_{22} & \cdots & g_{2J} \\ \vdots & \vdots & \ddots & \vdots \\ g_{I1} & g_{I2} & \cdots & g_{IJ} \end{bmatrix}. \tag{2.73}$$

**2.** More generally, *nonzero sum interactions* represent interactions between systems, $\mathcal{S}_1$ and $\mathcal{S}_2$, for simplicity, such that each system's objectives are defined by a dedicated *interaction matrix*. Hence, the two systems $\mathcal{S}_1$ and $\mathcal{S}_2$ have two associated interaction matrices $\hat{G}_1$ and $\hat{G}_2$, respectively. However, the columns of $\hat{G}_1$ and the rows of $\hat{G}_2$ represent a so-called fictive opponent (e. g., nature) and not the choice of actions available to the opposing system. ◇

The previous Definition 2.8.4 immediately raises the question of how to find the *reasonable actions* for system/player $\mathcal{S}_1$, often also referred to as *optimal actions or strategies*. Naturally, system $\mathcal{S}_1$ aims at maximizing its gains (or equivalently, minimizing its losses), and, obviously, the same holds for system $\mathcal{S}_2$. At first, system $\mathcal{S}_1$ might probably just look for the maximum payoff/gain in each single interaction. However, this can frequently lead to suboptimal results, which appear due to the fact that the opposing system $\mathcal{S}_2$ might be able to choose an action $b_j$, which results in a much lower gain $m_{ij}$ for $\mathcal{S}_1$, whose payoff will still be higher for an alternative, less lucrative action $a_i$. For instance, look at the subsequent Example 2.8.5.

**Example 2.8.5** (Loss in maximum profit strategies). Consider the interaction matrix,

$$\hat{G} := \begin{bmatrix} 4 & -3 \\ 2 & 1 \end{bmatrix}. \tag{2.74}$$

The *Maximum Profit Strategy (MPS)* implies for $\mathcal{S}_1$ to choose action $i = 1$ with the anticipated profit 4. However, the interaction matrix also tells us that there is no guarantee since system/player $\mathcal{S}_2$ has also the option of choosing action $j = 2$ resulting in the maximum loss $g_{12} = -3$ for system $\mathcal{S}_1$. In fact, this system interaction process governed by $\hat{G}$ even implies that *MPS* can lead to a *Maximum Loss Strategy (MLS)*. ◇

Example 2.8.5 urges for a systematic way to avoid situations resulting in an unwanted MLS. In fact, there is a central principle from nature that guides us here in finding the appropriate concept for a systematic framework providing us with the reasonable/ideal[12] action/strategy. This natural principle is related to the *second law of thermodynamics*, which expresses the fact that systems tend toward *(thermodynamic) equi-*

12 Note that we prefer here labels such as *reasonable* and *ideal* in difference to *optimal* to account for the fact that it is not a MPS.

*librium* (by *maximizing entropy*).[13] This translates for our situation of the two systems $\mathcal{S}_1$ and $\mathcal{S}_2$ interacting under (2.74) into choosing actions that minimize the losses for the parties involved, while still providing an as favorable as possible outcome to all of them. This means that the minimum of the maximal opponent's loss should be equal to the maximum of the minimal gains available, i. e., the opposing systems $\mathcal{S}_1$ and $\mathcal{S}_1$ are as close to equilibrium as the fairness of the interaction allows. Based on Example 2.8.5, this leads to the following consideration.

**Example 2.8.6** (Equilibrium by minimizing differences). This extension of the previous Example 2.8.5 aims to establish reasonable/ideal actions for both systems $\mathcal{S}_1$ and $\mathcal{S}_2$ by taking into account the fact of high losses under an *MPS* as well as the natural strive of systems to tend toward equilibrium, i. e., minimizing differences/gradients between the profits of participating systems. For the opposing system $\mathcal{S}_2$, we propose to look in a first step for actions that lead to the highest possible loss (which is sometimes just the smallest gain) and then balance inequality/reduce differences between the two opposing systems by only going for the smallest of the highest losses available. We correspondingly repeat the same steps for the gains by row actions of the system $\mathcal{S}_1$. That is, we first look for the minimal gains (possibly being even a real loss) row by row and then select the largest of these minimal gains. We can visualize this for the interaction matrix Ĝ as follows,

$$\begin{array}{cc|cc} 4 & -3 & -3 & \\ 2 & 1 & 1 & \mathbf{1} \\ \hline 4 & 1 & & \\ & \mathbf{1} & & \end{array} \,. \tag{2.75}$$

On the upper left square, we have the original interaction matrix Ĝ. The first row on the upper right square represents the first step of looking for the minimal gains for system $\mathcal{S}_1$ (i. e., for the minimal values in each row), and the second column of this right square then states the second step of selecting the largest of these identified maximal gains for system $\mathcal{S}_1$, which amounts to taking the maximum of the values obtained from the first step. In the same way, we proceed for actions/strategies of system $\mathcal{S}_2$, that is, we first look for the highest possible payoffs to $\mathcal{S}_1$ in each column on the first row of the left square on the bottom. The second row then represents the second step by stating the minimum of these gains obtained in the first step. Note that this interaction system for $\mathcal{S}_1$ and $\mathcal{S}_2$ shows the special case of an equilibrium system, in which the resulting payoffs obtained by the above motivated two-step procedure leads to the common value 1. This value is referred to as a saddle point, and the associated actions/strategies are called *pure actions/strategies*, which in fact represent *non-autonomous interactions*. ◇

13 Here, "increasing entropy" can be understood as forgetting about the fact that there would exist an alternative with a potential higher payoff.

The previous Example 2.8.6 motivates two essential building blocks, that is, a formal definition of *pure actions/non-autonomous actions* and a method for detecting it. More rigorously, we have the following definition.

**Pure Actions/Strategies**

**Definition 2.8.7.** Let $\hat{\mathrm{G}} \in \mathbb{R}^{I \times J}$ be a given interaction matrix for $I, J \in \mathbb{N}$. We call the common payoff value $g_{kl}$ for $1 \leq k \leq I$ and $1 \leq l \leq J$ to be a *saddle point*, if it holds that,

$$\max_{i} \min_{j} g_{ij} = g_{kl} = \min_{i} \max_{j} g_{ij} . \tag{2.76}$$

The actions $a_k$ of system $\mathcal{S}_1$ and $b_l$ of system $\mathcal{S}_2$ associated with this *saddle point* $g_{kl}$ are referred to as *pure/non-autonomous actions*.

In the literature, the pure/non-autonomous action introduced in Definition 2.8.7 is frequently referred to as *pure Nash equilibrium*, see [117], for instance.

**Remark 2.8.8** (Pure actions/strategies are ideal). We note that in case there exists a saddle point to an interaction system governed by an interaction matrix $\hat{\mathrm{G}}$, the associated interacting systems $\mathcal{S}_1$ and $\mathcal{S}_2$ should not deviate from their associated pure/non-autonomous strategies. ◇

These considerations immediately allow us to state the following method.

---

**Saddle Point Check**

For a given interaction matrix $\hat{\mathrm{G}}$, the saddle point check consists of computing the left- and right-hand sides of (2.76) and comparing the hereby obtained results for a possible common value $g_{kl}$. Applications of this method are given in Examples 2.8.6 and 2.8.9.

---

At the same time, we should also say that interacting CHeSs do not necessarily have the saddle point property, see the next Example 2.8.9.

**Example 2.8.9** (Lack of a saddle point). Consider the interaction matrix,

$$\hat{\mathrm{G}} := \begin{bmatrix} 4 & -3 \\ -1 & 2 \end{bmatrix} . \tag{2.77}$$

Applying the Saddle Point Check (2.76) gives,

$$\begin{array}{cc|cc} 4 & -3 & -3 & \\ -1 & 2 & -1 & \mathbf{-1} \\ \hline 4 & 2 & & \\ & \mathbf{2} & & \end{array} , \tag{2.78}$$

which clearly shows that there is no common value $g_{22}$, and hence, there does not exist a saddle point that would define for each of the two systems a unique and reasonable action. ◇

The previous example immediately leads to the central question of how we can obtain ideal/reasonable actions/strategies in situations where we do not have the saddle point property. A natural first thought might be to try to relax the concept of pure actions/strategies associated with saddle points. This immediately means that we would allow for choosing more than a single action at a time to be a favorable strategy. As a consequence, we end up with some degree of autonomy. Hence, in the case of just two possible actions, how can we decide about when to prefer one strategy over the other? In an attempt to answer this question and by taking into account our previous observation about how an MPS can turn into an MLS as well as nature's tendency for *equilibrium*, we know that we should look for a way to *counterbalance/weight* high-profit strategies toward a relaxed and sustainable gain-oriented approach. In [163], the authors formulate these observations in the form of two principles, i. e., one about systems tending toward maximum security and the other about systems tending toward equilibrium. In this context, let us consider the following simple example.

**Example 2.8.10** (Counterweighting as an action principle). Let us again consider the simple interaction of two systems $\mathcal{S}_1$ and $\mathcal{S}_2$ defined by the interaction matrix,

$$\hat{\mathrm{G}} := \begin{bmatrix} 4 & -3 \\ -1 & 2 \end{bmatrix}. \tag{2.79}$$

From the *Saddle Point Check* in Example 2.8.9, we know that we do not have pure actions for our interacting systems $\mathcal{S}_1$ and $\mathcal{S}_2$. We first focus on the actions of $\mathcal{S}_1$ and, in particular, on the differences in the gain of $\mathcal{S}_1$ resulting from the two possible actions of the opposing system $\mathcal{S}_2$. A large difference can represent in extremal circumstances a large gain with a small loss or a small gain with a large loss. Naturally, a chosen action whose result can show a large deviation (difference) from the desired one by the independent action of the opposing system represents a high risk and hence should be carefully picked. Therefore, a straightforward way for picking an action with such a discussed high difference $d_h$ is to increase the likelihood of picking the alternative action with probability $\frac{|d_h|}{|d_h|+|d_s|}$, where $d_s$ is the small difference of the alternative action. As a consequence, the action with the large difference in gain $d_h$ should be picked with probability $\frac{|d_s|}{|d_h|+|d_s|}$. Hence, if we apply this reasoning to our interaction matrix $\hat{\mathrm{G}}$, then we have that,

$$\begin{array}{cc|cc} 4 & -3 & 7 & \frac{3}{10} \\ -1 & 2 & -3 & \frac{7}{10} \\ \hline 5 & -5 & & \\ \frac{5}{10} & \frac{5}{10} & & \end{array}. \tag{2.80}$$

As a result, we obtain the mixed/weighted strategy $\mathbf{s}_1 = [0.3, 0.7]$ for system $\mathcal{S}_1$ and $\mathbf{s}_2 = [0.5, 0.5]$ for system $\mathcal{S}_2$.

Let us briefly investigate what happens if we apply these mixed strategies for $\mathcal{S}_1$ and $\mathcal{S}_2$. We get,

$$\begin{aligned}\hat{G}^T \mathbf{s}_1 &= \left[\frac{12}{10} - \frac{7}{10}, \frac{-9}{10} + \frac{14}{10}\right]' \\ &= [0.5, 0.5]', \\ \hat{G}\mathbf{s}_2 &= \left[2 - \frac{3}{2}, -\frac{1}{2} + 1\right]' \\ &= [0.5, 0.5]'.\end{aligned} \tag{2.81}$$

We will see later that the unique value 0.5 represents the value of the game, that is, this value represents the gain that both systems $\mathcal{S}_1$ and $\mathcal{S}_2$ *can expect on average* by applying their corresponding mixed actions/strategies. ◇

This immediately leads to the following well-established probabilistic or statistical action/strategy concept, see [163, 170], for instance.

**Mixed Actions/Strategies**

**Definition 2.8.11.** For a given interaction matrix $\hat{G}$ governing the interactions between two CHeSs $\mathcal{S}_1$ and $\mathcal{S}_2$, we call vectors of the form,

$$\mathbf{s}_k := \left\{ \mathbf{p} := [p_1, p_2, \dots, p_{d_k}]' \in \mathbb{R}^{d_k} \;\middle|\; \sum_{l=1}^{d_k} p_l = 1,\, p_l \geq 0 \right\}, \quad k = 1, 2, \tag{2.82}$$

*mixed actions/strategies*, where $d_1 = I \in \mathbb{N}$ and $d_2 = J \in \mathbb{N}$ denote the number of actions/strategies of systems $\mathcal{S}_1$ and $\mathcal{S}_2$, respectively. Mixed strategies provide a natural degree of autonomy.

**Remark 2.8.12** (Mixed actions). The above Definition 2.8.11 of mixed actions shows that a *pure action* represents a particular case of a *mixed action*, that is, the case where all the mass is put into a single action. As a consequence, the availability of mixed actions between interacting systems indicate a certain autonomy and hence autonomous interactions. ◇

Note that the action $\mathbf{s}_k$ represents the probability of system $\mathcal{S}_k$ applying action $a_l \in A_k$ from the set of possible actions $A_k$ of $\mathcal{S}_k$. The discussion in Example 2.8.10 establishes the following method, see also [175], which provides also a rich source for examples.

---

**!**

**Computing Reasonable Mixed Actions (RMAs)**

Suppose that we are given a $2 \times 2$-interaction matrix $\hat{G}$ that does not have a *saddle point*, see the *Saddle Point Check*. The computation of the *Reasonable Mixed Action (RMA)* belonging to the associated two CHeSs $\mathcal{S}_1$ and $\mathcal{S}_2$ is as follows:

**Step 1.** Compute the payoff differences for $\mathcal{S}_1$ by subtracting column two from column one of $\hat{G}$,

$$\begin{bmatrix} g_{11} - g_{12} \\ g_{21} - g_{22} \end{bmatrix} =: \begin{bmatrix} d_{11} \\ d_{12} \end{bmatrix}, \tag{2.83}$$

and similarly for $\mathcal{S}_2$, that is,

$$\begin{bmatrix} g_{11} - g_{21}, & g_{12} - g_{22} \end{bmatrix} =: \begin{bmatrix} d_{21}, & d_{22} \end{bmatrix}. \tag{2.84}$$

**Step 2.** Determine the mixed strategy $\mathbf{s}_1$ for $\mathcal{S}_1$ by calculating the fractions,

$$\mathbf{s}_1 = \left[\frac{|d_{12}|}{|d_{11}| + |d_{12}|}, \frac{|d_{11}|}{|d_{11}| + |d_{12}|}\right]'.$$

In the same way, we obtain the mixed strategy $\mathbf{s}_2$ for $\mathbf{S}_2$,

$$\mathbf{s}_2 = \left[\frac{|d_{22}|}{|d_{21}| + |d_{22}|}, \frac{|d_{21}|}{|d_{21}| + |d_{22}|}\right]'.$$

---

**Expected Interaction Gains (EIG)**
**Definition 2.8.13.** Let $\hat{G} = \{g_{ij}\}_{i,j=1}^{m,n}$ be a given interaction matrix between two systems $\mathcal{S}_1$ and $\mathcal{S}_2$, where the elements $m_{ij}$ represent the gains to system $\mathcal{S}_1$ and at the same time the losses of system $\mathcal{S}_2$. That is, we look at zero-sum interactions. Moreover, let $\mathbf{s}_1^i := \mathbf{e}_i^1 \in \mathbb{R}^m$ and $\mathbf{s}_2^j := \mathbf{e}_j^2 \in \mathbb{R}^n$ be pure actions of the systems $\mathcal{S}_1$ and $\mathcal{S}_2$, respectively, and let $\mathbf{s}_1 \in \mathbb{R}^m$ and $\mathbf{s}_2 \in \mathbb{R}^n$ be reasonable mixed actions (RMAs) of systems $\mathcal{S}_1$ and $\mathcal{S}_2$, respectively, where $\mathbf{e}_i^1 \in \mathbb{R}^m$ and $\mathbf{e}_i^2 \in \mathbb{R}^n$ are the canonical bases elements. The *pure Interaction Gains (pIG)* $v_1^i$ and $v_2^j$ for $\mathcal{S}_1$ and $\mathcal{S}_2$, respectively, are then defined by,

$$\begin{cases} v_1^i := (\mathbf{s}_1^i)'\hat{G}\mathbf{s}_2 = \sum_{j=1}^{n} g_{ij}s_{2,j}, \\ v_2^j := \mathbf{s}_1'\hat{G}\mathbf{s}_2^j = \sum_{i=1}^{m} g_{ij}s_{1,i}. \end{cases} \tag{2.85}$$

Similarly, the *Expected Interaction Gain (EIG)* $\overline{v}_1$ of system $\mathcal{S}_1$ is defined by,

$$\overline{v}_1 := \mathbf{s}_1'\hat{G}\mathbf{s}_2. \tag{2.86}$$

Definition 2.8.13 and Example 2.8.10 motivate us to extend our investigation to finding out what is the minimal payoff that we can expect for mixed actions under system interactions governed by (2.79).

**Example 2.8.14** (Lower and upper possible Payoffs (LPP) and (UPP), respectively). First, let us start with the arbitrary mixed action $\mathbf{s}_1 = [0.5, 0.5]'$ for system $\mathcal{S}_1$, i. e., we on purpose avoid a RMA, whereas keeping $\mathcal{S}_2$'s mixed action open, that is, we set $\mathbf{s}_2 = [p_1, p_2]'$. Starting with these inputs to the EIG, we immediately can establish the following lower bound called LPP,

$$\begin{aligned} \mathbf{s}_1'\hat{G}\mathbf{s}_2 &= [3/2, -0.5][p_1, p_2]' = 1.5p_1 - 0.5p_2 \\ &\geq -0.5(p_1 + p_2) \geq -0.5. \end{aligned} \tag{2.87}$$

In the same way, we get the following upper bound referred to as UPP,

$$\begin{aligned} \mathbf{s}_1'\hat{G}\mathbf{s}_2 &= [3/2, -0.5][p_1, p_2]' = 1.5p_1 - 0.5p_2 \\ &\leq 1.5(p_1 + p_2) \leq 1.5. \end{aligned} \tag{2.88}$$

By exploiting the *reasonable mixed action* for system $\mathcal{S}_1$ identified in Example 2.8.10, i. e., $\mathbf{s}_1 = [0.3, 0.7]'$, we want to recompute the *LPP* and *MPP* for $\mathcal{S}_1$ in this case,

$$\begin{aligned} \mathbf{s}_1 \hat{\mathrm{G}} \mathbf{s}_2' &= [0.5, 0.5][p_1, p_2]' = 0.5(p_1 + p_2) \\ &= 0.5\,. \end{aligned} \tag{2.89}$$

Hence, we have that LPP = MPP = 0.5 being exactly between the LPP = −0.5 and MPP = 1.5 for the chosen $\mathbf{s}_1 = [0.5, 0.5]$. This motivates Definition 2.8.13 coining *reasonable actions* as an *EIG*. ◇

**Example 2.8.15** (Even/odd bets in a fair roulette). We look at the subinteraction $\mathcal{S}_2$ of the *European roulette*, where an entity $\mathcal{S}_1$ (e. g., human or machine) only chooses between/bets on *even* and *odd* numbers as the roulette's upcoming result. To this end, we first recall some necessary *basics of roulette*:

a. *The total number of nonnegative integers on the roulette's wheel are* $N = 37$, *which includes the number* zero.
b. *The payoff for the even-odd-subinteraction is one-to-one, i. e.,* $\mathcal{S}_1$ *receives double its input.*
c. *For completeness, note that betting on the appearance of zero gives the payoff amounting to* $N - 1 = 36$ *times the initial input.*

This immediately leads to the following *even/odd-subsystem*'s interaction matrix,

$$\hat{\mathrm{G}} := \begin{bmatrix} 2 & -1 & -1 \\ -1 & 2 & -1 \end{bmatrix}. \tag{2.90}$$

The innovative step is to apply the fixed *mixed action* of the roulette $\mathcal{S}_2$ by accounting for the likelihoods of outcomes to happen, supposing a fair system $\mathcal{S}_2$. Therefore, we have that,

$$\mathbf{s}_2 := \left[\frac{N-1}{2N}, \frac{N-1}{2N}, \frac{1}{N}\right] = \left[\frac{18}{37}, \frac{18}{37}, \frac{1}{37}\right]. \tag{2.91}$$

Let us compute the *MIG* based on (2.91) and (2.90), where we keep $\mathcal{S}_1$'s mixed action open, i. e., $\mathbf{s}_1 = [p_1, p_2]$. Hence, we obtain,

$$[p_1, p_2]\hat{\mathrm{G}}\mathbf{s}_2' = [p_1, p_2]\begin{bmatrix} \frac{17}{37} \\ \frac{17}{37} \end{bmatrix} = p_1\frac{17}{37} + (1 - p_1)\frac{17}{37} = \frac{17}{37}, \tag{2.92}$$

where we applied the mixedness property. Unexpectedly, the *MIG* is entirely independent from our choice of action, which must be a matter of symmetry, i. e., absolute fairness between even and odd, after another look at the problem setup. ◇

The rather surprising fact of our analysis in Example 2.8.15 tells us that our expected loss in interacting in an even/odd-subsystem of the European roulette is larger than 0.5. Naturally, this raises the question of *how a roulette needs to be modified (by changing the fairness level)* for us to be interested in interacting with it at all? This is the topic of the following example.

**Example 2.8.16** (Roulette's fairness and reasonable bets). We look at the even/odd interaction system from Example 2.8.15. First, we introduce a mechanism allowing us to apply various fairness levels. The considerations in Example 2.8.15 motivate that we should find a way to break the intrinsic interaction symmetry given by $\mathbf{s}_2$ and $\hat{G}$. To this end, we define the *fairness parameter* $\varphi$ as follows,

$$\begin{cases} \varphi > 0\,, & \text{favoring even} = a_1\text{-bets,} \\ \varphi = 0\,, & \text{fair bets (not favoring any bet),} \\ \varphi < 0\,, & \text{favoring odd} = a_2\text{-bets,} \end{cases} \tag{2.93}$$

such that the roulette's mixed action reads,

$$\mathbf{s}_2 := \left[\frac{18+\varphi}{37}, \frac{18-\varphi}{37}, \frac{1}{37}\right]' . \tag{2.94}$$

In an attempt to maximize system $\mathcal{S}_1$'s *EIG*, we compute the *EIG* for an open mixed action $\mathbf{s}_2$,

$$\begin{aligned} \overline{v}_1(p_1, \varphi) := [p_1, p_2]\hat{G}\mathbf{s}_2' = [p_1, p_2]\begin{bmatrix} \frac{17+3\varphi}{37} \\ \frac{17-3\varphi}{37} \end{bmatrix} &= p_1\frac{17+3\varphi}{37} + (1-p_1)\frac{17-3\varphi}{37} \\ &= \frac{6\varphi}{37}p_1 + \frac{17-3\varphi}{37} . \end{aligned} \tag{2.95}$$

We can immediately identify the local extremum,

$$\nabla\overline{v}_1(p_1, \varphi) = \left[\frac{\partial\overline{v}_1}{\partial p_1}, \frac{\partial\overline{v}_1}{\partial\varphi}\right]' = \left[\frac{6\varphi}{37}, \frac{6p_1}{37} - \frac{3}{37}\right]' \stackrel{!}{=} \mathbf{0}\,,$$

as the saddle point $(p_1, \varphi) = (0.5, 0)$ belonging to a fair roulette with the associated EIG, $\overline{v}_1(0.5, 0) = 17/37$. Studying $\overline{v}_1(p_1, \varphi)$ for different parameters $p_1$ and $\varphi$ in Figure 2.16 on the top tells us that we need to turn to an unfair game by choosing a fairness parameter $|\varphi| > \frac{20}{3}$ to have an EIG larger than 1, that is, we obtain more than just the amount we bet. For instance, this amounts to increasing the appearance of *even numbers* (i. e., $a_1$) by 7/37, i. e., by 18.9 %. ◇

Next, we present the *Fundamental Theorem of Game Theory (FTGT)* (see [163], for instance), since its proof relies on a central concept referred to as *Linear Programming (LP)*. LP allows us to find reasonable interactions for general CHeSs. To this end, we first need to introduce additional definitions.

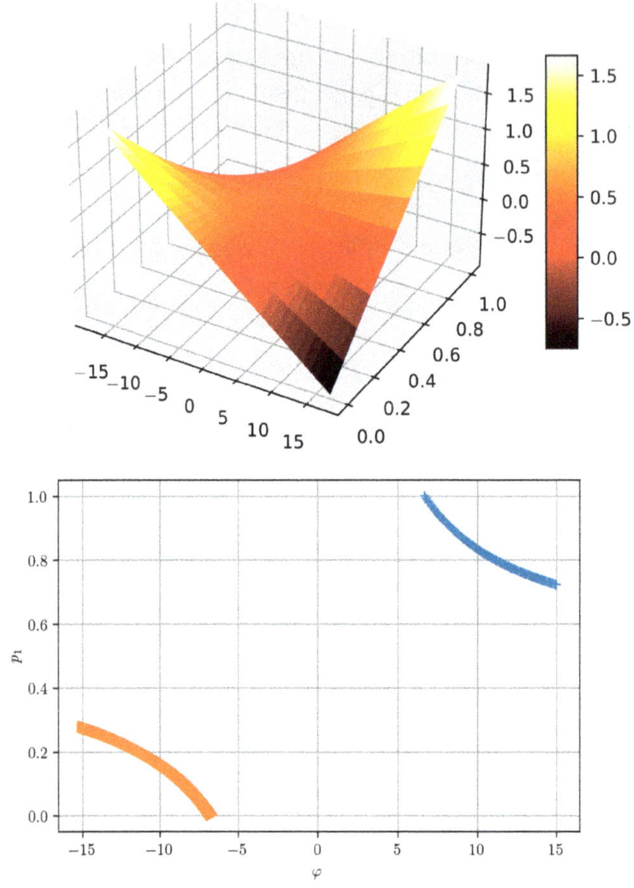


**Figure 2.16:** *Mean Interaction Gain (MIG). Top:* Plotted in dependence of fairness $\varphi \in [-15, 15]$ and probability $p_1 \in [0, 1]$ of choosing action/bet $a_1$ = even. The local maximal EIGs are given by $V(0, -15) = 62/37$ (choosing and favoring odd $= a_2$) and $V(1, 15) = 62/37$ (choosing and favoring even $= a_1$). Finally, the saddle point is obtained by $V(0.5, 0) = 17/37$. *Bottom:* $\pm\frac{20}{3}$ are the lower limits of the fairness $\varphi$ providing lower bounds (on the left) and upper bounds (on the right) on $p_1$ for achieving EIGs larger than 1 (i. e., $\varphi$ making the roulette interesting).

**Interaction Security**

**Definition 2.8.17.** Suppose that we have two interacting systems $\mathcal{S}_1$ and $\mathcal{S}_2$ governed by the interaction matrix $\hat{G} := \{g_{ij}\}_{i,j=1}^{I,J}$. Assuming $\mathbf{s}_1 \in A_1$ and $\mathbf{s}_2 \in A_2$ being mixed actions of system $\mathcal{S}_1$ and $\mathcal{S}_2$, respectively, their associated *Interaction Securities* are given by,

$$\begin{cases} \min_{\mathbf{s}_2 \in A_2} \mathbf{s}_1 \hat{G} \mathbf{s}_2' = \min_{1\le j\le J} \mathbf{s}_1 \hat{G}^j \,, \\ \max_{\mathbf{s}_1 \in A_1} \mathbf{s}_1 \hat{G} \mathbf{s}_2' = \max_{1\le i\le I} \hat{G}_i \mathbf{s}_2 \,, \end{cases} \tag{2.96}$$

where $\hat{G}^j$ and $\hat{G}_i$ denote the $j$-th column and the $i$-th row of $\hat{G}$, respectively.

We can immediately extend the previous security terms as follows.

**Optimal Interaction Security**

**Definition 2.8.18.** Suppose that we have two interacting systems $\mathcal{S}_1$ and $\mathcal{S}_2$ governed by the interaction matrix $\hat{G} := \{g_{ij}\}_{i,j=1}^{I,J}$. Assuming $\mathbf{s}_1 \in A_1$ and $\mathbf{s}_2 \in A_2$ being mixed actions of system $\mathcal{S}_1$ and $\mathcal{S}_2$, respectively, their associated *Optimal Interaction Securities (OIS)* are given by,

$$\begin{cases} v_1 = \max_{\mathbf{s}_1 \in A_1} \min_{\mathbf{s}_2 \in A_2} \mathbf{s}_1 \hat{G} \mathbf{s}_2' = \max_{\mathbf{s}_1 \in A_1} \min_{1 \le j \le J} \mathbf{s}_1 \hat{G}^j , \\ v_2 = \min_{\mathbf{s}_2 \in A_2} \max_{\mathbf{s}_1 \in A_1} \mathbf{s}_1 \hat{G} \mathbf{s}_2' = \min_{\mathbf{s}_2 \in A_2} \max_{1 \le i \le I} \hat{G}_i \mathbf{s}_2 , \end{cases} \tag{2.97}$$

where $\hat{G}^j$ and $\hat{G}_i$ denote the $j$-th column and the $i$-th row of $\hat{G}$, respectively.

We have now all the basic ingredients to state the following central result, going back to von Neumann in 1928, by establishing the existence of an equilibrium for interactions where no saddle point exists and hence where the concept of mixed actions provides an alternative solution.

**Fundamental Theorem of Game Theory**

**Theorem 2.8.19.** *Let* $\hat{G} \in \mathbb{R}^{I \times J}$ *be an arbitrary* Interaction Matrix *governing the interaction between two systems* $\mathcal{S}_1$ *and* $\mathcal{S}_2$. *There exist actions* $\bar{\mathbf{s}}_1 \in A_1$ *of system* $\mathcal{S}_1$ *and* $\underline{\mathbf{s}}_2 \in A_2$ *of system* $\mathcal{S}_2$ *such that,*

$$\begin{cases} v_1 = \max_{\mathbf{s}_1 \in A_1} \min_{1 \le j \le J} \mathbf{s}_1 \hat{G}^j = \min_{1 \le j \le J} \bar{\mathbf{s}}_1 \hat{G}^j , \\ v_2 = \min_{\mathbf{s}_2 \in A_2} \max_{1 \le i \le I} \hat{G}^i \mathbf{s}_2' = \max_{1 \le i \le I} \hat{G}^i \underline{\mathbf{s}}_2' , \end{cases} \tag{2.98}$$

*as well as* $v_1 = v_2$.

Instead of proving this theorem, we restrict ourselves to presenting the main tool and idea for establishing Theorem 2.8.19 and refer the interested reader to the literature, e. g., [163]. The previously mentioned central and powerful tool is the so-called *Linear Programming*. In fact, the following property is the key in the proof of Theorem 2.8.19 and of interest in applications.

**Equivalence between IM and LP** !

Assume that we know the IM $\hat{G} \in \mathbb{R}^{I \times J}$ between two CHeSs $\mathcal{S}_1$ and $\mathcal{S}_2$. For simplicity, we state the method for the particular case $I = J = 2$. We can rewrite the interaction problem governed by IM into an LP problem for each of the two systems involved as follows:

**(LP1)** *Minimize* $w_1 = L(\tilde{\mathbf{s}}_1) := \tilde{p}_1 + \tilde{p}_2$ *subject to,*

$$\begin{cases} g_{11}\tilde{p}_1 + g_{21}\tilde{p}_2 & \ge 1 , \\ g_{12}\tilde{p}_1 + g_{22}\tilde{p}_2 & \ge 1 , \\ \tilde{p}_1, \tilde{p}_2 & \ge 0 . \end{cases} \tag{2.99}$$

**(LP2)** *Maximize* $w_2 = L(\tilde{\mathbf{s}}_2) = \tilde{q}_1 + \tilde{q}_2$ *subject to,*

$$\begin{cases} g_{11}\tilde{q}_1 + g_{12}\tilde{q}_2 & \le 1 , \\ g_{21}\tilde{q}_1 + g_{22}\tilde{q}_2 & \le 1 , \\ \tilde{q}_1, \tilde{q}_2 & \ge 0 . \end{cases} \tag{2.100}$$

Finally, since the solutions $\tilde{\mathbf{s}}_1$ and $\tilde{\mathbf{s}}_2$ of **(LP1)** and **(LP2)** above are not normalized, we can do that at the end thanks to the definition of the objective function $L$, that is,

$$\mathbf{s}_i = \frac{\tilde{\mathbf{s}}_i}{w_i} = \frac{\tilde{\mathbf{s}}_i}{\mathbf{c}'\tilde{\mathbf{s}}_i} \quad \text{for } i = 1, 2 . \tag{2.101}$$

Before looking at solving such formulations, we first need basic terms characterizing such problems.

**LP Problem in Standard Form**

**Definition 2.8.20.** We say that an LP problem is in *standard form* if it satisfies the following properties:

**(LP1)** All variables of the LP problem show nonnegative constraints.

**(LP2)** Apart from **(LP1)**, we are left with equations of the form,

$$\hat{A}_{SF}\mathbf{x}_{SF} = \mathbf{b}_{SF} .$$

**(LP3)** The right-hand side $\mathbf{b}_{SF}$ of the equation from **(LP2)** is nonnegative.

The central question arising from the above equivalence is how to find solutions to LP problems. As an answer to this question, we present *Dikin's algorithm* based on the following example.

**Example LP Problem (ELPP).** *Minimize for* $\mathbf{c} := [-2, -1]'$ *and* $\tilde{\mathbf{s}}_1 := [p_1, p_2]'$ *the cost/goal function,*

$$w = L(\tilde{\mathbf{s}}_1) := \mathbf{c}'\tilde{\mathbf{s}}_1 , \tag{2.102}$$

*subject to,*

$$\begin{cases} \tilde{p}_1 + \tilde{p}_2 \leq 5 , \\ 2\tilde{p}_1 + 3\tilde{p}_2 \leq 12 , \\ \tilde{p}_1, \tilde{p}_2 \geq 0 . \end{cases} \tag{2.103}$$

◇

Note that the above **(ELPP)** is stated with the auxiliary action $\tilde{\mathbf{s}}_1$ of system $\mathcal{S}_1$ as in the IM-to-LP equivalence, since it is not normalized as required for *mixed actions* $\mathbf{s}_1 \in A_1$. By definition, the weight $w$ for the normalization is given by the goal function $L$ such that,

$$\mathbf{s}_1 = \frac{\tilde{\mathbf{s}}_1}{w} = \frac{\tilde{\mathbf{s}}_1}{\mathbf{c}'\tilde{\mathbf{s}}_1} . \tag{2.104}$$

**Remark 2.8.21** (Initial feasible interior solution). We note that an *initial feasible interior solution* $\tilde{\mathbf{s}}_1^0$ of **(ELPP)** is,

$$\tilde{\mathbf{s}}_1^0 := [1, 2] ,$$

for instance. ◇

Unfortunately, the **(ELPP)** needs a reformulation with the help of so-called *slack variables* before our subsequently presented LP-algorithm can be applied. By introducing the slack variables $\tilde{p}_3$ and $\tilde{p}_4$, we can rewrite (2.103) in *standard form*:

**Slack-(ELPP).** *Minimize the cost,*

$$w = L_{sl}(\tilde{\mathbf{s}}_1) = -2\tilde{p}_1 - \tilde{p}_2 + 0\tilde{p}_3 + 0\tilde{p}_4 , \tag{2.105}$$

*subject to,*

$$\begin{cases} \tilde{p}_1 + \tilde{p}_2 + \tilde{p}_3 = 5 , \\ 2\tilde{p}_1 + 3\tilde{p}_2 + \tilde{p}_4 = 12 , \\ \tilde{p}_1, \tilde{p}_2, \tilde{p}_3, \tilde{p}_4 \geq 0 . \end{cases} \tag{2.106}$$

◇

Note that an interior feasible starting point for the **Slack-(ELPP)** is,

$$\tilde{\mathbf{s}}_1^0 = [1, 2, 2, 4]' .$$

Moreover, this slack-reformulation of **(ELPP)** can be written in matrix form as follows,

$$\hat{G}_{sl} := \begin{bmatrix} 1 & 1 & 1 & 0 \\ 2 & 3 & 0 & 1 \end{bmatrix} , \tag{2.107}$$

with the cost vector $\mathbf{c} = [-2, -1, 0, 0]'$. We are now in the position to state Dikin's algorithm, which is applicable to this **Slack-(ELPP)**, i. e., the slack-reformulation of **(ELPP)**, see [39].

---

**Dikin's Method**

!

Let $\tilde{\mathbf{s}}^0$ denote an initial feasible interior solution that fulfills the following equation,

$$\hat{G}_{SF}\mathbf{s}^0 = \mathbf{b}_{SF} . \tag{2.108}$$

Assume that the objective function is to minimize (otherwise, multiply the coefficient vector of the objective function by −1 and then minimize). The iterative computation of a solution to an LP problem in standard form consists of the following steps:

1. Start counter: Set $k = 0$.
2. Define the matrix,

$$\hat{D} := \text{diag}\left(\tilde{\mathbf{s}}^0\right) = \begin{bmatrix} 1 & 0 & 0 & 0 \\ 0 & 2 & 0 & 0 \\ 0 & 0 & 2 & 0 \\ 0 & 0 & 0 & 4 \end{bmatrix} . \tag{2.109}$$

3. Compute the auxiliary matrices,

$$\hat{B} = \hat{G}\hat{D} = \begin{bmatrix} 1 & 1 & 1 & 0 \\ 2 & 3 & 0 & 1 \end{bmatrix} \begin{bmatrix} 1 & 0 & 0 & 0 \\ 0 & 2 & 0 & 0 \\ 0 & 0 & 2 & 0 \\ 0 & 0 & 0 & 4 \end{bmatrix} = \begin{bmatrix} 1 & 2 & 2 & 0 \\ 2 & 6 & 0 & 4 \end{bmatrix} , \tag{2.110}$$

$$\mathbf{d} = \hat{D}\mathbf{c} = \begin{bmatrix} 1 & 0 & 0 & 0 \\ 0 & 2 & 0 & 0 \\ 0 & 0 & 2 & 0 \\ 0 & 0 & 0 & 4 \end{bmatrix} \begin{bmatrix} -2 \\ -1 \\ 0 \\ 0 \end{bmatrix} = \begin{bmatrix} -2 \\ -2 \\ 0 \\ 0 \end{bmatrix} . \tag{2.111}$$

4. Compute the projection matrix,

$$\hat{\mathrm{P}} := \hat{\mathrm{I}} - \hat{\mathrm{B}}'\left(\hat{\mathrm{B}}\hat{\mathrm{B}}'\right)^{-1}\hat{\mathrm{B}}\,. \tag{2.112}$$

5. Compute the steepest descent direction,

$$\boldsymbol{\pi} = -\hat{\mathrm{P}}\mathbf{d}\,. \tag{2.113}$$

6. Set $\theta = -\min_j \pi_j$.
7. Test for unbounded objective: *If $\theta \leq 0$, then the objective is unbounded, and hence stop the algorithm!*
8. Compute the update for $0 < \alpha < 1$ (in practice, $0.9 \leq \alpha \leq 0.95$), that means,

$$\begin{cases} \tilde{\mathbf{s}}_1^{k+1} = \hat{\mathrm{I}} + \frac{\alpha}{\theta}\boldsymbol{\pi}\,, \\ \mathbf{s}^{k+1} = \hat{\mathrm{D}}\tilde{\mathbf{s}}^{k+1}\,. \end{cases} \tag{2.114}$$

9. Termination check for $\epsilon > 0$: Stop, if it holds that,

$$\left\|\mathbf{s}^{k+1} - \mathbf{s}^k\right\| < \epsilon\,. \tag{2.115}$$

---

**Remark 2.8.22** (Dikin's algorithm versus the simplex method). **1.** We note that for very large LP problems formulated for practical problems, the interior-point methods such as Dikin's algorithm are often faster than the Simplex Method. **2.** Moreover, the above algorithm and the subsequent Python implementation follow the intuitive presentation from [39], that is, the matrix $\hat{\mathrm{P}}$ is explicitly computed instead of applying a factorization such as the QR-decomposition. ◇

**Dikin's Method**, as stated above, can be implemented by the following lines of Python code:

```
def dikinsLP(lmbd,L0,n):
    # INPUT:
    # x0: n-dimensional numpy array representing an initial feasible
    #        guess
    # A: (m,n)-system numpy matrix of the LP problem transformed into
    #    standard form via slack variables
    # c: n-dimensional numpy array representing coefficients of
    #      linear cost/goal function, i.e., w=c^Tx
    #
    # OUTPUT:
    # LP solution x

    # Parameter of algorithm:
    alph = 0.9  # choice in most codes

    # Start loop
```

```
#
k = 0
xOld = x0
err = 1.0e19   # initial error satisfying while-condition
while err > eps:

    # increase counter.
    k += 1

    # Step 1: Create matrix D = diag(x^k)
    D = np.diag(xOld)

    # Step 2: Compute centering transformations
    B = np.matmul(A,D)
    d = np.matmul(D,c)

    # Step 3: Compute the projection matrix P
    P = np.eye(np.shape(c)[0])-np.matmul(B.T, \
        np.matmul(np.linalg.inv(np.matmul(B,B.T)),B))

    # Step 4: Compute the steepest descent direction
    p = -np.matmul(P,d)

    # Step 5: Check boundedness of objective
    theta = -min(p)
    if theta <= 0.0:
        print("Attention: objective is unbounded  => stopping!")
        return

    # Step 6: Compute update
    xTmp = np.ones(np.shape(c)[0]) + alph/theta*p
    # transform back
    x = np.matmul(D,xTmp)

    # Step 7: Termination check
    err = np.linalg.norm(x-xOld)
    if err 0<= eps:
        print("Success: Dikin's algorithm finished with " \
              +str(k)+" iterations!")
        return x
    elif k >= maxItr:
```

```
            print("Attention: algorithm did not converge for " \
               +str(maxItr)+" iterations!")
            return x

        # Step 8: Print & reset variables for next iteration
        print("Iteration k =",k)
        print("Cost/goal value w =",np.matmul(c,x))
        print("L2-err =",err)
        print("--------------------------------------")
        xOld = x

    return x
```

We consider an example of solving an LP problem with the help of Dikin's algorithm.

**Example 2.8.23** (Application of Dikin's algorithm). Consider interactions between systems $\mathcal{S}_1$ and $\mathcal{S}_2$ governed by the following interaction matrix,

$$\hat{\mathrm{G}} := \begin{bmatrix} 1 & 3 \\ 4 & 0 \end{bmatrix} . \tag{2.116}$$

Before we are able to apply **Dikin's Method**, we need to cast the interacting CHeS into *standard form*. In a first step, this involves to rewrite (2.116) as an LP problem for system $\mathcal{S}_1$, that is,

$$\text{Minimize } w = L(\tilde{\mathbf{s}}_1) = \tilde{p}_1 + \tilde{p}_2 \,, \tag{2.117}$$

*subject to,*

$$\begin{cases} \tilde{p}_1 + 4\tilde{p}_2 & \geq 1\,, \\ 3\tilde{p}_1 & \geq 1\,, \\ \tilde{p}_1, \tilde{p}_2 & \geq 0\,. \end{cases} \tag{2.118}$$

Hence, with the help of the additional *surplus variables* $\tilde{p}_3 \geq 0$ and $\tilde{p}_4 \geq 0$, the *standard form LP problem* reads as follows,

$$\text{Minimize } w = L_{\text{sp}}(\tilde{\mathbf{s}}_{1,\text{sp}}) = \mathbf{c}'\tilde{\mathbf{s}}_{1,\text{sp}} = \tilde{p}_1 + \tilde{p}_2 + \tilde{p}_3 + \tilde{p}_4 \,, \tag{2.119}$$

*subject to,*

$$\begin{cases} \tilde{p}_1 + 4\tilde{p}_2 - \tilde{p}_3 & = 1\,, \\ 3\tilde{p}_3 - \tilde{p}_4 & = 1\,, \\ \tilde{p}_1, \tilde{p}_2, \tilde{p}_3, \tilde{p}_4 & \geq 0\,. \end{cases} \tag{2.120}$$

We can solve the latter problem with **Dikin's Method** in Python with the help of the previously stated Python version of Dikin's algorithm by running the following lines of code:

```
import numpy as np

Gsp = -np.array([[-1,-4,1,0],[-3,0,0,1]]) # surplus formulation
c = np.array([1,1,0,0])
x0 = np.array([0.5,0.25,0.5,0.5])

maxItr = 200  # maximal number of iterations in Dikin's algorithm
eps = 0.1e-6   # threshold parameter stopping the iteration
x = dikinsLP(x0,Gsp,c,eps,maxItr)
print("x=",x)
```

The output obtained during runtime of the Dikin's method is shown in Figure 2.17.

The mixed action/solution to the original interaction system governed by (2.116) finally reads as follows,

$$\mathbf{s}_1 = \frac{\tilde{\mathbf{s}}_1}{w} = \frac{\tilde{\mathbf{s}}_1}{\mathbf{c}'\tilde{\mathbf{s}}_1}, \tag{2.121}$$

where $\tilde{\mathbf{s}}_1$ denotes the solution obtained by **Dikin's Method** applied to **Slack-(RLPP)**. We can realize this last step in Python as follows:

```
xSol = 1/np.dot(c,x)*x
print("Optimal mixed action for player 1:",xSol[0:2])
```

with the resulting final solution/mixed action depicted in Figure 2.18.

For completeness, we also look at the optimal/reasonable action for system $\mathcal{S}_2$, whose associated LP problem reads as follows:

*Maximize* $L^*(\mathbf{s}_2') = \mathbf{c}^T\mathbf{s}_2' := y_1' + y_2'$ *subject to,*

$$\begin{cases} y_1' + 3y_2' & \leq 1\,, \\ 4y_1' & \leq 1\,, \\ y_1', y_2' & \geq 0\,. \end{cases} \tag{2.122}$$

As for $\mathcal{S}_1$, we can rewrite this maximization problem (2.122) with the help of *slack variables* by the following *standard LP problem:*

```
Iteration k = 1
Cost/goal value z = 0.534375
L2-err = 0.632030767150619
-----------------------------------
Iteration k = 2
Cost/goal value z = 0.50816638227442
L2-err = 0.08523083399662765
-----------------------------------
Iteration k = 3
Cost/goal value z = 0.5020644407991554
L2-err = 0.022895943838080926
-----------------------------------
Iteration k = 4
Cost/goal value z = 0.500525149801885
L2-err = 0.005503107907543225
-----------------------------------
Iteration k = 5
Cost/goal value z = 0.5001333624775012
L2-err = 0.0014521225527726602
-----------------------------------
Iteration k = 6
Cost/goal value z = 0.5000339018270685
L2-err = 0.0003579193139119047
-----------------------------------
Iteration k = 7
Cost/goal value z = 0.5000086145743419
L2-err = 9.346564038008757e-05
-----------------------------------
Iteration k = 8
Cost/goal value z = 0.5000021894211409
L2-err = 2.3150962384584642e-05
-----------------------------------
Iteration k = 9
Cost/goal value z = 0.5000005564020235
L2-err = 6.032422954908445e-06
-----------------------------------
Iteration k = 10
Cost/goal value z = 0.5000001414311066
L2-err = 1.4956820757834087e-06
-----------------------------------
Iteration k = 11
Cost/goal value z = 0.5000000357145153
L2-err = 3.896088218223215e-07
-----------------------------------
Success: Dikin's algorithm ordinarily finished with 12 iterations!
x= [3.33333343e-01 1.66666667e-01 1.03152477e-08 2.62156840e-08]
```

Figure 2.17: *Output of Dikin's Method during runtime.*

```
Optimal mixed action for player 1: [0.66666667 0.33333333]
```

Figure 2.18: *Normalized LP solution:* $\mathcal{S}_1$'s reasonable mixed action $\mathbf{s}_1 = [2/3, 1/3]'$.

$$\textit{Minimize } L^*_{\mathrm{sl}}(\mathbf{s}'_{2,\mathrm{sl}}) = -\mathbf{c}^T\mathbf{s}_{2,\mathrm{sl}} := y'_1 + y'_2 + y'_3 + y'_4 = w \textit{ subject to,}$$

$$\begin{cases} y'_1 + 3y'_2 + y'_3 & = 1, \\ 4y'_1 + y'_4 & = 1, \\ y'_1, y'_2, y'_3, y'_4 & \geq 0, \end{cases} \tag{2.123}$$

where $\mathbf{c}^T = [1, 1, 0, 0]$.

Let us solve this LP problem with the help of Dikin's algorithm in Python:

```
import numpy as np

A = -np.array([[-1,-3,-1,0],[-4,0,0,-1]])
c = -np.array([1,1,0,0])
y0 = np.array([0.125,0.25,0.125,0.5])

maxItr = 200  # maximal number of iterations
eps = 0.1e-6  # threshold parameter stopping the iteration
y = dikinsLP(y0,G,c,eps,maxItr)
print("y=",y)
```

Finally, we obtain the reasonable action $\mathbf{s}_2$ via the corresponding equation (2.121) albeit neglecting the sign of the cost by the following associated code lines:

```
ySol = 1/np.abs(np.dot(c,y))*y
print("Optimal mixed action for player 1:",ySol[0:2])
```

The application of the first part of the above code lines leads to the output shown in Figure 2.19. The second part of the code just prints $\mathcal{S}_2$'s final solution as can be seen in Figure 2.20.

As a final validation step, we also provide the analytical derivation of the above numerical result. First, we perform the Saddle Point Check (2.76) as follows,

$$\begin{array}{cc|cc} 1 & 3 & 1 & \mathbf{1} \\ 4 & 0 & 0 & \\ \hline 4 & 3 & & \\ & \mathbf{3} & & \end{array} . \tag{2.124}$$

As expected, this interaction problem does not have a pure action as an equilibrium strategy. This requires us to compute RMAs by hand,

$$\begin{array}{cc|cc} 1 & 3 & -2 & \frac{4}{6} \\ 4 & 0 & 4 & \frac{2}{6} \\ \hline -3 & 3 & & \\ \frac{3}{6} & \frac{3}{6} & & \end{array} , \tag{2.125}$$

so that the RMAs for $\mathcal{S}_1$ and $\mathcal{S}_2$ read,

$$\begin{cases} \mathbf{s}_1 := [2/3, 1/3]' , \\ \mathbf{s}_2 := [1/2, 1/2]' . \end{cases} \tag{2.126}$$

◇

```
Iteration k = 1
Cost/goal value z = -0.48750000000000004
L2-err = 0.4772970773009196
----------------------------------
Iteration k = 2
Cost/goal value z = -0.49688097266720577
L2-err = 0.046764284306976156
----------------------------------
Iteration k = 3
Cost/goal value z = -0.49921156277856443
L2-err = 0.006680884964974156
----------------------------------
Iteration k = 4
Cost/goal value z = -0.4997994283295333
L2-err = 0.0031310087349738615
----------------------------------
Iteration k = 5
Cost/goal value z = -0.4999490662341331
L2-err = 0.00042596565776995476
----------------------------------
Iteration k = 6
Cost/goal value z = -0.49998705208330474
L2-err = 0.00020392813556111185
----------------------------------
Iteration k = 7
Cost/goal value z = -0.49999670990821726
L2-err = 2.7453018541592882e-05
----------------------------------
Iteration k = 8
Cost/goal value z = -0.49999916381484794
L2-err = 1.3194060332381309e-05
----------------------------------
Iteration k = 9
Cost/goal value z = -0.49999978752880464
L2-err = 1.7723610198036612e-06
----------------------------------
Iteration k = 10
Cost/goal value z = -0.49999994611422727
L2-err = 8.52453503502969e-07
----------------------------------
Iteration k = 11
Cost/goal value z = -0.49999998619365604
L2-err = 1.1446060589160938e-07
----------------------------------
Success: Dikin's algorithm ordinarily finished with 12 iterations!
y= [2.49999998e-01 2.49999999e-01 7.50926789e-09 5.90942980e-09]
```

Figure 2.19: *Output of Dikin's Method during runtime:* Solving $\mathcal{S}_2$'s LP problem.

```
Optimal mixed strategy for player 1: [0.5 0.5]
```

Figure 2.20: *Normalized LP solution:* $\mathcal{S}_2$'s reasonable mixed action $\mathbf{s}_2 = [1/2, 1/2]'$.

### 2.8.1 Learning interaction strategies: the fictitious play

Let us first look at specific aspects of the *interacting CHeSs' theory* presented so far, such as assumptions made and solutions gained. In particular, a careful and critical investigation will help us to increase the practical relevance of this theory. This will allow us to choose the correct approach and formulation and to extract meaningful and statistically relevant information from results gained hereby.

Naturally, capturing decision relevant information in any interaction framework requires a careful and systematic approximation taking the key/dominant characteristics into account. To this end, assumptions on the interactions of interest provide a rigorous tool for selecting relevant properties and for neglecting contributions showing diminishing influence on decisions/actions of involved systems. At the same time, we need also to be careful with the correct interpretation and application of results computed with interaction frameworks as presented in the previous sections. For instance, the probabilities of mixed actions/strategies do not represent an element of information that allows us to make the right choice in a single action/draw, but rather over the course of multiple actions/draws. This means that we should select actions such that the normalized frequencies of the chosen actions correspond to the computed mixed action/strategy in the end.

Here, our primary focus is the fact that we cannot be sure that the opposing system acts reasonably with respect to our choice of utility/cost function, i. e., it does not need to act according to the (optimal/equilibrium) mixed action due to complexity inferring lack of detail/knowledge in the utility, for instance. A historical example is the refinement of Bernoulli's *Utility Concept* [15] toward the *Prospect Theory* by Kahneman and Tversky [81] by adding a *Reference Point* and a *Loss Aversion*. Since the assumption about a rationally acting opposing system is crucial to many results rigorously developed in game theory (e. g., [170]) and hence of relevance for interacting CHeSs, we believe that it will serve useful in applications to look at approaches that extend existing rigorous statements toward situations where the "rationality" assumption is questionable due to uncertainties in the utility/cost function describing the scenario of interest. Recall that rationality is directly linked to the underlying objective function. It is exactly this scenario that we want to have a closer look at in this section. We propose a straightforward data-based approach by *learning the opposing system strategy* over several (fictitious) interactions.

The concept of *fictitious interactions*[14] provides a simple and intuitive learning strategy, and hence we look at it in more detail. This learning method seems to go back to a result of an iterative solution concept proposed in [30] and which has been rigorously proved in [128]. Before we state the approach, we describe the context of its relevance. Consider two systems $\mathcal{S}_1$ and $\mathcal{S}_2$, whose interactions are governed by an interaction matrix $\hat{\mathrm{G}} \in \mathbb{R}^{m_1 \times m_2}$, where $m_1, m_2 \in \mathbb{N}_{\geq 0}$. Moreover, let $\mathbf{s}_1 \in \mathbb{R}^{m_1}$ and $\mathbf{s}_2 \in \mathbb{R}^{m_2}$ denote the equilibrium/optimal strategies for the systems $\mathcal{S}_1$ and $\mathcal{S}_2$, respectively.

---

**Fictitious Interaction Learning Method**

Assume that we know $K \in \mathbb{N}_{\geq 0}$ previous actions chosen by system $\mathcal{S}_2$, that is, $\mathbf{s}_2^1, \mathbf{s}_2^2, \ldots, \mathbf{s}_2^K \in \mathbb{R}^{m_2}$. Then we can estimate the following *average mixed action* for system $\mathcal{S}_2$,

$$\tilde{\mathbf{s}}_2 = \frac{1}{K} \sum_{k=1}^{K} \mathbf{s}_2^k, \tag{2.127}$$

and therefore, we obtain the deviation of $\mathcal{S}_2$ by $\mathbf{d}_2 := \tilde{\mathbf{s}}_2 - \mathbf{s}_2$ from its equilibrium/optimal interaction choice $\mathbf{s}_2$. Hence, system $\mathcal{S}_1$'s adapted/learned equilibrium/optimal reaction reads,

---

14 Fictitious interactions are referred to as *fictitious play* in game theory.

$$\bar{\mathbf{s}}_1 = \max_{\mathbf{s} \in \mathbb{R}^{m_1}_{\geq 0}} \mathbf{s}' \hat{G} \tilde{\mathbf{s}}_2 , \tag{2.128}$$

where $\mathbf{s} \in \mathbb{R}^{m_1}_{\geq 0}$ chosen in the minimization is considered to be a mixed action.

---

We will demonstrate the use of this method in the context of *monitoring Cyber Physical Systems (CPSs)*, see the subsequent Example 2.8.24.

**Example 2.8.24** (Learning from cyber intrusions). We make the following assumptions for our CPS of interest:

**(A1)** The CPS consists of $R$ critical(/valuable) assets. The value of the $i$-th asset is $i \in \mathbb{N}$, where $1 \leq i \leq R$. The CPS's owner has a surveillance/monitoring capacity of one asset per check, that is, there are $m_1 := \binom{R}{1}$ possible choices per check.[15]

**(A2)** Attackers are interested in the $R$ assets of the CPS but can only attack two assets at the same time, that is, they have to pick $m_2 := \binom{R}{2}$ specific assets per attack.

**(A3)** Using information **(A1)** and **(A2)**, we can assign the following cost function governing associated interactions,

$$g(\mathbf{s}_{1,i}, \mathbf{s}_{2,j}) := \sum_{k=1}^{R} k s_{1,i,k} s_{2,j,k} + k s_{2,j,k} (s_{1,i,k} - 1) . \tag{2.129}$$

These assumptions allow us to define interactions with the help of the interaction matrix $\hat{G} := \{g(\mathbf{s}_{1,i}, \mathbf{s}_{2,j})\}_{i,j=1}^{m_1,m_2}$, such that we have,

$$\hat{G} := \begin{bmatrix} g(\mathbf{s}_{1,1}, \mathbf{s}_{2,1}) & g(\mathbf{s}_{1,1}, \mathbf{s}_{2,2}) & \dots & g(\mathbf{s}_{1,1}, \mathbf{s}_{2,m_2}) \\ g(\mathbf{s}_{1,2}, \mathbf{s}_{2,1}) & g(\mathbf{s}_{1,2}, \mathbf{s}_{2,2}) & \dots & g(\mathbf{s}_{1,2}, \mathbf{s}_{2,m_2}) \\ \vdots & & \ddots & \vdots \\ g(\mathbf{s}_{1,m_1}, \mathbf{s}_{2,1}) & g(\mathbf{s}_{1,m_1}, \mathbf{s}_{2,2}) & \dots & g(\mathbf{s}_{1,m_1}, \mathbf{s}_{2,m_2}) \end{bmatrix} . \tag{2.130}$$

For simplicity, we restrict ourselves to the case of $R = 3$ assets in total. Using Assumption **(A3)**, we immediately can identify the following interaction table, i. e.,

| $\mathbf{s}_1 \setminus \mathbf{s}_2$ | $(1,1,0)$ | $(0,1,1)$ | $(1,0,1)$ |
|---|---|---|---|
| $(1,0,0)$ | $-1$ | $-5$ | $-2$ |
| $(0,1,0)$ | $1$ | $-1$ | $-4$ |
| $(0,0,1)$ | $-3$ | $1$ | $2$ |

, (2.131)

which immediately gives the interaction matrix by,

$$\hat{G} := \begin{bmatrix} -1 & -5 & -2 \\ 1 & -1 & -4 \\ -3 & 1 & 2 \end{bmatrix} . \tag{2.132}$$

---

15 Recall that $\binom{N}{n} = \frac{N!}{n!(N-n)!}$.

To numerically compute reasonable actions for $\mathcal{S}_1$ and $\mathcal{S}_2$, we first need to apply the *equivalence between IM and LP*, i. e., identify (2.99) and (2.100) and their reformulation in *standard form*.

Let us first compute the equilibrium actions for both systems, i. e., the CPS monitoring system $\mathcal{S}_1$ and the attacking system $\mathcal{S}_2$, with the help of Dikin's algorithm. To this end, we add the constant $\kappa = 1 - \min_{i,j} g_{ij}$ to all elements of the interaction matrix $\hat{\mathrm{G}}$, such that we end up with the following positive interaction matrix,

$$\hat{\mathrm{G}} := \begin{bmatrix} 5 & 1 & 4 \\ 7 & 5 & 2 \\ 3 & 7 & 8 \end{bmatrix}, \tag{2.133}$$

whose variable we keep the same as of its original matrix (2.132). Note that this constant shift of the components in the interaction matrix does not change anything regarding reasonable/equilibrium actions, but it shifts the fairness level of the interaction.

For $\mathcal{S}_1$, we first need to compute a feasible solution representing an initial guess. We perform this directly in Python with the subsequent lines of code:

```
import numpy as np

# interaction matrix
C = np.array([[5,1,4],[7,5,2],[3,7,8]]).T

x_4 = 0.05 # surplus variable
x_5 = 0.20 # surplus variable
x_6 = 0.075 # surplus variable
b = np.array([[1-x_4],[1-x_5],[1-x_6]])

x_00 = np.linalg.solve(C,b)
x0 = np.append(x_00,[x_4,x_5,x_6])
x0
```

```
array([0.07476852, 0.05462963, 0.06458333, 0.05      , 0.2       ,
       0.075     ])
```

Figure 2.21: *Feasible solution:* representing an initial guess for Dikin's method.

The output of the previous code lines gives a feasible solution as shown in Figure 2.21. Next, we apply Dikin's algorithm to a surplus-variable-reformulated LP problem as shown by the following Python code:

```
Gsp = np.array([[5,7,3,1,0,0],[1,5,7,0,1,0],[4,2,8,0,0,1]]
c = np.array([1,1,1,0,0,0])  # coefficients of the linear cost
```

```
maxItr = 200  # maximal number of iterations in Dikin's algorithm
eps = 0.1e-6   # threshold parameter stopping the iteration
x = dikinsLP(x0,Gsp,c,eps,maxItr)
print("x=",x)

xSol = 1/np.dot(c,x)*x
print("Optimal mixed strategy for player 1:",xSol[0:3])
```

The result obtained from Dikin's algorithm called in the above lines of codes reads as depicted in Figure 2.22, and its associated mixed strategy follows in Figure 2.23.

Similarly, we can compute the reasonable equilibrium action for $\mathcal{S}_2$. We first compute a feasible solution (initial guess) $\mathbf{s}_2^0$ for $\mathcal{S}_2$ by the corresponding lines of code:

```
import numpy as np

C = np.array([5,1,4],[7,5,2],[3,7,8]])  # interaction matrix

y_4 = 0.05 # slack variable
y_5 = 0.20 # slack variable
y_6 = 0.075 # slack variable
b = np.array([[1-y_4],[1-y_5],[1-y_6]])

y_00 = np.linalg.solve(C,b)
y0 = np.append(y_00,[y_4,y_5,y_6])
y0
```

The execution of the above Python instructions leads to the result printed in Figure 2.24.

This allows us to apply Dikin's algorithm to a slack-variable-reformulated LP problem as realized in the following Python code:

```
Asl = np.array([[5,1,4,1,0,0],[7,5,2,0,1,0],[3,7,8,0,0,1]])
c =- np.array([1,1,1,0,0,0])  # coefficients of the linear cost

maxItr = 200  # maximal number of iterations in Dikin's algorithm
eps = 0.1e-6   # threshold parameter stopping the iteration
y = dikinsLP(y0,Asl,c,eps,maxItr)
print("y=",y)

ySol = 1/np.abs(np.dot(c,y))*y
print("Optimal mixed strategy for player 1:",ySol[0:3])
```

```
Iteration k = 1
Cost/goal value z = 0.16030647001240583
L2-err = 0.2495884278291641
-----------------------------------
Iteration k = 2
Cost/goal value z = 0.11464268083145315
L2-err = 0.38882490620203314
-----------------------------------
Iteration k = 3
Cost/goal value z = 0.055523803865769605
L2-err = 0.5375538552648231
-----------------------------------
Iteration k = 4
Cost/goal value z = 0.009346890837780737
L2-err = 0.5102640348990016
-----------------------------------
Iteration k = 5
Cost/goal value z = 0.0024484760452320813
L2-err = 0.06712021187321061
-----------------------------------
Iteration k = 6
Cost/goal value z = 0.0008213747535716189
L2-err = 0.013537665394461454
-----------------------------------
Iteration k = 7
Cost/goal value z = 0.00017863488700832898
L2-err = 0.005175011452464272
-----------------------------------
Iteration k = 8
Cost/goal value z = 5.562141098419328e-05
L2-err = 0.0010546639043007861
-----------------------------------
Iteration k = 9
Cost/goal value z = 1.1796273028716828e-05
L2-err = 0.0003660036021741205
-----------------------------------
Iteration k = 10
Cost/goal value z = 3.754718935685096e-06
L2-err = 6.534071851689051e-05
-----------------------------------
Iteration k = 11
Cost/goal value z = 8.083356518182431e-07
L2-err = 2.3758867579734144e-05
-----------------------------------
Iteration k = 12
Cost/goal value z = 2.5274740558181406e-07
L2-err = 4.756788411431966e-06
-----------------------------------
Iteration k = 13
Cost/goal value z = 5.345761081952855e-08
L2-err = 1.663813156931287e-06
-----------------------------------
Iteration k = 14
Cost/goal value z = 1.7043669660338304e-08
L2-err = 2.960752614412582e-07
-----------------------------------
Iteration k = 15
Cost/goal value z = 3.6689119804812545e-09
L2-err = 1.0785174874965605e-07
-----------------------------------
Success: Dikin's algorithm ordinarily finished with 16 iterations!
x= [4.68716808e-10 1.81111903e-10 4.97245412e-10 9.99999995e-01
 9.99999995e-01 9.99999994e-01]
```

Figure 2.22: *LP solution:* Computed with Dikin's algorithm.

```
Optimal mixed strategy for player 1: [0.40861946 0.15789032 0.43349022]
```

Figure 2.23: *Mixed action:* $\mathbf{s}_1 := [0.41, 0.16, 0.43]$ represents the frequencies with respect to which checks of the $R = 3$ assets have to be performed.

```
array([1.30554630e-01, 2.77648148e-03, 3.61125926e-02, 2.00000000e-01,
       1.00000000e-05, 3.00000000e-01])
```

Figure 2.24: *Feasible solution for $\mathcal{S}_2$:* Representing an initial guess for Dikin's method.

```
Iteration k = 1
Cost/goal value z = -0.19642227951100505
L2-err = 0.28948106849064326
----------------------------------
Iteration k = 2
Cost/goal value z = -0.19921503827451112
L2-err = 0.029734378997251676
----------------------------------
Iteration k = 3
Cost/goal value z = -0.19981998257417793
L2-err = 0.010585708587711845
----------------------------------
Iteration k = 4
Cost/goal value z = -0.19995368329505017
L2-err = 0.0014255894198536949
----------------------------------
Iteration k = 5
Cost/goal value z = -0.19998750776588953
L2-err = 0.0006747144161471782
----------------------------------
Iteration k = 6
Cost/goal value z = -0.19999620896891873
L2-err = 9.222745788112964e-05
----------------------------------
Iteration k = 7
Cost/goal value z = -0.19999871219707038
L2-err = 4.5656548385338724e-05
----------------------------------
Iteration k = 8
Cost/goal value z = -0.19999974774464882
L2-err = 9.145709686624093e-06
----------------------------------
Iteration k = 9
Cost/goal value z = -0.199999914324624
L2-err = 2.9943113561751126e-06
----------------------------------
Iteration k = 10
Cost/goal value z = -0.1999999812783381
L2-err = 5.934507611230377e-07
----------------------------------
Iteration k = 11
Cost/goal value z = -0.19999998277334335
L2-err = 2.0233751898540408e-07
----------------------------------
Success: Dikin's algorithm ordinarily finished with 12 iterations!
y= [1.19999949e-01 3.06921720e-09 7.99999798e-02 8.00000131e-02
 2.52172192e-09 3.90530104e-09]
```

Figure 2.25: *LP solution for $\mathcal{S}_2$:* Computed with Dikin's algorithm.

The result obtained with the above Python lines calling Dikin's algorithm is shown in Figure 2.25, and the output of the final reasonable mixed action $\mathbf{s}_2$, which requires an appropriate normalization as written in the second last lines of code, reads as printed in Figure 2.26.

```
Optimal mixed strategy for player 1: [5.99999950e-01 1.53460912e-08 4.00000035e-01]
```

Figure 2.26: *Reasonable mixed action for $\mathcal{S}_2$:* Attack the $R = 3$ assets with frequencies as given in the following mixed action $\mathbf{s}_2 = [0.6, 0, 0.4]$.

So far, we have not applied the **Fictitious Interaction Learning Method (FILM)** and only analyzed the ideal situation where both, CHeSs $\mathcal{S}_1$ and $\mathcal{S}_2$, are considered to be rational. To account for deviations in rationality, we depend on additional information such as historical data on actions chosen by the opposing CHeS $\mathcal{S}_2$. To this end, we introduce an additional assumption:

**(A4)** *We know $K = 10$ past attacks/actions chosen by $\mathcal{S}_2$, that is,*

$$A_2^{\text{hist}} := \{[1,0,1],[0,1,1],[1,0,1],[1,0,1],[0,1,1],[1,1,0],[1,0,1],\\ [1,1,0],[1,0,1],[1,0,1]\}\,. \tag{2.134}$$

Assumption **(A4)** allows us to compute the average mixed action of the attacker $\mathcal{S}_2$, which reads,

$$\tilde{\mathbf{s}}_2 := [0.4, 0.2, 0.4]'\,. \tag{2.135}$$

Herewith, we can determine the so-called adapted/learned equilibrium/reasonable action via (2.128), that is, the reasonable strategy for $\mathcal{S}_1$ amounts to solving the following problem,

$$\bar{\mathbf{s}}_1 = \max_{\mathbf{s}\in\mathbb{R}_{\geq 0}^{m_1}} \mathbf{s}'\hat{\mathbf{G}}\tilde{\mathbf{s}}_2\,. \tag{2.136}$$

Applying (2.135) in (2.136) finally leads to the *pure action* $\bar{\mathbf{s}}_1 := [0,0,1]'$ for $\mathcal{S}_1$,

$$\begin{aligned}\bar{\mathbf{s}}_1 &= \max_{\mathbf{s}\in\mathbb{R}_{\geq 0}^{m_1}} 3.8s_{1,1} + 4.6s_{1,2} + 5.8(1 - s_{1,1} - s_{1,2})\\ &= \max_{\mathbf{s}\in\mathbb{R}_{\geq 0}^{m_1}} 5.8 - 2.0s_{1,1} - 1.2s_{1,2}\\ &= [0,0,1]'\,,\end{aligned} \tag{2.137}$$

which implies on average a gain of 5.8 (but recall the previously applied shift in the fairness level). ◇

The result obtained by the FILM might seem to be obvious after having worked out the solution: *always go for the most valuable asset*! However, we motivate that this is generally not the case as we can see by the following attack-defense interaction (e. g., see [175]) between two systems/countries $\mathcal{S}_1$ and $\mathcal{S}_2$,

$$\hat{\mathbf{G}} := \begin{bmatrix} 4 & 1 \\ 3 & 4 \end{bmatrix}. \tag{2.138}$$

The background information for the above interaction matrix $\hat{G}$ is that $\mathcal{S}_1$ only has the ability to protect one of two assets $a_1$ and $a_2$. The values of assets $a_1$ and $a_2$ are 1 and 3, respectively. Similarly, $\mathcal{S}_2$ can only attack one asset at a time. Applying the previously established analytical method for computing reasonable actions for both systems/parties $\mathcal{S}_1$ and $\mathcal{S}_2$, that is, first performing the *Saddle Point Check* and then computing RMAs due to the absence of pure actions (i. e., no saddle point) will immediately demonstrate that $\mathcal{S}_1$ should not only protect the most valuable asset $a_2$ (which corresponds to a pure action/strategy), but also to defend both assets according to the following RMA,

$$\mathbf{s}_1 = [1/4, 3/4]' . \tag{2.139}$$

At least, our intuition is satisfied by the fact that $\mathcal{S}_1$ should defend the more valuable asset more often. Similarly, $\mathcal{S}_2$'s attack strategy should not be a pure action but the following RMA,

$$\mathbf{s}_1 = [3/4, 1/4]' , \tag{2.140}$$

which amounts to attacking the less valuable asset $a_1$ more often than the more valuable asset $a_2$.

As a consequence, we motivate that a system $\mathcal{S}_1$ can learn the level of rationality of its opposing system $\mathcal{S}_2$ by wisely triggering actions $\mathbf{s}_2$ in a seemingly unintentional interaction between $\mathcal{S}_1$ and $\mathcal{S}_2$. In the cyberspace, for this purpose, so-called honey pots are placed, which are nonvaluable assets in a system with valuable assets to distract attackers. These honeypots can be used to learn how an attacker circumvents security measures. Naturally, such a *seemingly unintentional interaction*/honeypot should be designed in such a way that it does not raise suspicion of $\mathcal{S}_2$.

To conclude, the framework for *smart interacting CHeSs* represents the *intelligence* for choosing the right action in a particular scenario, and the *FILM* serves as a straightforward way for learning from previous experiences.

### 2.8.2 Framework for decision dynamics: kinematic evolution and utility functions

Let us extend the previously developed interaction framework so far, to account for decision dynamics based on specific state evolutions of the systems $\mathcal{S}_1$ and $\mathcal{S}_2$ and to allow for each system to choose a specific control in time on its state evolution.

**Systems' evolution: state dynamics.** For simplicity, we present the fundamental ideas based on kinematic state evolutions associated with the systems $\mathcal{S}_1$ and $\mathcal{S}_2$ and described by the coordinates $\mathbf{x}^n \in \mathbb{R}^2$ and $\mathbf{y}^n \in \mathbb{R}^2$, respectively, where $n \in \mathbb{N}$ denotes a discrete time step $t_n \in \mathbb{R}_{\geq 0}$ such that,

$$\mathbf{x}^{n+1} = \mathbf{x}^n + \alpha \Delta t \mathbf{v}_i^n , \tag{2.141}$$

$$\mathbf{y}^{n+1} = \mathbf{y}^n + \alpha\Delta t \mathbf{w}_j^n\,, \tag{2.142}$$

where $\mathbf{v}_i^n$ and $\mathbf{w}_j^n$ admit the following fixed velocities,

$$\mathbf{v}_i^n = \mathbf{w}_i^n := \begin{cases} [1,1]\,, & i = 1\,,\\ [1,0]\,, & i = 2\,,\\ [1,-1]\,, & i = 3\,,\\ [0,-1]\,, & i = 4\,,\\ [-1,-1]\,, & i = 5\,,\\ [-1,0]\,, & i = 6\,,\\ [-1,1]\,, & i = 7\,,\\ [0,1]\,, & i = 8\,, \end{cases} \tag{2.143}$$

and $\alpha$ denotes a parameter depending on the distance between the systems $\mathcal{S}_1$ and $\mathcal{S}_2$. From a physical point of view, the parameter $\alpha$ can be chosen to be a distance-dependent velocity. For instance, by setting $\alpha := \|\mathbf{x}^n - \mathbf{y}^n\|$ (for time 1), it increases with growing distance between $\mathcal{S}_1$ and $\mathcal{S}_2$.

**Objective functions: decisions and path planning.** There are two naturally arising objective/payoff functions associated with the above systems $\mathcal{S}_1$ and $\mathcal{S}_2$.

**1. Kinematic interaction.** The first and simpler payoff function models so-called kinematic interactions based on the distance between the two systems. That is, larger distances favor approaching movements, whereas shorter distances might lead to oscillating behavior due to possible cancelation effects enabled by the opposite sign of summands. Hence, the described payoff function reads as follows,

$$\begin{aligned} g_{ij} &:= g(\mathbf{v}_i^n, \mathbf{w}_j^n) \\ &:= \frac{1}{2}\|\mathbf{x}^n - \mathbf{y}^n - \alpha\Delta t \cdot \mathbf{w}_j^n\|^2 - \frac{1}{2}\|\mathbf{y}^n - \mathbf{x}^n - \alpha\Delta t \cdot \mathbf{v}_i^n\|^2\,, \end{aligned} \tag{2.144}$$

where $1 \le i,j \le 8$, see (2.143). The components defined by (2.144) allow us to define the following *(kinematic) interaction matrix*,

$$\hat{\mathrm{G}} := \begin{bmatrix} g_{11} & g_{12} & \cdots & g_{18} \\ g_{21} & g_{22} & \cdots & g_{28} \\ \vdots & \vdots & \ddots & \vdots \\ g_{81} & g_{82} & \cdots & g_{88} \end{bmatrix}. \tag{2.145}$$

Since (2.145) denotes a zero-sum interaction matrix, we can apply Dikin's algorithm to find the optimal actions/strategies for the systems $\mathcal{S}_1$ and $\mathcal{S}_2$.

**2. Chaser–Evader interaction.** Simple objective functions $g_{ij}^1$ and $g_{ij}^2$ making system $\mathcal{S}_1$ a chaser and system $\mathcal{S}_2$ an evader, respectively, can be defined as follows,

$$\begin{aligned} g_{ij}^1 &:= g(\mathbf{v}_i^n, \mathbf{w}_j^n) \\ &:= -\beta \|\mathbf{y}^n + \alpha\Delta t \mathbf{w}_j^n - \mathbf{x}^n - \alpha\Delta t \cdot \mathbf{v}_i^n\|^2, \end{aligned} \tag{2.146}$$

and,

$$\begin{aligned} g_{ij}^2 &:= g(\mathbf{v}_i^n, \mathbf{w}_j^n) \\ &:= -\beta \|\mathbf{x}^n + \alpha\Delta t \mathbf{v}_i^n - \mathbf{y}^n - \alpha\Delta t \cdot \mathbf{w}_j^n\|^2. \end{aligned} \tag{2.147}$$

Note that the payoff components defined in (2.146) and (2.147) define the interaction matrices $\hat{\mathrm{G}}^1 := \{g_{ij}^1\}_{i,j=1}^{8,8}$ and $\hat{\mathrm{G}}^2 := \{g_{ij}^2\}_{i,j=1}^{8,8}$, where the column actions of $\hat{\mathrm{G}}^1$ and the row actions of $\hat{\mathrm{G}}^2$ are associated with nature as it does not directly represent the corresponding opponent.[16]

**Remark 2.8.25** (Lennard–Jones interactions). To demonstrate dynamic decision making based on the previous interacting systems theory, we restricted ourselves to primarily geometric payoff functions. However, we motivate that we can apply more sophisticated interaction potentials known from physical and chemical studies of media such as fluids and solids. In this context, the so-called Lennard–Jones potential serves as a natural payoff function to model interactions between equal particles/species. ◇

Of course, velocities (2.143) provide limited control to the dynamics of these single systems. Hence cat-and-mouse games based on the above framework are not expected to lead to a real catch but demonstrate that cats will in a reasonable fashion run after the mices, which themselves show to generally take rational decisions about their escape paths. Nevertheless, this simplicity allows a coarse validation of the here presented strategic decision framework, as we will motivate in the subsequent examples.

**Example 2.8.26** (Kinematic interactions). We consider the evolution of systems $\mathcal{S}_1$ and $\mathcal{S}_2$ based on equations (2.141) and (2.142), respectively. We investigate six different cases, which are characterized by the following initial conditions and time step sizes:

**Case 1.** $\mathbf{x}^0 = [-5, 5]$, $\mathbf{y}^0 = [5, 0]$, and $\Delta t = 5.0$.
**Case 2.** $\mathbf{x}^0 = [-4.5, 5]$, $\mathbf{y}^0 = [5, -5]$, and $\Delta t = 1.9$.
**Case 3.** $\mathbf{x}^0 = [1.5, -5]$, $\mathbf{y}^0 = [-5, 2.5]$, and $\Delta t = 4.575$.
**Case 4.** $\mathbf{x}^0 = [-5, 2.5]$, $\mathbf{y}^0 = [5, -3.5]$, and $\Delta t = 4.3$.
**Case 5.** $\mathbf{x}^0 = [-5, 2.5]$, $\mathbf{y}^0 = [4.5, -3.5]$, and $\Delta t = 2.4$.
**Case 6.** $\mathbf{x}^0 = [-4.5, 3.5]$, $\mathbf{y}^0 = [3.5, -3.5]$, and $\Delta t = 1.525$.

The computations of these six cases lead to the kinematics depicted in Figure 2.27. Since both, the blue and red systems, do not follow the optimal paths belonging to a corre-

16 As for the well-known Prisoner's Dilemma, such interactions are not zero-sum interactions anymore. Hence we introduces nature as an opponent such that we still can apply the zero-sum methods.

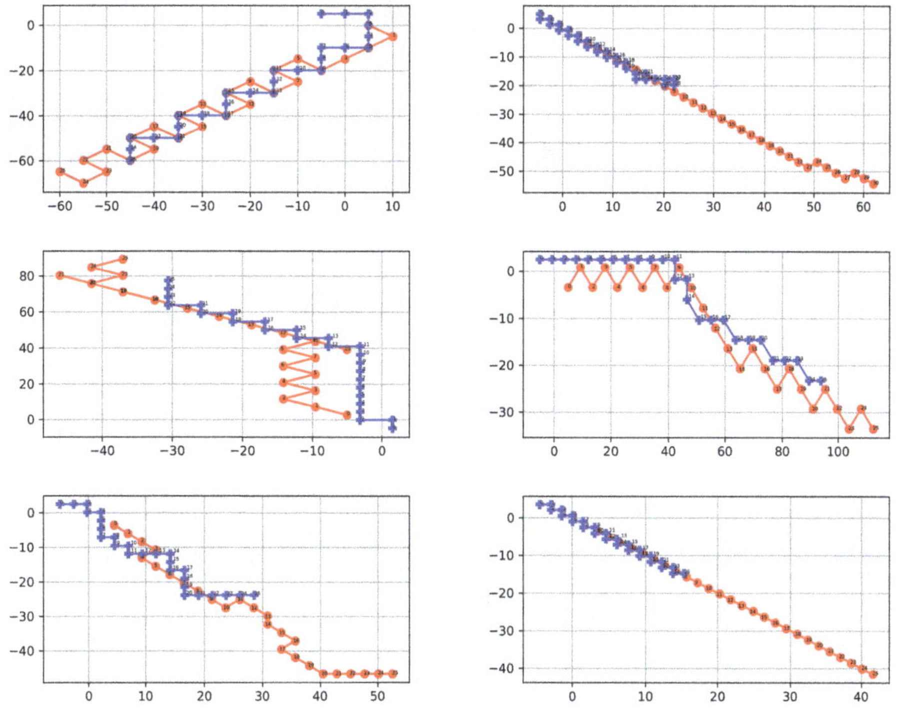


Figure 2.27: *Top left:* Case 1 with the dynamics starting at the top right corner and then turns into a stable oscillating pattern. *Top right:* Case 2 starts in the top left corner with mainly oscillations from the chaser/blue $\mathbf{x}^n$. *Middle left:* Case 3 sets off in the bottom right corner with oscillations alternating between blue and red. *Middle right:* Case 4 begins in the top left corner with diagonally oscillating patterns at the end. *Bottom left:* Case 5 starts at the top left corner, and blue follows red in changing patterns. *Bottom right:* Case 6 shows blue and red moving from the top left to the bottom right corner. Blue has a fixed oscillating movement, whereas red runs off on a straight line.

sponding chaser and evader situation, respectively, we refer to this blue–red system governed by the goal/gain function (2.144) as a kinematic interaction system. ◇

The second geometric payoff example is the chaser–runner interactions presented in the next example.

**Example 2.8.27** (Chaser–Evader interactions). We refine the payoff function into the two specific payoff functions (2.146) and (2.147), which define a so-called chaser and evader, respectively. As before, we look again at six different initial conditions and time steps, that is:

**Case 1.** $\mathbf{x}^0 = [-5, 5]$, $\mathbf{y}^0 = [5, 0]$, and $\Delta t = 0.25$.

**Case 2.** $\mathbf{x}^0 = [-4.0, 5]$, $\mathbf{y}^0 = [5, -4]$, and $\Delta t = 0.08$.

**Case 3.** $\mathbf{x}^0 = [1.5, -5]$, $\mathbf{y}^0 = [-5, 2.5]$, and $\Delta t = 0.285$.

**Case 4.** $\mathbf{x}^0 = [-5, 2.5]$, $\mathbf{y}^0 = [5, -3.5]$, and $\Delta t = 0.1$.

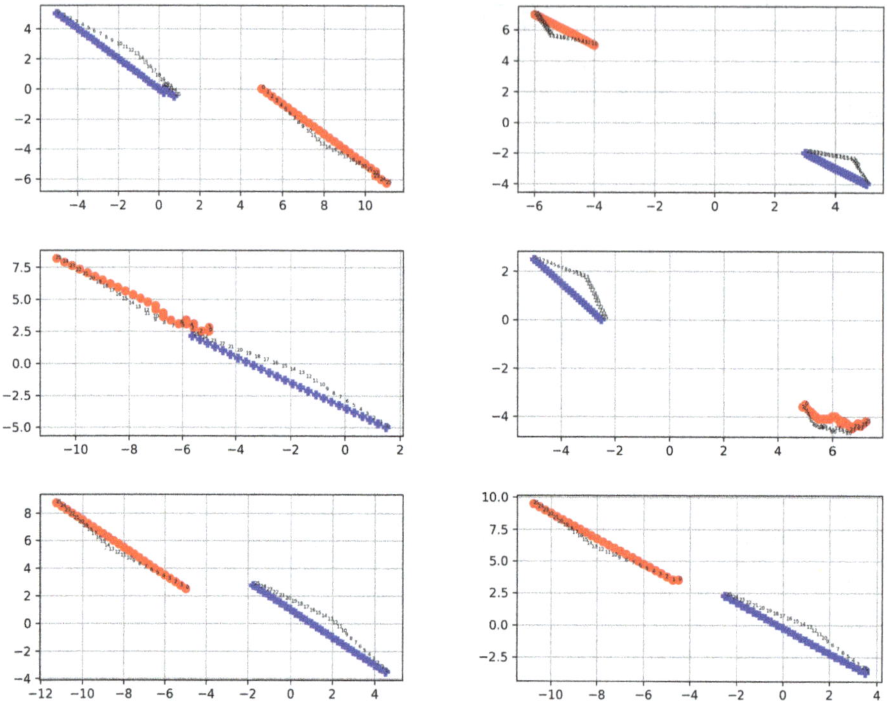

Figure 2.28: *Top left:* Case 1 chaser (blue) starts at the top left corner, and the runner (red) starts near the center. Their paths follow mainly straight lines along a diagonal as expected. *Top right:* Case 2 shows blue (chaser) starting in the bottom right corner and red (runner) on the top left corner. Both pick a straight diagonal path that amounts to the optimal choice in this context. *Middle left:* Case 3 shows the runner (red) evading from the center on a diagonal to the top left corner. The chaser (blue) runs after red on a corresponding path from the bottom right corner. *Middle right:* Case 4 shows roughly the same situation as the previous case except that red seems not to feel that threatened by the chaser blue. *Bottom left and Bottom right:* Cases 5 and 6, respectively, show the expected optimal runner–evader paths in the diagonal direction.

**Case 5.** $\mathbf{x}^0 = [-5, 2.5]$, $\mathbf{y}^0 = [4.5, -3.5]$, and $\Delta t = 0.25$.
**Case 6.** $\mathbf{x}^0 = [-4.5, 3.5]$, $\mathbf{y}^0 = [3.5, -3.5]$, and $\Delta t = 0.25$.

Taking into account the available velocity choices, which show the largest distance changes on the diagonal velocities, it becomes immediately obvious that the running–chasing interaction primarily happens along diagonal directions, see Figure 2.28. This is exactly what we would expect, if blue is the chaser and red is the evader. ◇

The presented dynamic decision framework can be used to perform simulations with rational agents. The rational decision framework of each agent can be explicitly defined with the help of the payoff/(personal) goal functions. Moreover, uncertainties in the agents decisions are naturally included by the *mixed actions* systematically computed from the corresponding interaction matrices with the help of linear programming.

Nevertheless, the reliability and usefulness of the results obtained hereby still require rigorous mathematical analysis. For instance, the important and not immediately obvious role of the time step size in the decision dynamics and the subsequent solvability of the LP problems represent important open questions. Hence, we believe that this framework provides meaningful and practical extensions for applications and promising research directions.

# 3 Examples of CHeSs: toward reliable and systematic descriptions

## 3.1 Nonequilibrium CHeSs: transport in solids

An application of the nonequilibrium thermodynamic framework, GENERIC (see Section 2.7 and [65, 121, 119]), to a phase-separating binary system described by the state vector $\mathbf{z} := \{\rho, \phi, \mathbf{M}, u\}$ with the following general total energy and entropy has been proposed in [77],

$$\begin{aligned} U(\mathbf{z}) &:= \int_D \left( \frac{\mathbf{M}^2}{2\rho} + u + \frac{\lambda_U}{2} |\nabla\phi|^2 \right) d\mathbf{x}, \\ S(\mathbf{z}) &:= \int_D \left( s(\rho, \phi, u) + \frac{\lambda_S}{2} |\nabla\phi|^2 \right) d\mathbf{x}, \end{aligned} \tag{3.1}$$

where $s$ and $u$ are problem specific entropy and internal energy densities, respectively. The variables $\rho$, $\phi$, $\mathbf{M}$, and $u$ denote the *total mass density*, *scalar phase field*, *total momentum density*, and *internal energy density*, respectively.

As motivated already by van der Waals [168], we use the square gradient penalty for an energetic contribution and an entropic contribution with the corresponding coefficients $\lambda_U$ and $\lambda_S$, respectively. Looking at the usual Helmholtz free energy,[1] that is, $F(\mathbf{z}) = U(\mathbf{z}) - k_B T S(\mathbf{z})$, for the internal energy $U$, allows us to write $\lambda = \lambda_U - k_B T \lambda_S$. This is the classical phase field regularization parameter $\lambda$.

Based on (3.1), the symmetric velocity gradient $\mathbf{S} := \nabla\mathbf{v} + (\nabla\mathbf{v})^T$, and $\boldsymbol{\tau} = \boldsymbol{\Pi} - p\mathbf{I}$ for the identity matrix $\mathbf{I}$ and the total pressure tensor $\boldsymbol{\Pi}$ composed of energetic and entropic parts, i. e., $\boldsymbol{\Pi} = \boldsymbol{\Pi}_E + \boldsymbol{\Pi}_S$, Jelić [77] shows that the following generalized Cahn–Hilliard-based binary mixture formulation is consistent with the GENERIC framework (2.65),

$$\begin{cases} \rho_t = -\operatorname{div}(\rho\mathbf{v}), \\ \phi_t = -\mathbf{v} \cdot \nabla\phi + \operatorname{div}\Big(MT\nabla\Big(\frac{1}{T}(\mu^* - (\lambda_E - T\lambda_S)\Delta\phi)\Big)\Big), \\ \mathbf{M}_t = -\operatorname{div}(\mathbf{v} \otimes \mathbf{M}) - \operatorname{div}(\boldsymbol{\Pi} + \boldsymbol{\tau}), \\ u_t = -\operatorname{div}(u\mathbf{v}) - \frac{1}{2}\boldsymbol{\Pi}_S : \mathbf{S} - \boldsymbol{\tau} : (\nabla\mathbf{v})^T - \operatorname{div}(\mathbf{j}^q) \\ \qquad + \kappa_E \Delta\phi \operatorname{div}\Big(MT\nabla\Big(\frac{1}{T}(\mu' - (\kappa_E - T\kappa_S)\Delta\phi)\Big)\Big). \end{cases} \tag{3.2}$$

Previous mathematical investigations verified thermodynamic dissipation properties of entropic/irreversible processes in binary mixtures, see, e. g., [101]. Almost at the same time as [77], a more approximate model formulation, which does not account for the

1 Helmholtz free energy describes maximum amount of work at constant volume and temperature.

https://doi.org/10.1515/9783110579543-003

intrinsic reversible–irreversible couplings as (3.2), has been proposed in [1]. Therein, a link to a corresponding sharp interface model is advocated. Similarly, in [98, 75], the authors propose a slightly different variational approach to obtain both, reversible and irreversible parts, in the associated evolution equations.

**System (3.2) fullfils GENERIC:** We will subsequently define all the operators appearing in the GENERIC framework and hence verify the validity of the associated equations. Gibbs'[2] fundamental equation of thermodynamics [119, e. g., p. 9] allows us to identify the variational derivatives from (3.1). To this end, we state it here in differential form,

$$dU = -pdV + \mu dN + TdS\,. \tag{3.3}$$

Equation (3.3) follows from the first and second laws of thermodynamics with respect to the fundamental equilibrium concept of thermodynamics, which reads, $dU = dW + dQ$, for work $dW$ and heat $dQ$.

Following the intuitive and physical motivation in [119], we accordingly employ the concept of local equilibrium. Hereby, we are able to divide a large nonequilibrium system with *nonuniform state variables* into small systems for which we can identify *local state densities*. Relying on the description of nonequilibrium systems of volume $V$ with state densities such as,

$$\rho(\mathbf{x},t) := \frac{mN(\mathbf{x},t)}{V}\,, \quad u(\mathbf{x},t) := \frac{U(\mathbf{x},t)}{V}\,, \quad \text{and} \quad s(\mathbf{x},t) := \frac{S(\mathbf{x},t)}{V}\,, \tag{3.4}$$

has proven to be more instructive instead of using the associated extensive variables $N$, $U$, and $S$ defined with respect to a small (equilibrium) volume element. In fact, the volume $V$ in (3.4) can represent a reference volume in the form of a small volume element being in *local thermodynamic equilibrium* or even the whole volume of the system in which we are interested in.

Unlike the state variable $\mathbf{z}$ consisting of order parameters, which allow us to systematically describe phase transitions, we also introduce the state variable $\mathbf{z}^* = \{n_\alpha, n_\beta, \mathbf{M}, u\}$ of independent variables for the description of binary fluids composed of species $i \in \{\alpha, \beta\}$ with number densities $n_i := \frac{N_i}{V}$, total momentum $\mathbf{M}(\mathbf{x},t)$, and the internal energy $u(\mathbf{x},t)$. This gives $s^*(n_\alpha(\rho,\phi), n_\beta(\rho,\phi), u) = s(\rho(n_\alpha, n_\beta), \phi(n_\alpha, n_\beta), u)$, and, moreover, (3.3) reads as follows,

$$du = -pdv + \mu_\alpha dn_\alpha + \mu_\beta dn_\beta + Tds^*\,. \tag{3.5}$$

We then immediately get the following equation for the pressure,

$$p = \mu_\alpha n_\alpha + \mu_\beta n_\beta + Ts^* - u\,, \tag{3.6}$$

2 Gibbs free energy describes the maximum amount of work at constant pressure and temperature.

derived by integrating over a small volume $v$ in local equilibrium and exploiting the property of constant chemical potentials $\mu_i$ = const. in $v$, $i \in \{\alpha, \beta\}$.

Similarly, (3.5) implies the following definitions of chemical potentials $\mu_i$, $i \in \{\alpha, \beta\}$, and temperature $T$,

$$-\frac{\mu_i}{T} := \frac{\partial s^*}{\partial n_i}, \quad i \in \{\alpha, \beta\}, \quad \text{and} \quad \frac{1}{T} := \frac{\partial s^*}{\partial u}. \tag{3.7}$$

For the mass $m_i$ of species $i \in \{\alpha, \beta\}$, we finally get with the relations inverse to,

$$\rho(\mathbf{x}, t) := m_\alpha n_\alpha + m_\beta n_\beta, \tag{3.8}$$

and,

$$\phi(\mathbf{x}, t) := \frac{m_\alpha n_\alpha}{m_\alpha n_\alpha + m_\beta n_\beta}, \tag{3.9}$$

and with $\mu = \mu_\alpha \phi / m_\alpha + \mu_\beta (1 - \phi)/m_\beta$ and $\mu^* = (\mu_\alpha / m_\alpha - \mu_\beta / m_\beta)\rho$, the following partial derivatives,

$$\frac{\partial s(\rho, \phi, u)}{\partial \rho} = -\frac{\mu}{T}, \quad \frac{\partial s(\rho, \phi, u)}{\partial \phi} = -\frac{\mu^*}{T}, \quad \text{and} \quad \frac{\partial s(\rho, \phi, u)}{\partial u} = 1/T. \tag{3.10}$$

Thanks to (3.10), the variational derivatives of the total energy $E$ and entropy $S$ read as stated in [77],

$$\begin{cases} \frac{\delta E}{\delta \mathbf{z}} = \{-\mathbf{v}^2(\mathbf{x}, t)/2, \mathbf{v}(\mathbf{x}, t), 1, -\lambda_E \Delta\phi(\mathbf{x}, t)\}, \\ \frac{\delta S}{\delta \mathbf{z}} = \{-\mu(\mathbf{x}, t)/T(\mathbf{x}, t), \mathbf{0}, 1/T(\mathbf{x}, t), -\mu^*(\mathbf{x}, t)/T(\mathbf{x}, t) - \lambda_S \Delta\phi(\mathbf{x}, t)\}. \end{cases} \tag{3.11}$$

Also in [77], the following Poisson matrix accounting for the reversible/convective behavior has been obtained,

$$[\mathcal{L}(\mathbf{z})](\mathbf{x}) := -\begin{bmatrix} 0 & \nabla\rho & 0 & 0 \\ \rho\nabla & [\nabla\mathbf{M} + \mathbf{M}\nabla]^T & u\nabla + \operatorname{div}\mathbf{\Pi}_S & -\nabla\phi \\ 0 & \nabla u + \mathbf{\Pi}_S \cdot \nabla & 0 & 0 \\ 0 & \nabla\phi & 0 & 0 \end{bmatrix}. \tag{3.12}$$

Recall the decomposition $\mathbf{\Pi} = \mathbf{\Pi}_E + \mathbf{\Pi}_S$ of the pressure tensor in energetic and entropic contributions, which leads to,

$$-\operatorname{div}\mathbf{\Pi} = -\nabla p + \left(\frac{\delta E_\phi}{\delta\phi} - T\frac{\delta S_\phi}{\delta\phi}\right), \tag{3.13}$$

where,

$$E_\phi := \frac{\lambda_E}{2}\int_D |\nabla\phi|^2\, d\mathbf{x}, \quad \text{and} \quad S_\phi := \frac{\lambda_S}{2}\int_D |\nabla\phi|^2\, d\mathbf{x}. \tag{3.14}$$

Finally, we need to consider irreversible and additive quantities such as viscosity, diffusion, and heat conduction, which all can be accounted for by the friction matrix $\mathcal{M} = \mathcal{M}^H + \mathcal{M}^D$. Using the thermal conductivity $\lambda^q$, the viscosity $\eta$, the dilatational viscosity $\kappa$ (i. e., $\hat{\kappa} := \kappa - \frac{2}{3}\eta$), the symmetric velocity gradient $\mathbf{S} := \nabla\mathbf{v} + (\nabla\mathbf{v})^T$, and the components,

$$\begin{aligned}
m^H_{\mathbf{MM}} &:= -(\nabla(\eta T)\nabla\,\cdot + (\mathbf{I}\ \mathrm{div})\eta T\nabla\,\cdot)^T - \nabla\hat{\kappa}T\nabla\,\cdot\,, \\
m^H_{\mathbf{M}u} &:= \mathrm{div}(\eta T\mathbf{S}\,\cdot) + \nabla\left(\frac{\hat{\kappa}}{2}\,\mathrm{tr}\,\mathbf{S}\,\cdot\right), \\
m^H_{u\mathbf{M}} &:= -\eta T\mathbf{S}\nabla\,\cdot - \frac{\hat{\kappa}T}{2}\,\mathrm{tr}\,\mathbf{S}\nabla\cdot\,, \\
m^H_{uu} &:= \frac{\eta T}{2}\mathbf{S} : \mathbf{S}\,\cdot + \frac{\hat{\kappa}}{4}(\mathrm{tr}\,\mathbf{S})^2 - \mathrm{div}(\lambda^q T^2\nabla\,\cdot)\,,
\end{aligned} \tag{3.15}$$

as defined in [77, 119], we can write the general hydrodynamic friction matrix as follows,

$$\mathcal{M}^H := \begin{bmatrix} 0 & 0 & 0 & 0 \\ 0 & m^H_{\mathbf{MM}} & m^H_{\mathbf{M}u} & 0 \\ 0 & m^H_{u\mathbf{M}} & m^H_{uu} & 0 \\ 0 & 0 & 0 & 0 \end{bmatrix}. \tag{3.16}$$

Similarly, the diffusive contribution [77] follows from the symmetry and degeneracy requirements,

$$\mathcal{M}^D := \begin{bmatrix} 0 & 0 & 0 & 0 \\ 0 & 0 & 0 & 0 \\ 0 & 0 & m^D_{uu} & m^D_{u\phi} \\ 0 & 0 & m^D_{\phi u} & m^D_{\phi\phi} \end{bmatrix}, \tag{3.17}$$

where,

$$\begin{aligned}
m^D_{uu} &:= -\kappa_E\Delta\phi\,\mathrm{div}(MT\nabla(\kappa_E\Delta\phi\,\cdot))\,, \\
m^D_{u\phi} &:= -\kappa_E\Delta\phi\,\mathrm{div}(MT\nabla\,\cdot)\,, \\
m^D_{\phi u} &:= -\mathrm{div}(MT\nabla(\kappa_E\Delta\phi\,\cdot))\,, \\
m^D_{\phi\phi} &:= -\mathrm{div}(MT\nabla\,\cdot)\,.
\end{aligned} \tag{3.18}$$

The above derivation of the GENERIC operators verifies that the hereby resulting phase field equations represent a consistent nonequilibrium thermodynamic system of equations.

### 3.1.1 Applications and limits of phase field formulations

The well-known Cahn–Hilliard phase field equation, i. e., $(3.2)_2$ for $\mathbf{v} = \mathbf{0}$, is going back to the work [31] by Cahn and Hilliard in 1958 and hence has a long history. Its validity has been demonstrated repeatedly, both experimentally and theoretically. For instance, it coincides with the time-dependent Ginzburg–Landau (GL) theory for a conserved order parameter, i. e., the 4th-order partial differential equation obtained as the $H^1$-gradient flow of the classical GL excess free energy,

$$F(\phi) = \int w(\phi(\mathbf{x})) + \lambda|\nabla\phi|^2\, d\mathbf{x}\,, \tag{3.19}$$

where $w(s) = (s^2 - 1)^2$ represents a simple prototype to describe phase transitions. In fact, the parameter $\lambda$ is connected to the underlying microscopic interaction potential via its second moment. Moreover, the Cahn–Hilliard description can be obtained from a corresponding stochastic Ising model and reliably captures coarsening rates of sharp interface formulations describing dynamics of interfaces governed by the energy of surfaces [57, 58, 127]. That is, the growth rates of a specific length, governed by the so-called Mullins–Sekerka and the surface diffusion formulations, have been computationally and rigorously (i. e., in a time-averaged sense) validated, see [86, 169] and Section 3.2, for instance.

Hence, the Cahn–Hilliard *diffuse interface* formulation is a widespread and often used computational formulation in applications such as image impainting [16], electrowetting [102, 47], contact line behavior [155, 156], batteries [138], and fuel cells [126], for instance. Very recently, these equations have been formally and rigorously upscaled toward *porous media phase field equations* in [146, 143] and computationally validated in [147, 169].

Next to this wide range of very important, experimentally validated, and practically and computationally relevant applications, we believe that it is valuable to recall that the free energy (3.19) clearly shows limits in the level of detail of the thermodynamic information provided on the interface. An investigation dedicated to this aspect is [85], where thermodynamic consistency is analyzed with the help of asymptotic expansions and GENERIC. Such consistency studies can provide a valuable insight into the applicability in cases where the thermodynamic state of the interface plays a central role. In fact, the Cahn–Hilliard specific energy and entropy contributions, i. e., $U$ and $S$ in (3.1) for $\mathbf{M} = \mathbf{0}$ and $u = 0$, only define the interface mathematically by a penalty on the interfacial extension governed by the term $\lambda|\nabla\phi|^2$ in (3.19). It does not include any specific free energy describing the transition region of the interface.

A simple extension, which approximates a specific physical structure of the interfacial region, exploits the intrinsic connection between the Cahn–Hilliard free energy and the surface tension as rigorously established by Modica and Mortola using Gamma-convergence [110], see [142].

## 3.2 Universal CHeSs: coarsening in heterogeneous media

We continue our discussion on the dynamics of interfaces separating two phases in a mixture by recalling theoretical and computational results from the work [169]. Therein, the authors investigate the radial dependence of the coarsening/ripening process in the context of heterogeneous media. The observation of this process goes back to Ostwald in 1900. This radial growth phenomenon plays a central role in many applications such as materials science and hence happens in almost all phase transition processes. Therefore, it governs the morphology of the resulting microstructure whose self-similarity can be observed after a long enough coarsening time.

The general driving force for the Ostwald ripening process is the system's goal to minimize its interfacial area/interfacial energy. Coarsening can generally be observed in systems where a single large particle can have much lower interfacial area than many small particles. In fact, this ripening/coarsening can be identified with respect to different characteristic length scales, e. g., the particle distance, particle radius, or the inverse of the interfacial area per volume,

$$L(t) = |D|/E_{\text{mix}}\,,$$

where $\frac{1}{|D|}E_{\text{mix}}$ is the volume-averaged interfacial area. In fact, the Gibbs free energy of mixing $E_{\text{mix}}$ can be realized with the Cahn–Hilliard free energy density $f_r$ defining the free energy functional $F_r(\phi)$ for the order parameter $\phi$, see (8.43), and its use in the Cahn–Hilliard equation (8.49). Therefore, we will subsequently focus on the length $L = 1/F_r(\phi)$.

About 60 years later since Ostwald's discovery of this growth phenomenon, Lifshitz and Slyozov [97] and Wagner [171] proposed a mean field equation whose solution gives the number of droplets of a particular radius $r$ at time $t$. The coarsening rate,

$$L(t) \le Ct^{1/3}\,, \tag{3.20}$$

has been validated experimentally and computationally in [178]. So far, a rigorous proof of (3.20) has only been obtained for a time-averaged version in [86],

$$\frac{1}{T}\int_0^T E_{\text{mix}}^2\,dt \ge \frac{C}{T}\int_0^T (t^{-1/3})^2\,dt\,. \tag{3.21}$$

### 3.2.1 Interfacial dynamics in heterogeneous media: coarsening rates

The authors of [169] investigate the influence of heterogeneities on the coarsening rate. For example, these heterogeneities can be porous media showing porosity gradients, see Figure 3.1. The well-known coarsening rate (3.20) for homogeneous media has been

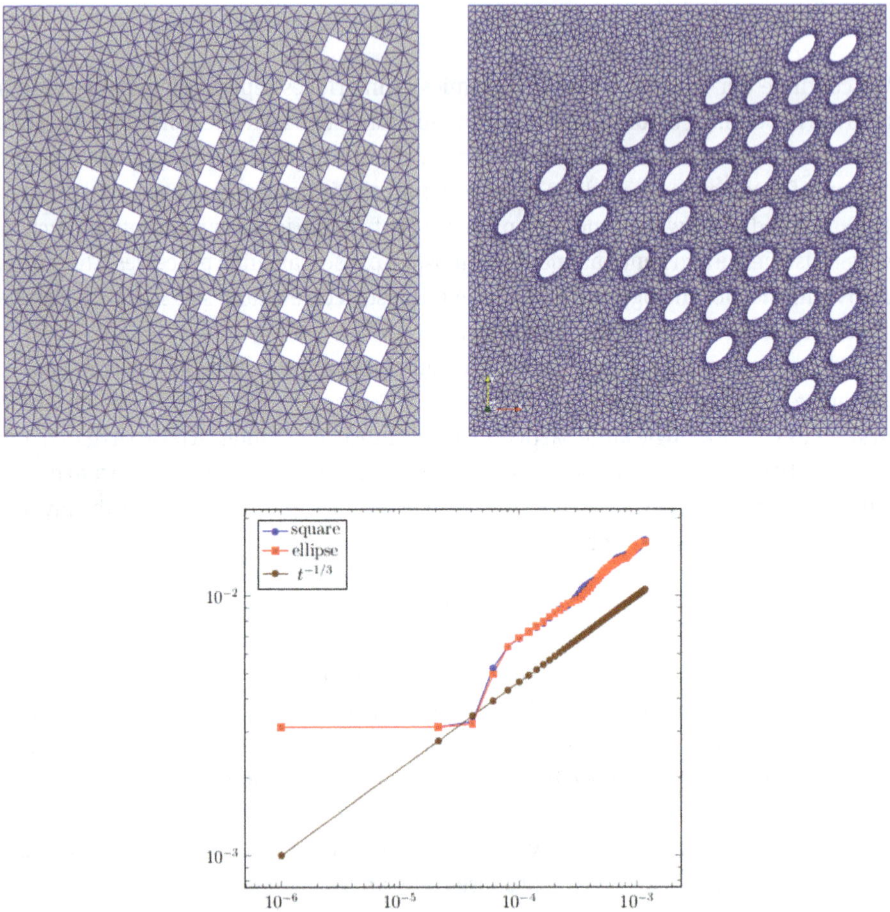


Figure 3.1: *Top Left and Right:* Different geometries and porosity gradients along the horizontal direction from the left to right side, i. e., the porosity increases. *Bottom:* Coarsening rate seems to be universal, i. e., independent of pore geometries and porosity gradients.

recovered in [144] in the context of porous media having neutral wetting properties (i. e., contact angle of 90°). As already motivated in [144], this fact provides a strong argument for the exponent 1/3 in (3.20) to represent a so-called *universal* coarsening rate, see Figure 3.2.

#### 3.2.1.1 Fluctuations: Langevin dynamics ($L^\epsilon$)

Let $(\Omega, \mathcal{F}, \mathrm{P})$ be a standard probability space and consider the random force $\zeta^\epsilon(\mathbf{x}, t; \omega) : D^\epsilon \times (0, T) \times \Omega \to \mathbb{R}$ defined by,

$$\begin{aligned} \langle \zeta^\epsilon(\mathbf{x}, t), \zeta^\epsilon(\mathbf{x}', t') \rangle &= \delta(\mathbf{x} - \mathbf{x}')\delta(t - t'), \\ \langle \zeta^\epsilon(\mathbf{x}, t) \rangle &= 0, \end{aligned} \tag{3.22}$$

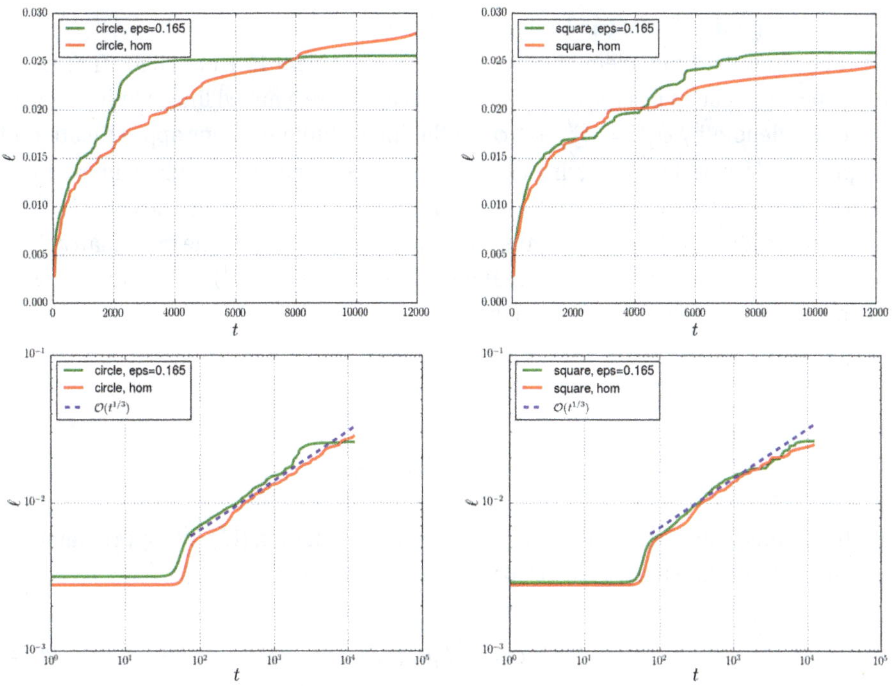


Figure 3.2: Influence of porosity on the coarsening rate. The universal growth rate $\mathcal{O}(\epsilon^{1/3})$ is recovered in the porous media setting [129, 51, 86].

where $D^\epsilon$ denotes the microscopic porous domain fully resolving the pores. Taking into account statistical properties such as the first and second moments is a general approach in coarse graining of microscopic descriptions due to a loss of information, see [129].

Moreover, the well-accepted fluctuation-dissipation theorem [90] allows us to include thermal fluctuations by adding the product of the parameter $\sigma(\theta) := \sqrt{2k_B\theta \mathrm{M}} = \sqrt{2\mathrm{D}}$ with $\zeta^\epsilon$ to the deterministic problem at hand, i. e., (8.49), where $k_B$ is the Boltzmann constant, $\theta$ is the temperature, M is the mobility parameter, D is the diffusion parameter, and $\zeta^\epsilon$ is Gaussian and uncorrelated (white) noise. As a result, we obtain the conserved Langevin dynamics of the Cahn–Hilliard equation (8.49), that is,

$$(\mathbf{L}^\epsilon) \qquad \begin{cases} \partial_t \phi^\epsilon = \operatorname{div}(\mathrm{M}\nabla(f_r'(\phi^\epsilon) - \lambda\Delta\phi^\epsilon)) + \sigma(\theta)\zeta^\epsilon & \text{in } (D^\epsilon)_T\,, \\ \mathbf{n}\cdot\nabla\phi^\epsilon = 0 & \text{on } (\partial D^\epsilon)_T\,, \\ \mathbf{n}\cdot\nabla\Delta\phi^\epsilon = 0 & \text{on } (\partial D^\epsilon)_T\,, \end{cases} \tag{3.23}$$

which also represents the so-called *Ginzburg–Landau equation* for identifying the phase transition of a material to a superconducting state, see [14]. Recall the reliable and consistent nonequilibrium thermodynamic derivation of the Cahn–Hilliard phase field formulation based on GENERIC in Section 3.1.

#### 3.2.1.1.1 Discretization of the Langevin equation ($L^\epsilon$)

We first recall the finite element discretization of problem $\mathbf{L}^\epsilon$ from [169]. That is, we apply the $\theta$-method to discretize in time. In space, we use a mixed linear finite element space. We denote by $\phi_n^{\epsilon,l} := \sum_{j=1}^J \phi_j^{n,l}\varphi_j(\mathbf{x})$ the linear finite element approximation of the phase field variable $\phi^\epsilon$ solving (3.23), where $\varphi_j(\mathbf{x})$ denotes a standard linear finite element basis on the domain of interest discretized with a given mesh size $h > 0$.

Next, we identify the discrete finite element operators such as the mass matrix $\hat{\mathrm{M}} := \{m_{ij}\}_{i,j=1}^J$, stiffness matrix $\hat{\mathrm{S}} := \{s_{ij}\}_{i,j=1}^J$, and nonlinear tensor $\hat{\mathrm{K}}(\phi_n^{\epsilon,l}) := \{k_{ij}(\phi_n^{\epsilon,l})\}_{i,j=1}^J$ with the following associated matrix elements,

$$
\begin{aligned}
m_{ij} &:= (\varphi_i, \varphi_j)\,, \\
s_{ij} &:= (\nabla\varphi_i, \nabla\varphi_j)\,, \\
k_{ij}(\phi_n^{\epsilon,l}) &:= ((\phi_n^{\epsilon,l})^2\varphi_i, \varphi_j)\,.
\end{aligned}
\tag{3.24}
$$

It remains to discretize the additive white noise term $\zeta^\epsilon$ in ($\mathbf{L}^\epsilon$). We approximate $\zeta^\epsilon$ as a space-time white noise (Wiener process) (see, e. g., [100]),

$$
\zeta(\mathbf{x}, t) = \sum_{j=1}^{\infty} \beta_j(t_n)\varphi_j(\mathbf{x})\,, \tag{3.25}
$$

where $\beta_j(t_n) = \beta_j(t_{n-1}) + Z$, $Z \sim \mathcal{N}(0, k)$, represent *i. i. d. Brownian motions*, and $k > 0$ denotes the time step size. The above white noise approximation is possible, since $\zeta^\epsilon$ is uncorrelated. Introducing the vector $\boldsymbol{\beta}_n := \{\beta_{j,n}\}_{j=1}^J$ allows us to discretize the nonlinear Langevin equation (3.23) by iterates $(\boldsymbol{\phi}_{n+1}^{\epsilon,l+1}, \boldsymbol{\mu}_{n+1}^{\epsilon,l+1})$ solving the following equations,

$$
(\textbf{Scheme } \mathbf{L}_{h,k}^{\epsilon,l}) \qquad
\begin{cases}
\mathbb{M}\boldsymbol{\phi}_{n+1}^{\epsilon,l+1} + km\mathbb{S}\boldsymbol{\mu}_{n+\theta}^{\epsilon,l+1} = \mathbb{M}\boldsymbol{\phi}_n^{\epsilon} + \sigma(\tilde{\theta})\mathbb{M}\boldsymbol{\beta}_n\,, \\
\mathbb{M}\boldsymbol{\mu}_{n+\theta}^{\epsilon,l+1} + \mathbb{K}(\boldsymbol{\phi}_{n+1}^{\epsilon,l})\boldsymbol{\phi}_{n+1}^{\epsilon,l+1} - \frac{1}{2}\mathbb{M}\boldsymbol{\phi}_{n+1}^{\epsilon,l+1} - \lambda\mathbb{S}\boldsymbol{\phi}_{n+1}^{\epsilon,l+1} = 0\,.
\end{cases}
\tag{3.26}
$$

Thermal fluctuations are taken into account by $\sigma(\tilde{\theta})$, which therefore requires no discretization as the coefficient solely depends on a given temperature $\tilde{\theta}$. Due to the $\theta$-method already occupying the variable $\theta$ in the context of the time steps, we rewrite the temperature variable here by $\tilde{\theta}$ instead of remaining with an ambiguous $\theta$.

### 3.2.1.2 Influence of the pore geometry and porosity

Let us investigate how thermal fluctuations influence the coarsening rate in the regular solution formulation $\mathbf{L}^\epsilon$, that is, equation (3.26), with the help of numerical solutions $(\phi_{n+1}^{\epsilon,l+1}, \mu_{n+1}^{\epsilon,l+1})$ to the scheme $\mathbf{L}_{h,k}^{\epsilon,l}$. Our study relies on the following random initial conditions,

$$
\phi(\mathbf{x}, 0) = 0.5 + n_c(0.5 + r_{\mathbf{x}})\,. \tag{3.27}
$$

The variable $r_{\mathbf{x}} \in [0,1]$ imposes on locations $\mathbf{x} \in \mathcal{D} = (0,1)^2$ random fluctuations with zero mean. Moreover, we control the influence of the noise by $n_c > 0$, which we set to $n_c = 0.125$. With (3.27), we introduce the so-called *critical quenches* by setting the mean value to 0.5 such that the process goes through the *critical point*, which is exactly between the equilibrium limiting values 0 and 1 of the double-well free energy,

$$w(\xi) = a\xi^2(1-\xi)^2\,, \quad a > 0\,. \tag{3.28}$$

The double-well free energy $w(\xi)$ is well accepted due to its computational stability and its close qualitative characteristics to the *regular solution* free energy (8.43).

Despite the appearance of *percolating phases* with critical quenches, we still observe the universal scaling $\mathcal{O}(t^{1/3})$ of the coarsening phenomenon as depicted in Figure 3.3 with rather complex dynamics. Surprisingly, for long enough times, we observe a sudden decrease of the size (i. e., the characteristic size $L$) as shown in Figure 3.3 (right), and hence a process different from coarsening dominates the evolution. In a future investigation, it will be of interest to explore the underlying scientific/physical mechanisms responsible for this behavior.

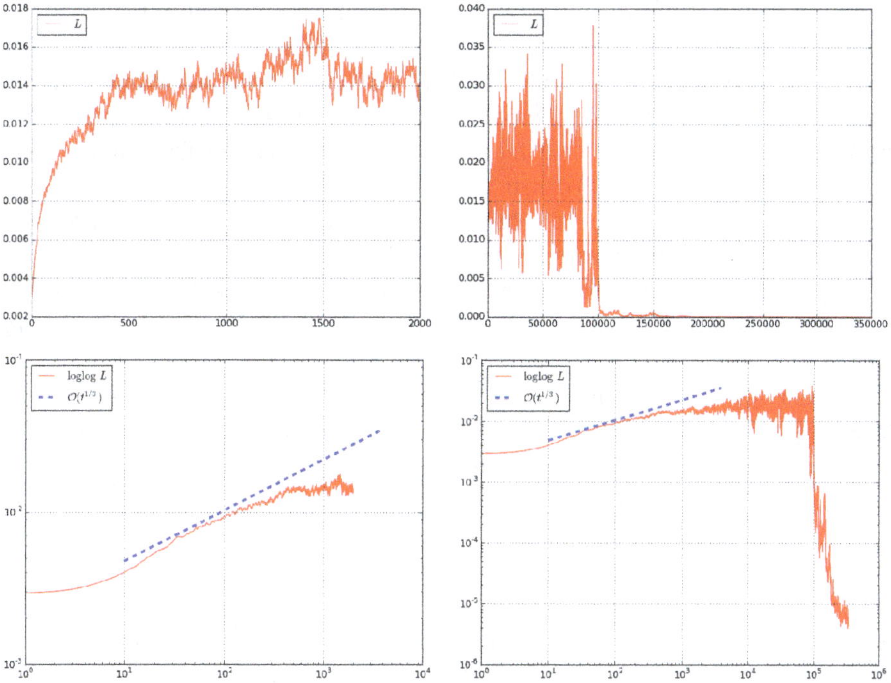


Figure 3.3: Coarsening in a circular porous medium for the Langevin formulation (3.26) for critical quenches. Characteristic length $L$ for the first 2000 time steps *(top left)* and according loglog plot *(bottom left)*. Characteristic length for 3500 time steps *(top right)* and according loglog plot *(bottom right)*.

Finally, note that a deeper investigation of multiscale effects such as heterogeneous media with spatially depending pore characteristics and the associated effective macroscopic/upscaled formulations have been performed in [169, 144]. In fact, the homogenization result, that is, the upscaling of the Cahn–Hilliard phase field equation for strongly heterogeneous media, has been rigorously achieved in [143] for the first time, to the best of our knowledge, and previously motivated by formally derived and validated equations in [146, 147].

## 3.3 Information/data processing systems as CHeSs: from PC over DC and HPC to QC

Let us first recall our motivation for thermodynamics as being a fundamental scientific branch driving first investigations of (complex) thermodynamic input (energy as heat) and output systems (mechanical energy plus entropy) and, in particular, the concepts of entropy and information; see Szilard's engine in Section 2.5.1. In fact, Szilard's engine represents a single-bit processing system and hence builds an elementary entity of multibit processing systems such as the *Turing machine*. Therefore, it is only natural that we discuss *information/data processing systems* such as a classical *Desktop/Personal Computer (PC)*, a *Data Center (DC)*, a *High Performance Computer (HPC)*, and a *Quantum Computer (QC)* as information/data processing CHeSs. Already a PC alone represents unlimited many complex CHeSs solely by the unlimited number of possible virtual CHeSs running in the form of programs/simulations on a PC, for instance. An often neglected but frequent reason for unexpected system behavior (which can be considered as emerging (unexpected) dynamics and hence representing a CHeS) is induced by human operators interacting via a *Human–Machine Interface (HMI)* for application or management and control purposes.

In the spirit of Szilard's engine (see again Section 2.5.1), we emphasize that temperature plays a crucial role in the design of information/data processing systems starting with PCs up to HPCs. Therefore, we give a basic overview of the four mentioned CHeSs, i. e., PC and DC in Figure 3.4 and HPC and QC in Figure 3.5. Finally, we qualitatively visualize the primary usage scenarios for PCs, DCs, HPCs, and QCs in relation to a primarily data or computational intense application or even both, see Figure 3.6.

### 3.3.1 PC as a CHeS

We consider here a PC(/Laptop) as the most elementary commercially (off-the-shelf) available and affordable information/data processing unit as depicted on the top left in Figure 3.4. It becomes immediately clear that once a link between elements such as CPU, storage, and software (e. g., OS and application) has been made, we obtain as a result a linked system of systems, in which each contributing system has its own internal

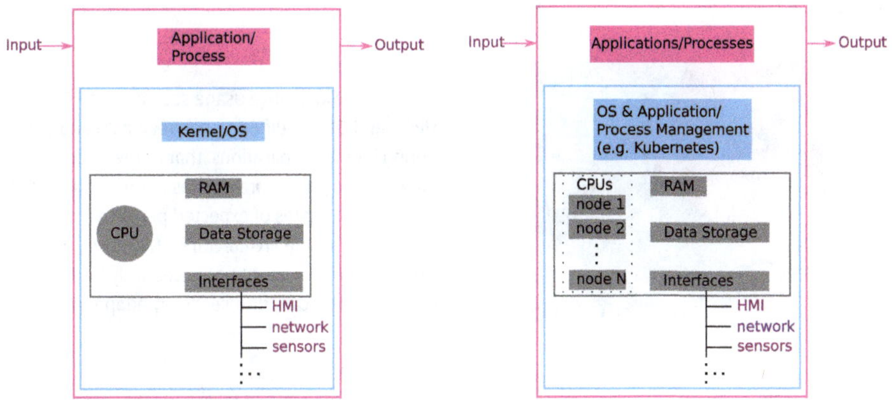


Figure 3.4: *Left:* PC as a CHeS, that is, a simple representation of a classical PC containing the basic elements of a Turing machine, see Section 2.3.1. *Right:* DC as a CHeS, which can be understood as a classical PC where the role of the function of the OS is extended by an application management such as Kubernetes (K8s), for instance.

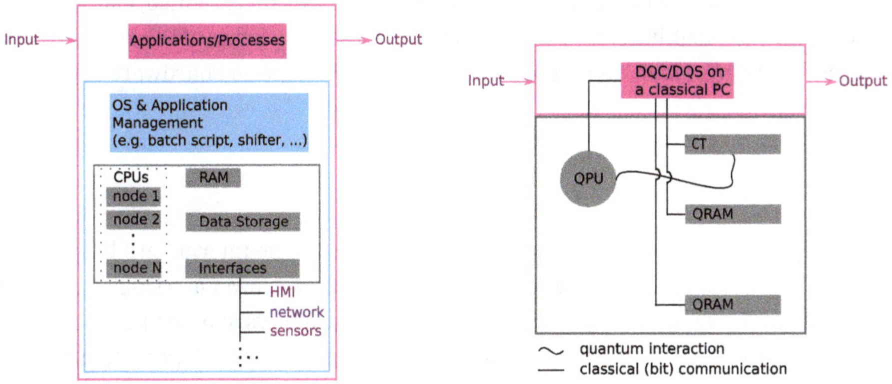


Figure 3.5: *Left:* HPC as a CHeS, where there currently seems to be no main stream approach how to manage applications/jobs. Approaches are classical batch scripts. A tool in development is, for instance, the shifter idea; see Section 3.3.3. *Right:* QC as a CHeS, which consists of a classical PC part running the *Dynamic Quantum Compiler (DQC)/Dynamic Quantum Scheduler (DQS)* that controls the execution of a quantum program on the *Quantum Processing Unit (QPU)* and the read/write operations to the *Quantum RAM (QRAM)*.

state and update rules. If we also take into account the infinite amount of possible inputs, then it immediately becomes clear that there exists a likelihood for an unexpected combination of inputs to a current internal system state, which leads to an unexpected or unforseen output. Exactly this possibility motivates us here to label a PC together with its user as a CHeS.

Naturally, it is the interest of software companies to eliminate unforseen output/behavior of their *software*. To this end, there exist increasingly systematic principles and

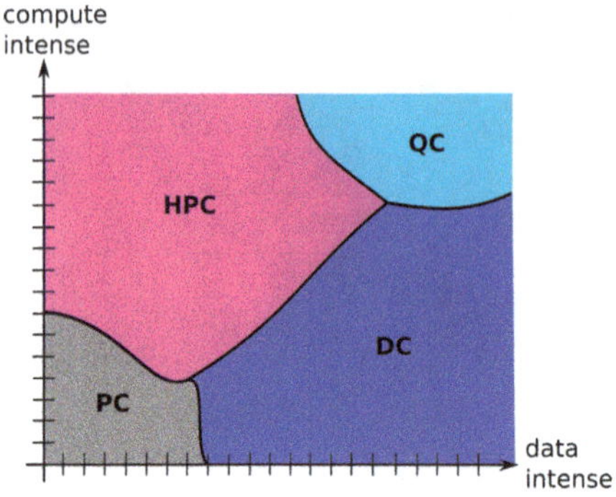


Figure 3.6: Qualitative usage scenarios of PCs, DCs, HPCs, and QCs by differing between data and computing intense applications, that is, the interfaces between the phases are not based on real data, but rather on estimates of expected performances of currently available architectures. Moreover, the transitions/interfaces might not necessarily be sharp, i. e., could be locally diffuse/slowly adapting for instance.

methods to ensure that a software works as required, e. g., *exception handling*. Such principles and methods are collected in the so-called *Software Assurance Process*, see also Section 5.5.1 on software reliability. Finally, we note that the subsequent extensions toward scalability in parallel load (data intense, i. e., DC) processing and parallel computing (computing intense, i. e., HPC) can be realized in its most primitive form by suitably connecting PCs to a cluster with an appropriate, i. e., goal oriented, management software for this resulting cluster. In fact, these were the early forms of DCs realized with low-cost hardware, albeit under the additional risk of unexpected hardware failures.

### 3.3.2 DC as a CHeS

A primary driver for DCs is the fact that user requests on the internet are handled by web services. Naturally, such services require computational power as a single PC makes available, but for an often unpredictable number of customers at a certain instance in time. Therefore, the main advantage of a DC over a single PC is that it provides scalability and a high enough degree of automation (to limit maintenance work and) to cover varying customer demand in the (public) cloud, see Figure 3.7.

A possible DC architecture from Cisco is depicted in Figure 3.7 and consists of the following layers (given here in the exact wording from the following website www.cisco.com visited on 19 January 2023):

**Virtual Compute.** The virtual compute farm contains two UCS 5108 chassis with 16 UCS B200 servers (dual quad-core Intel Xeon X5570 CPU at 2.93 GHz, and 72 GB RAM) with 10 GE Menlo-Emulex converged network adapters (CNAs) organized into a VMware ESX cluster; and 32 servers (2 clusters) within a Compute Point-of-Delivery (PoD). Each server has two CNAs dual-attached to the UCS 6100 Fabric Interconnect. The CNAs provide LAN and SAN connectivity to the servers, which run VMware ESX4.0 hypervisor. The CNAs provide LAN and SAN services to the hypervisor.

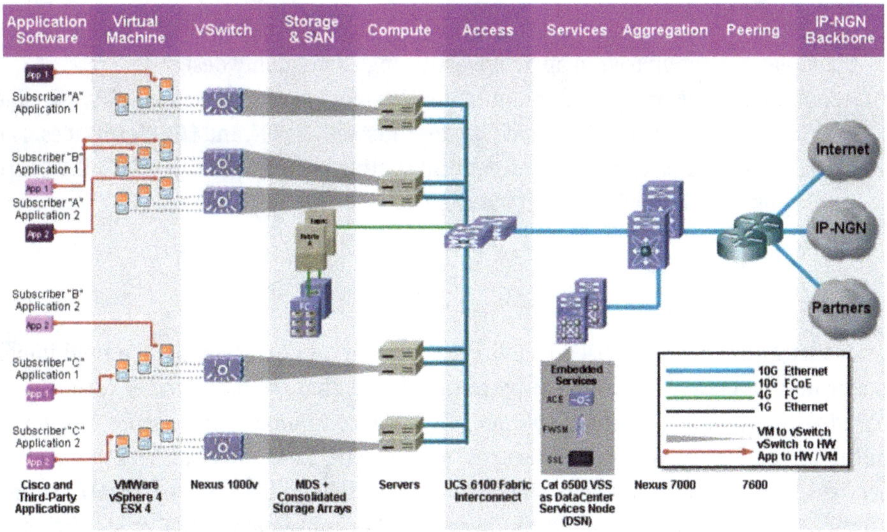


Figure 3.7: *Cisco multitenant DC as a CHeS.* (Source of image: www.cisco.com. Date of access: Sep. 24, 2023.)

**Storage Array Network (SAN).** This consists of storage arrays that support Fiber Channel (FC) and information(/data) lifecycle management (ILM) services. The storage arrays connect through MDS SAN switches to the UCS 6120 switches in the access layer.

**VM Virtual Access Layer.** Cisco Nexus 1000V DVS acts as the virtual access layer for the virtual machines (VMs). Edge LAN policies such as QoS marking and vNIC ACLs are implemented at this layer in Nexus 1000V port-profiles. There is one Nexus 1000V virtual supervisor module (VSM) per ESX cluster. Each ESX server runs an instance of the Nexus 1000V Virtual Ethernet Module (VEM).

**Access Layer.** In the Layer 2 access layer, redundant pairs of Cisco UCS 6120 switches aggregate VLANs from the Nexus 1000V DVS. FCoE SAN traffic from VMs are handed off as FC traffic to a pair of MDS SAN switches and then to a pair of storage array controllers. FC expansion modules in the UCS 6120 switch provides SAN interconnects to dual SAN fabrics. The UCS 6120 switches are in N Port virtualization (NPV) mode to interoperate with the SAN fabric.

**Aggregation.** Redundant Cisco Nexus 7010 switches provide Layer 2 switching between compute nodes and PODs. The core supports Layer 2 multipathing to the access and services layers through virtual port-channels (vPCs). The Nexus 7010 switches serve as collapsed Core/Aggregation Layer 2 devices in this design.

**Services Layer.** A Data Center Service Node (DSN) virtual switching system (VSS) – a pair of Cisco Catalyst 6500 chassis with the VSS supervisors – provides L3 GW services and security services for the hosts. The Cisco Application Control Engine (ACE-20) and Cisco Firewall Services Module (FWSM) on the Catalyst 6500-VSS provide virtual firewall and server load-balancing services to the VMs. Dual FWSM and ACE modules are configured in an active/active high availability design.[3]

**DC/WAN Edge or Peering.** Redundant Cisco 7600 Series routers act as DC/WAN edge routers and provide 10GE connectivity for Internet, L3VPN, and L2VPN services. The 7600 Series router runs IS-IS and mpBGP routing protocols to the WAN/MPLS core and runs OSPF and BGP toward the cloud.[4]

### 3.3.3 HPC as a CHeS

An investigation of cloud-based HPC architectures shows that this is still a field under major development. The principal reason for this is that the main driver for a cloud-based redesign of conventional HPC models is the continuously increasing popularity of DC approaches such as container (i. e., deployment, execution, and maintenance of services/functionalities in a lightweight and easily deliverable software form) and microservices (e. g., container-based web services). As a consequence, HPC centers such as NERSC and CSCS follow an approach based on the shifter framework depicted in Figure 3.8 and described in [95].

In short, the main complexity in HPC is the demand in computational processing power for a single application, which either cannot be efficiently (network/hardware latency) scaled further by distributing subtasks or whose algorithms simply cannot be divided into additional distributable subpieces for parallel computing. Hence, let us briefly collect immediate differences between DCs and HPCs.

In general, customers of DC infrastructures have rather steady demands on a DC's service level, which either slowly increases due to an increasing number of users consuming the service or due to imposed software updates that suddenly require more computational power or storage. On the other hand, HPC infrastructures are primarily for a less steady environment such as research and development departments, universities, etc. That summarizes into the following central aspect:

**HPC versus DC**
The number and type of applications require HPCs to change over time and hence should not simply be fixed via Service Level Requirements (SLRs) and Service Level Agreements (SLAs) as for customers of DCs. Such SLRs and SLAs generally prevent or even reduce the creation of valuable innovation and useful information in the form of new knowledge.

3 DSN and Virtual Switching System (VSS) are used interchangeably. A DSN refers to a Catalyst 6500 with installed FWSM and ACE services modules.

4 Note that we use DC Edge Router, Peering Router, and WAN Router interchangeably to refer to the Cisco 7600 router in this solution architecture.

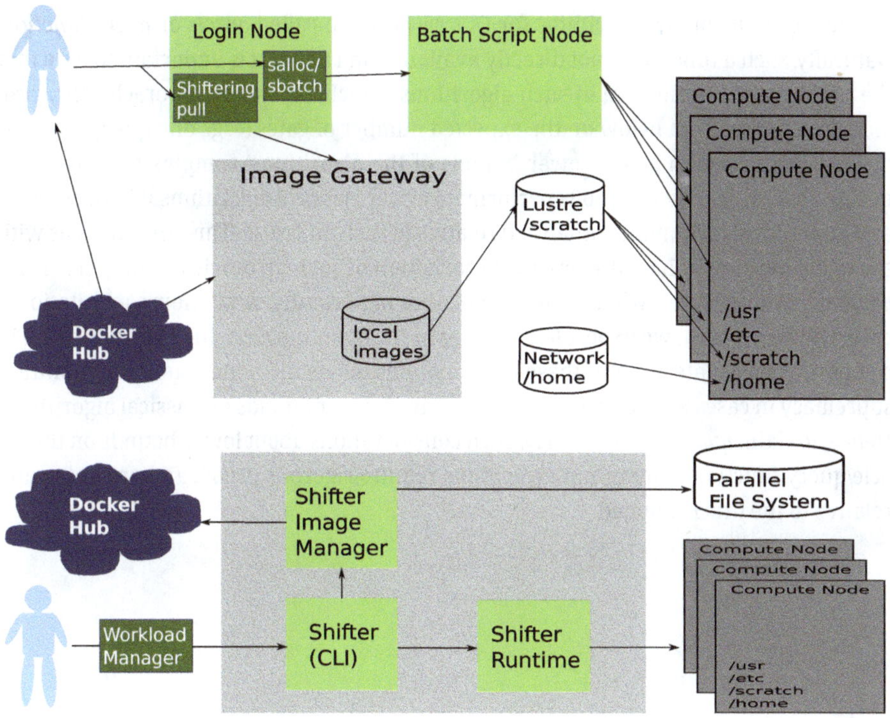


Figure 3.8: *HPC as CHeS:* Research Scientific Computer Center (NERSC) shifter architecture (*top*) and Swiss National Supercomputing Center (CSCS) shifter architecture (*bottom*), see [95].

### 3.3.4 QC as a CHeS

The frequent news over the last decade about successful increases in the number of qubits of Quantum Computers (QCs) motivate us, on one hand, to look into the basic working principles and concepts for programming QCs; on the other hand, it also shows us that a QC is still primarily a research intense technology without well-established industrial standards such as they are available for PCs. Therefore, we discuss a rather general design of a QC as proposed in [118], see Figure 3.5 on the right-hand side.

The main components of a QC as proposed in [118] are a so-called *Quantum Processing Unit (QPU)* (also called quantum Arithmetic Logic Unit (ALU)), a *Quantum RAM (QRAM)*, a *Code Teleporter (CT)*, and a *Dynamic Quantum Compiler (DQC)/Dynamic Quantum Scheduler (DQS)*. The QRAM itself requires a CT and a *Qubit Refresher (QR)*, which prevents qubits from decohering over time. Without going further into the interesting physical details of these single components, we refer the reader interested in the basic elements leading to quantum algorithms and (software) code running on QCs to Chapter 7.

To conclude, many algorithms for QCs rely on a so-called *oracle* or *query function*. Naturally, such a function is not directly available, and hence it is important to note that the performance evaluation of such algorithms, which depend on an oracle, only provide an estimate from below on the expected number of gates (e. g., one query requires at least one quantum gate). Lower bounds of the algorithms' complexity allow us to decide about their advantageous performance over classical algorithms if both classical and quantum algorithms are analyzed relative to such an oracle. This still leaves us with the open question of how the practical realization of such an oracle in both, the classical and quantum cases, will look like and, more importantly, what their computational costs will be. At least, we expect less queries in the quantum case due to qubits being in a superimposed state. Finally, these lower bounds tell us the nonexistence of quantum supremacy in cases where they are above derived upper bounds of classical algorithms. Hence, to claim *quantum supremacy*, such considerations about lower bounds on the oracle/query function in quantum algorithms require rigorous proofs and are only valid relative to the oracle applied.

# 4 CHeSs and measurements: state and parameter estimation

## 4.1 CHeS identification and description: stochastic models and filters

The identification and proper understanding of CHeSs relies on carefully and systematically performed observations, see Figure 2.2 for a general setup, whereas Figure 4.1 represents a classical and passive observation form (observer does not "illuminate" the CHeS). Hence, a primary concern for observations is the loss of information, which can have various origins such as *channel deterioration* (e. g., interference and noise), *lack of data transmission capacity* (e. g., video of high-frequency events), and *chaotic dynamics of CHeSs (and/or Observer)*.

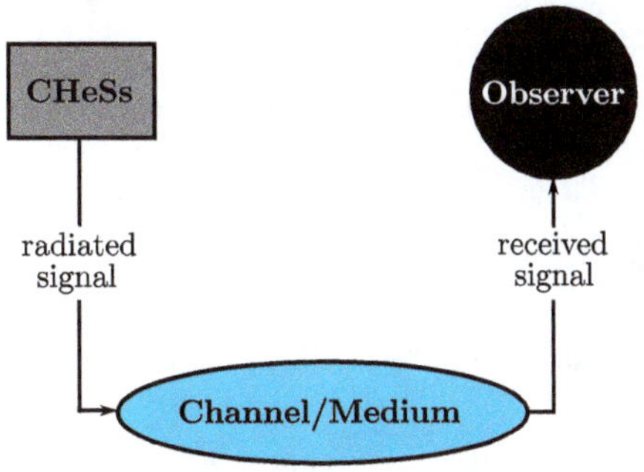


Figure 4.1: *Top left:* The system radiates, reflects, or transports information/material in an optical/electromagnetic/radioactive/energetic/material manner (or by a mix of these transport modes). *Center at the bottom:* Depending on the transport medium, the information/material transported is also convected (fluid) by the environment or dispersed due to interference or possible collisions with obstacles or the same or other particles, for instance. *Top right:* The observer/sensor receives/measures the signals/materials.

As a consequence, the signal/information/material received by the Observer is distorted and noisy, see Definition 2.1.3 about observable systems. Therefore, the main task of the Observer is to estimate/filter the original, undistorted information content.

### 4.1.1 CHeSs as stochastic input/output models

Before we introduce basic and for CHeSs relevant stochastic models, we briefly recall useful and important characterizations of discrete-time stochastic processes. Note that the label "stochastic" refers here to the problem of erroneous measurements/perception/observation and also the possible incomplete scientific description of the problem at hand. In fact, characterizations of random processes are generally the best we can do, since it is impossible to determine the "real" joint probability distribution in most realistic settings. Real-world scenarios generally only allow us to exploit the first (expectation/mean) and second moments (variance/standard deviation).

https://doi.org/10.1515/9783110579543-004

Let $U(n)$ be an arbitrary discrete-time stochastic process, and let $u(n)$ be its realization/sample. For simplicity, we subsequently denote time series with an observation window of $M$ past events,

$$U(n), U(n-1), \dots, U(n-M+1) \in \mathbb{C}, \tag{4.1}$$

simply by $U_M(n)$. Herewith, we can directly list the main quantities/functions characterizing *stochastic processes/time series*: for $k = 0, \pm 1, \pm 2, \dots$, we have,

$$\begin{aligned} &\text{(mean-value)} && \mu(n) = \mathbb{E}[U(n)], \\ &\text{(autocorrelation)} && r(n, n-k) = \mathbb{E}[U(n)U^*(n-k)], \\ &\text{(autocovariance)} && c(n, n-k) = \\ & && \mathbb{E}[(U(n) - \mu(n-k))(U(n-k) - \mu(n-k))^*], \end{aligned}$$

where $U^*$ denotes the complex conjugate of $U$.

**Remark 4.1.1** (Relation between characteristics of $U$ and stationarity). 1. We recall that the mean-value ($\mu(n)$), the autocorrelation ($r(n, n-k)$), and the autocovariance ($c(n, n-k)$) satisfy the following relation,

$$c(n, n-k) = r(n, n-k) - \mu(n)\mu^*(n-k). \tag{4.2}$$

Hence, the autocovariance and autocorrelation are the same in the case of zero mean ($\mu = 0$).

2. For *strictly stationary processes*,[1] i. e., $U(n) = U$, the above characterizations simplify for all $n$ to,

$$\begin{aligned} &\text{(mean-value)} && \mu(n) = \mu, \\ &\text{(autocorrelation)} && r(n, n-k) = r(k), \\ &\text{(autocovariance)} && c(n, n-k) = c(k). \end{aligned}$$

Finally, this implies for $k = 0$ the following identities,

$$\begin{cases} r(0) = \mathbb{E}[|U|^2], \\ c(0) = \mathbb{E}[|U - \mu|^2] = \operatorname{Var}[U] = \sigma_U^2. \end{cases} \tag{4.3}$$

◇

A very simple and standard nonstationary stochastic process is obtained by adding noise to a sine wave as shown in the following example.

1 Recall that the distributions of a strictly stationary process $\{U(n)\}_n$, $n \in \mathbb{N}_{\geq 0}$, are the same over all corresponding time steps $t_n$, $n \in \mathbb{N}$.

**Example 4.1.2** (Correlation matrix). The mathematical formulation,

$$U(n) = A\sin(\omega n) + v(n),\ n \geq 0, \tag{4.4}$$

finds a lot of applications in signal processing, where it could represent a receiver for a specific input signal with frequency $\omega$, amplitude $A$, and thermal noise $v(n)$ with zero mean and the following autocorrelation,

$$r_v(k) = \mathbb{E}\big[v(n)v^*(n-k)\big] = \begin{cases} \sigma_v^2\,, & k = 0\,, \\ 0\,, & k \neq 0\,. \end{cases} \tag{4.5}$$

Moreover, we use the notation $i = \sqrt{-1}$ and canonically write $U(n)$ as a complex signal still denoted by $U(n)$. In general, signal and noise are independent, so that the autocorrelation function for the composed signal $U$ reads,

$$r_U(k) = \begin{cases} |A|^2 + \sigma_v^2(n)\,, & k = 0\,, \\ |A|^2\exp(i\omega k)\,, & k \neq 0\,. \end{cases} \tag{4.6}$$

Hence, identifying by $\rho = \frac{|A|^2}{\sigma_v^2}$ the *signal-to-noise ratio* allows us to immediately write the correlation matrix $\hat{\mathrm{R}}_U = |A|^2\hat{\hat{\mathrm{R}}}_{U_M}$ with the help of (4.6) as follows,

$$\hat{\hat{\mathrm{R}}}_{U_M} := \begin{bmatrix} 1+\frac{1}{\rho} & \exp(i\omega) & \exp(i\omega 2) & \dots & \exp(i\omega(M-1)) \\ \exp(-i\omega) & 1+\frac{1}{\rho} & \exp(i\omega) & \dots & \exp(i\omega(M-2)) \\ \vdots & \ddots & \ddots & \ddots & \vdots \\ & & & & \exp(i\omega) \\ \exp(-i\omega(M-1)) & & \dots & \exp(-i\omega) & 1+\frac{1}{\rho} \end{bmatrix}, \tag{4.7}$$

for given samples $u(n), u(n-1), \dots, u(n-M+1)$. ◇

**Remark 4.1.3** (Singular autocorrelation matrix). We note that the particular case of vanishing noise and hence infinite signal-to-noise ratio, i. e., $\rho = \infty$, and for simplicity, an observation window of size $M = 2$ easily show that the resulting autocorrelation matrix,

$$\hat{\mathrm{R}}_{U_2} = |A|^2 \begin{bmatrix} 1 & \exp(i\omega) \\ \exp(-i\omega) & 1 \end{bmatrix}, \tag{4.8}$$

is singular, i. e., $\det(\hat{\mathrm{R}}_{U_2}) = 0$. This implies a linear dependence inside the correlation matrix and hence a linear dependence in the signal $U(n)$ over time. ◇

In applications, three types of linear stochastic models are frequently identified, depending on whether we account for present and past input values up to present and past output values, see Figure 4.2. These three formulations are as follows:

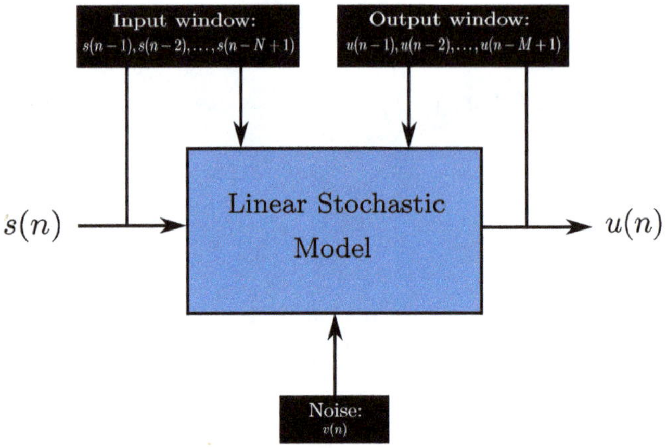


Figure 4.2: *Linear Stochastic Models:* The present input $s(n)$, present output $u(n)$, past input values $s(n-1)$, $s(n-2), \ldots, s(n-N+1)$ represent a so-called *input window*, and the past output values $u(n-1), u(n-2)$, $\ldots, u(n-M+1)$ form an *output window*. Hence, the input and output windows have the observation lengths $N$ and $M$, respectively. Linearity appears due to $N$ coefficients $\mathbf{b} := [b_0, b_1, \ldots, b_{N-1}]' \in \mathbb{C}^N$ weighting the input values and $M$ coefficients $\mathbf{a} := [a_0, a_1, \ldots, a_M]' \in \mathbb{C}^M$ with $a_0 = 1$ weighting the output values. The *noise process* $v(n)$ is zero mean and additive.

- **autoregressive (AR) models:** no past values of model inputs are used, but rather present and past model outputs;
- **moving-average (MA) models:** no past values of model outputs are used, but rather present and past values of inputs;
- **mixed AR and MA models, i. e., ARMA models:** both present and past input and output values are used.

Therefore, we can immediately state the following formal definitions.

**AR models**

**Definition 4.1.4.** Let $a_1, a_2, \ldots, a_{M-1} \in \mathbb{C}$ be constants, and let $v$ be a zero-mean white noise process. We call realizations $\{u(i)\}_{i=n-M+1}^{n}$ to be an *AR model of order* $M-1>0$, if it holds that,

$$u(n) + a_1^* u(n-1) + \cdots + a_{M-1}^* u(n-M+1) = v(n) . \tag{4.9}$$

We note that equation (4.9) can also be understood as a *discrete convolution of order* $M-1$ *(i. e., of length M)*, that is,

$$(a * u)_M(n) := \sum_{k=0}^{M-1} a_k^* u(n-k) = v(n) , \tag{4.10}$$

where we set $a_0 = 1$.

Similarly, the second stochastic model can be formally defined as follows.

**MA models**

**Definition 4.1.5.** Let $b_0, b_1, \dots, b_{N-1} \in \mathbb{C}$ be constants, and let $v$ be a zero-mean white noise process. Moreover, $\{s(i)\}_{i=n-N+1}^{n}$ denote present and past input values of order $N-1$ (i. e., of length $N > 0$). We call the realization $u(n)$ to be an *MA model*, if it holds that,

$$\begin{aligned} u(n) &= b_0^* s(n) + b_1^* s(n-1) + \dots + b_{M-1}^* s(n-M+1) + v(n) \\ &= (b * s)_{M-1}(n) + v(n) . \end{aligned} \tag{4.11}$$

Finally, the last model ARMA can be formally captured in the following definition.

**Mixed order $(M-1)$-AR and order $(N-1)$-MA models**

**Definition 4.1.6.** Let $a_0, a_1, \dots, a_{M-1} \in \mathbb{C}$ and $b_0, b_1, \dots, b_{N-1} \in \mathbb{C}$ be constants, the so-called *ARMA parameters*, and let $v$ be a zero-mean white noise process. As in Definition 4.1.5, let $\{s(i)\}_{i=n-N+1}^{n}$ represent present and past input values of order $N-1$ (i. e., of length $N > 0$). We call realizations $\{u(i)\}_{i=n-M+1}^{n}$ to be a *mixed order $(M-1)$-AR and order $(N-1)$-MA process*, if it holds that,

$$(a * u)_{M-1}(n) = (b * s)_{N-1}(n) + v(n) , \tag{4.12}$$

where $a_0 = 1$.

For many analytical and practical aspects, the concepts of the *transfer/response function* and the z-transform are very helpful. Therefore we briefly recall these technical and useful tools.

**$z$-transform, $\mathcal{Z}$, and Region of Convergence (RoC), $\zeta$**

**Definition 4.1.7.** For a time series $\{u(n)\}_{n=-\infty}^{\infty} \in \mathbb{C}$, we define the *z-transform* $\mathcal{Z}(\cdot)$ by,

$$U(z) := \mathcal{Z}\big(u(n)\big) := \sum_{n=-\infty}^{\infty} u(n) z^{-n} , \tag{4.13}$$

where $z \in \mathbb{C}$.

Moreover, we call the set $\zeta \subset \mathbb{C}$ the *Region of Convergence (RoC)*, if for all $z \in \zeta$, the $z$-transform $U(z) = \mathcal{Z}(u(n))$ is finite.

**Remark 4.1.8** (Unilateral z-transformation). In applications such as signal processing, a so-called *unilateral z-transformation* $\mathcal{Z}_+$ is generally applied,

$$\mathcal{Z}_+(u(n)) := \sum_{n=0}^{\infty} u(n) z^{-n} . \tag{4.14}$$

◇

We briefly summarize the central properties of this transformation.

**Properties of the $z$-transform $\mathcal{Z}$**

**Lemma 4.1.9.** *Let $\{u_1(n)\}$ and $\{u_2(n)\}$ be two sequences with associated RoCs $\zeta_1 \subset \mathbb{C}$ and $\zeta_2 \subset \mathbb{C}$, respectively. We denote by $U_1(z)$ and $U_2(z)$ the z-transforms of $\{u_1(n)\}$ and $\{u_2(n)\}$, respectively. Moreover, $a_1 \in \mathbb{C}$ and $a_2 \in \mathbb{C}$ are two complex constants. The z-transform $\mathcal{Z}$ has the following properties:*

$$\textbf{(Z1)}\quad \mathcal{Z}\big(a_1u_1(n) + a_2u_2(n)\big) = a_1U_1(z) + a_2U_2(z)\,, \quad z \in \zeta_1 \cap \zeta_2\,;$$

$$\textbf{(Z2)}\quad \mathcal{Z}\big(u_1(n-k)\big) = z^{-k}U_1(z)\,, \quad z \in \zeta_1 \setminus \mathcal{U}_k\,;$$

$$\textbf{(Z3)}\quad \mathcal{Z}\big(a_1^n u_1(n)\big) = U_1\big(a_1^{-1}z\big)\,, \quad z \in |a_1|\zeta_1\,,$$

$$\textbf{(Z4)}\quad \mathcal{Z}\big((u_1 * u_2)(n)\big) = U_1(z)U_2(z)\,, \quad z \in \zeta_1 \cap \zeta_2\,,$$

*where*

$$\mathcal{U}_k := \begin{cases} \{0\} & \text{for } k > 0\,, \\ \{\infty\} & \text{for } k < 0\,, \end{cases}$$

*and properties **(Z1)–(Z4)** hold at least in the RoC.*

*Proof.* **Linearity (Z1).** Starting with the left-hand side expression allows us to transform it with the help of the definition of the $z$-transform as follows,

$$\begin{aligned} \mathcal{Z}(a_1u_1(n) + a_2u_2(n)) &= \sum_{k=-\infty}^{\infty} (a_1u_1(k) + a_2u_2(k))z^{-k} \\ &= \sum_{k=-\infty}^{\infty} a_1u_1(k)z^{-k} + \sum_{k=-\infty}^{\infty} a_2u_2(k)z^{-k} \\ &= a_1\mathcal{Z}(u_1(n)) + a_2\mathcal{Z}(u_2(n)) \\ &= a_1U_1(z) + a_2U_2(z)\,, \end{aligned} \tag{4.15}$$

which proves **(Z1)**.

**Shifting (Z2).** Again, beginning with the left-hand side as above leads to,

$$\begin{aligned} \mathcal{Z}(u_1(n-k)) &= \sum_{n=-\infty}^{\infty} u_1(n-k)z^{-n} = z^{-k}\sum_{n=-\infty}^{\infty} u_1(n-k)z^{-(n-k)} \\ &= z^{-k}\mathcal{Z}(u_1(n)) = z^{-k}U_1(z)\,, \end{aligned} \tag{4.16}$$

which is the wanted statement.

**Multiplication (Z3).** We can directly rewrite the left-hand side as,

$$\begin{aligned} \mathcal{Z}(a_1^n u_1(n)) &= \sum_{n=-\infty}^{\infty} a_1^n u_1(n)z^{-n} = \sum_{n=-\infty}^{\infty} u_1(n)(z/a_1)^{-n} \\ &= \mathcal{Z}(u_1(n))(z/a_1) = U_1(z/a_1)\,, \end{aligned} \tag{4.17}$$

which represents the required equation.

**Convolution (Z4).** Again, starting with the left-hand side, the definition of the convolution operator immediately gives,

$$\begin{aligned}\mathcal{Z}((u_1 * u_2)(n)) &= \sum_{n=-\infty}^{\infty}\left(\sum_{l=-\infty}^{\infty} u_1(l)u_2(n-l)\right)z^{-n}\\ &= \sum_{n=-\infty}^{\infty}\left(\sum_{l=-\infty}^{\infty} u_1(l)u_2(n-l)\right)z^{-n}z^{-l}z^{l}\\ &= \sum_{l=-\infty}^{\infty} u_1(l)z^{-l}\sum_{n=-\infty}^{\infty} u_2(n-l)z^{-(n-l)}\\ &= \mathcal{Z}(u_1(n))\mathcal{Z}(u_2(n)) = U_1(z)U_2(z)\,,\end{aligned} \tag{4.18}$$

which proves that last statement. □

These stochastic models can be analyzed at once with the general stochastic ARMA model (4.12). The $z$-transformation provides a rather intuitive tool for this purpose, if we can systematically transform the problem into a noise-free form. For instance, a noise-free setting can be easily obtained by taking the expectation on both sides. Taking the $z$-transform on both sides of (4.12) and thanks to property **(Z4)** from Lemma 4.1.9, we immediately obtain,

$$A(z)U(z) = \mathcal{Z}((a * u)_{M-1}(n)) = \mathcal{Z}((b * s)_{N-1}(n)) = B(z)S(z)\,, \tag{4.19}$$

where

$$\begin{aligned}A(z) &= \sum_{k=0}^{M-1} a_k^* z^{-k}\,,\\ B(z) &= \sum_{k=0}^{N-1} b_k^* z^{-k}\,,\\ U(z) &= \sum_{n=-\infty}^{\infty} u(n)z^{-k}\,,\\ S(z) &= \sum_{n=-\infty}^{\infty} s(n)z^{-k}\,.\end{aligned}$$

This allows us to identify the $z$-transform of the *transfer/response function* $h$, i.e, $u(n) = h(w(n))$, as follows,

$$H(z) := \frac{B(z)}{A(Z)}\,, \tag{4.20}$$

so that $U(z) = H(z)S(z)$. In the next section, we consider how the transfer function $H$ can be used to investigate filters.

### 4.1.2 CHeSs as a filter

The rather general theory presented in Section 4.1.1 allows us to investigate certain aspects of filter theory, which is closely related to (state and parameter) estimation theory. The common goal of both theories is to infer from noisy/erroneous data/measurements the correct underlying information content.

In signal processing and, more generally, in system theory such as CHeSs, the *transfer/response function* $h$, i. e.,

$$u(n) = h(s(n)), \quad n \in \{0, 1, \dots, N_T\}, \tag{4.21}$$

plays a crucial role in the analysis and understanding of how inputs $s(n)$ are transformed into outputs $u(n)$ by a particular system $\mathcal{S}$ of interest; $N_T$ denotes the number of time steps. Hence, for the rather broad class of linear stochastic ARMA processes (4.12), we can immediately state the $z$-transform of the underlying transfer function thanks to (4.20),

$$H(z) = \frac{B(z)}{A(Z)} = \frac{\sum_{k=0}^{N-1} b_k^* z^{-k}}{\sum_{k=0}^{M-1} a_k^* z^{-k}} = \frac{\sum_{k=0}^{N-1} b_k^* z^{-k}}{1 + \sum_{k=1}^{M-1} a_k^* z^{-k}}, \tag{4.22}$$

where we used the fact that $a_0 = 1$. Equation (4.22) allows us to motivate the important class of *Finite Impulse Response (FIR)* filters. These correspond to the so-called *MA models*, since FIR filters do not show any feedback, i. e., $M = 1$,

$$\textbf{(FIR filter)} \quad \begin{cases} H(z) = \sum_{k=0}^{N-1} b_k^* z^{-k} & \text{($z$-transformed transfer function)}, \\ u(n) = (b * s)_{N-1}(n) & \text{(MA process)}. \end{cases} \tag{4.23}$$

Naturally, the question is how to identify the *FIR parameters* $\{b_k\}_{k=0}^{N-1}$? If we denote by $u_d(n)$ the desired system output, then a common strategy is to find $b_k$, $k = 0, 1, \dots, N-1$, by minimizing the following quadratic error,

$$\{b\}_{k=0}^{N-1} := \min_{\{\tilde{b}\}_{k=0}^{N-1}} E(\{\tilde{b}_{k=0}^{N-1}\}) := \min_{\{\tilde{b}\}_{k=0}^{N-1}} \|u_d(\cdot) - (\tilde{b} * s)_{N-1}(\cdot)\|^2_{\ell^2(0,N-1)}, \tag{4.24}$$

where $\mathbf{b} := [b_0, b_1, \dots, b_{N-1}]'$, and $\|\cdot\|_{\ell^2(0,N)}$ is the following quadratic norm on finite sequences,

$$\|u(\cdot)\|^2_{\ell^2(0,N)} := \sum_{n=0}^{N} |u(n)|^2. \tag{4.25}$$

The quadratic form of (4.24) immediately motivates to determine the parameters $\mathbf{b} := [b_0, b_1, \dots, b_{N-1}]'$ in the classical *data-driven/data-science* way, called the Least-Squares Method (LSM). For more technical details on LSM, we refer the interested reader to Section 4.2.1.

## 4.2 Concepts of parameter estimation

Approaches to *parameter estimation* naturally find increasing interest due to the growing number of sensorial capabilities (e. g., measurements of temperature, pressure, magnetic fields, voltage, current, weight, velocity/radar, location/radar, etc.). In fact, an intrinsic property of repeated sensorial measurements of the same quantity generally shows small discrepancies in the results, called *measurement errors*. Next to filtering out measurement errors, *parameter estimation* provides also the capability for solving *inverse problems*, e. g., inferring the permeability of porous media as observed in ground water flows [114], which represents an example of an important class of multiscale problems.

To establish the basic ideas and methods of parameter estimation, we first need to introduce basic terms and notations.

**Estimation, estimator, and estimation error**

**Definition 4.2.1.** Consider an arbitrary problem that depends on an $r$-dimensional vectorial parameter $\boldsymbol{\theta} \in \mathbb{R}^r, r \in \mathbb{N}_{>0}$. *Parameter estimation* represents the general problem of finding a map $\overline{\Theta} : \mathbb{R}^q \to \mathbb{R}^r$, i. e., $\overline{\Theta}(\mathbf{z}) = \tilde{\boldsymbol{\theta}}$, that associates $r$-dimensional *estimates* $\tilde{\boldsymbol{\theta}} \in \mathbb{R}^r, r \in \mathbb{N}_{>0}$, with $q$-dimensional *measurements* $\mathbf{z} \in \mathbb{R}^q, q \in \mathbb{N}_{>0}$. This map $\overline{\Theta}$ is referred to as *an estimator*.

This represents the so-called *deterministic estimation problem (DEP)*, since it aims at estimating a nonrandom parameter, and hence its estimator is noise-free. Herewith, we can define the following pointwise estimation error,

$$\mathbf{e}_{\theta} := \tilde{\boldsymbol{\theta}} - \overline{\boldsymbol{\theta}}, \tag{4.26}$$

which quantifies the error between the estimated and the true value/parameter $\overline{\boldsymbol{\theta}}$.

To be able to qualitatively characterize estimators, the number of measurements performed to estimate the parameter of interest is generally taken into account. Hence, let $\ell \in \mathbb{N}_{>0}$ denote the number of such measurements, which leads to a vector $\boldsymbol{Z}^{\ell}$ containing all measurements,

$$\boldsymbol{Z}^{\ell} := \begin{bmatrix} \mathbf{z}^1 \\ \mathbf{z}^2 \\ \dots \\ \mathbf{z}^{\ell} \end{bmatrix}, \tag{4.27}$$

where $\mathbf{z}^i, 1 \leq i \leq \ell$, denotes the vector of the $i$th measurement.

In general, parameter and state estimation relies on the following two maps,

$$\begin{cases} \mathbf{z}^i = h(i, \mathbf{s}) & \text{(measurement map), and} \\ \tilde{\theta}^i = \overline{\Theta}(i, \boldsymbol{Z}^i) & \text{(estimator/estimation map),} \end{cases} \tag{4.28}$$

respectively. The variable $\mathbf{s}$ describes the state of a stationary system $\mathcal{S}$, for which we are able to get $\ell$ measurement vectors collected in the vector $\mathbf{Z}^{\ell}$. Note that the *measurement map* will play a crucial role in the context of *Kalman filters*, which provides a so-called *dynamic estimator* as opposed to *static estimators* such as parameter estimation, see Section 4.3. In the former context, an additional time step index $n$ is introduced such that $\mathbf{z}^{i,n}$ and $\mathbf{s}^n$ indicate the dynamic character of the measurements and state, respectively.

We need a further property, which guarantees that with an increasing number of measurements, our estimator should become exact, i. e., the estimation error $\mathbf{e}_\theta$ should vanish.

**Unbiased and consistent estimators**

**Definition 4.2.2.** Let $\overline{\Theta}(i, \mathbf{Z}^i)$ be an estimator. We call $\overline{\Theta}$ to be *unbiased*, if it holds that,

$$\mathbb{E}\left[\overline{\Theta}\left(\ell, \mathbf{Z}^\ell\right)\right] = \overline{\boldsymbol{\theta}}, \tag{4.29}$$

where $\ell$ denotes the number of measurements.

Similarly, an estimator is called *consistent* if $\mathbf{e}_{\theta,\ell} := \tilde{\boldsymbol{\theta}}^\ell - \overline{\boldsymbol{\theta}^\ell}$ converges as $\ell \to \infty$ in some stochastic sense to zero.[a]

---

*a* For example, a possible form of convergence is $\lim_{\ell\to\infty} \mathbb{E}[\|\mathbf{e}_{\theta,\ell}\|_p^p] = 0$, where $\|\cdot\|_p$ denotes the $p$-norm for vectors, i. e., $\|\mathbf{u}\|_p := (\sum_{i=1}^m |u_i|^p)^p$ for $\mathbf{u} = [u_1, u_2, \ldots, u_m]$.

Finally, we note that the estimation of nonrandom parameters is referred to as the *non-Bayesian/Fisher method* whereas the estimation of random parameters leads to *Bayesian methods*. Let us first look at the non-Bayesian/Fisher method which can be applied in cases where we have access to a conditional probability $p(\mathbf{z}|\boldsymbol{\theta})$ on measurements $\mathbf{z}$ for a given parameter vector $\boldsymbol{\theta}$. We employ the following notation,

$$\begin{cases} L_\mathbf{z}(\boldsymbol{\theta}) := p(\mathbf{z}|\boldsymbol{\theta}) & \text{single measurement,} \\ L_\ell(\boldsymbol{\theta}) := p(\mathbf{Z}^\ell|\boldsymbol{\theta}) & \ell \text{ measurements,} \end{cases} \tag{4.30}$$

where $L_\cdot(\boldsymbol{\theta})$ is called a *likelihood function* and gives the likelihood of a parameter vector $\boldsymbol{\theta}$ for given ($\ell$-times repeated) measurements $\mathbf{Z}^\ell$ with $\mathbf{z} = \mathbf{Z}^1$, see Section (4.2.2).

To demonstrate the basic concepts for establishing suitable estimators, we introduce a simple and widely used approach in the next section.

### 4.2.1 Least Squares Estimator (LSE)

The key strategy of LSE is based on a *variational approach* by introducing a *cost/goal function* $J(\boldsymbol{\theta})$ in such a way, that the parameter/state of interest $\boldsymbol{\theta}$ represents a minimizer of $J$.

---

**Least Squares Method (LSM)**

Assume that we are given a measurement/data vector $\mathbf{z} \in \mathbb{R}^r$ and a symmetric positive definite tensor $\hat{\Sigma}^{-1} := \{\sigma_{ij}^{-1}\}_{i,j=1}^{\ell} \in \mathbb{R}^{r\times r}$, whose components are weights of the corresponding measurement errors. The LSM then consists in finding the parameter/state $\boldsymbol{\theta}_m \in \mathbb{R}^q$ that minimizes the following cost/goal function,

$$\begin{aligned} J(\boldsymbol{\theta}) &:= \frac{1}{2}\sum_{i,j}^{r}\left(z_i - h_i(\boldsymbol{\theta})\right)^T \sigma_{ij}^{-1}\left(z_j - h_j(\boldsymbol{\theta})\right), \\ &= \frac{1}{2}\left(\mathbf{z} - \mathbf{h}(\boldsymbol{\theta})\right)^T \hat{\Sigma}^{-1}\left(\mathbf{z} - \mathbf{h}(\boldsymbol{\theta})\right), \end{aligned} \tag{4.31}$$

that is,

$$\boldsymbol{\theta}_m := \underset{\boldsymbol{\theta}\in\mathbb{R}^q}{\operatorname{argmin}}\, J(\boldsymbol{\theta}). \tag{4.32}$$

---

**Remark 4.2.3** (LSM versus Maximum Likelihood Method (MLM)). We note that the weighting tensor $\hat{\Sigma}$ in the LSM plays the role of the covariance matrix of the measurement error $\mathbf{w}$. As we will see, the MLM leads to the same estimator as the LSM with weighting matrix $\hat{\Sigma}^{-1}$. Moreover, the estimator $\boldsymbol{\theta}_m$ obtained by the LSM becomes random if the measurement map $\mathbf{h}$ models measurement errors by an additive random process $\mathbf{w}$, i. e., $\mathbf{z} = \mathbf{h}(\boldsymbol{\theta}) + \mathbf{w}$. ◇

An extension toward multiple measurements $\mathbf{Z}^{\ell}$ (see (4.27)) relies on the straightforward generalization of (4.31) toward,

$$J(\boldsymbol{\theta}) = \frac{1}{2}\sum_{k=1}^{\ell}(\mathbf{z}^k - \mathbf{h}(\boldsymbol{\theta}))^T \hat{\Sigma}^{-1}(\mathbf{z}^k - \mathbf{h}(\boldsymbol{\theta})), \tag{4.33}$$

which is called Batch LSM (BLSM) for $\ell$-measurements $\mathbf{z}^k, 1 \le k \le \ell$, of the same constant parameter/state vector $\boldsymbol{\theta}$ whose estimator becomes random if the associated measurement errors show a random description.

Applying this method to a linear DEP, we immediately obtain the following result.

**Solution of linear DEP with LSM**

**Lemma 4.2.4.** *We assume that the measurement map* $\mathbf{h} : \boldsymbol{\theta} \in \mathbb{R}^q \mapsto \mathbf{z} \in \mathbb{R}^r$ *is linear, i. e.,* $\mathbf{z} = \mathbf{c}_0 + \hat{\mathrm{H}}\boldsymbol{\theta}$ *for a constant vector* $\mathbf{c}_0 \in \mathbb{R}^r$ *and a tensor* $\hat{\mathrm{H}} \in \mathbb{R}^{r\times q}$*. The solution of the linear DEP obtained by the LSM can be represented as,*

$$\boldsymbol{\theta} = \left(\hat{\mathrm{H}}^T\hat{\Sigma}^{-1}\hat{\mathrm{H}}\right)^{-1}\hat{\mathrm{H}}^T\hat{\Sigma}^{-1}(\mathbf{z} - \mathbf{c}_0). \tag{4.34}$$

*Proof.* A primary but not complete requirement for identifying a minimizer $\boldsymbol{\theta}_m \in \mathbb{R}^q$ of the cost/goal function $J(\boldsymbol{\theta})$ (see equation (4.31)) is to solve the following equation,

$$\begin{aligned} \mathbf{0} &= \nabla_{\boldsymbol{\theta}} J(\boldsymbol{\theta}) \\ &= \frac{1}{2}(-(\nabla_{\boldsymbol{\theta}}\mathbf{h}(\boldsymbol{\theta}))^T\hat{\Sigma}^{-1}(\mathbf{z} - \mathbf{h}(\boldsymbol{\theta})) + (\mathbf{z} - \mathbf{h}(\boldsymbol{\theta}))^T\hat{\Sigma}^{-1}(\nabla_{\boldsymbol{\theta}}\mathbf{h}(\boldsymbol{\theta}))) \end{aligned}$$

$$= -(\nabla_{\theta}\mathbf{h}(\boldsymbol{\theta}))^T\hat{\Sigma}^{-1}\mathbf{z} + (\mathbf{h}(\boldsymbol{\theta}))^T\hat{\Sigma}^{-1}(\nabla_{\theta}\mathbf{h}(\boldsymbol{\theta})). \tag{4.35}$$

Taking the linearity of $\mathbf{h}$ into account, i. e., $\nabla_{\theta}\mathbf{h}(\boldsymbol{\theta}) = \hat{\mathrm{H}}$, leads to the wanted expression (4.34). It remains to note that the parameters (stationary state) $\boldsymbol{\theta}$ defined in this way represents an actual minimizer, since $\hat{\Sigma}^{-1}$ is symmetric positive definite, so that the Hessian $\nabla^2_{\theta}J(\boldsymbol{\theta})$ is also positive definite, and hence (4.34) defines a minimizer. □

We can extend the previous method toward nonlinear measurement maps $\mathbf{h}$.

**Solution of nonlinear DEP with LSM**

**Lemma 4.2.5.** *Let* $\mathbf{h} : \boldsymbol{\theta} \in \mathbb{R}^q \mapsto \mathbf{z} \in \mathbb{R}^r$ *be a nonlinear map, and let* $\boldsymbol{\theta}_0 \in \mathbb{R}^q$ *be a state vector close to the true one. The least squares solution of this nonlinear DEP then reads,*

$$\boldsymbol{\theta} = \left(\left(\nabla_{\theta}\mathbf{h}(\boldsymbol{\theta}_0)\right)^T\Sigma^{-1}\nabla_{\theta}\mathbf{h}(\boldsymbol{\theta}_0)\right)^{-1}\left(\left(\nabla_{\theta}\mathbf{h}(\boldsymbol{\theta}_0)\right)^T\hat{\Sigma}^{-1}\left(\mathbf{z} - \mathbf{h}(\boldsymbol{\theta}_0) + \nabla_{\theta}\mathbf{h}(\boldsymbol{\theta}_0)\boldsymbol{\theta}_0\right)\right). \tag{4.36}$$

*Proof.* The proof follows the same ideas as those in Lemma 4.2.4 but by replacing $\mathbf{h}$ with its linear approximation, i. e., the following first-order Taylor approximation,

$$\tilde{\mathbf{h}}(\boldsymbol{\theta}) = \mathbf{h}(\boldsymbol{\theta}_0) + \nabla_{\theta}\mathbf{h}(\boldsymbol{\theta}_0)(\boldsymbol{\theta} - \boldsymbol{\theta}_0), \tag{4.37}$$

for fixed vector $\boldsymbol{\theta}_0$ close enough to the variable vector $\boldsymbol{\theta}$. □

**Example 4.2.6** (Nonlinear LSM). Consider the nonlinear map $F : \mathbb{R}^3 \to \mathbb{R}^4$ defined by,

$$F(\mathbf{x}) := \begin{bmatrix} a_0 + a_1x_1^2 + a_2x_2 + a_3x_3^2 \\ b_0 + b_1x_1^2 + b_2x_2 + b_3x_3^2 \\ c_0 + c_1x_1^2 + c_2x_2 + c_3x_3^2 \\ d_0 + d_1x_1^2 + d_2x_2 + d_3x_3^2 \end{bmatrix}, \tag{4.38}$$

where we fix the coefficients as follows,

$$\begin{aligned} &a_0 = 1.5, && a_1 = 0.5, && a_2 = -1.0, && a_3 = 2.0, \\ &b_0 = 0.0, && b_1 = -0.5, && b_2 = 1.0, && b_3 = -1.0, \\ &c_0 = 1.0, && c_1 = 1.5, && c_2 = 2.0, && c_3 = -2.0, \\ &d_0 = 0.5, && d_1 = -1.5, && d_2 = -2.0, && d_3 = 1.0. \end{aligned} \tag{4.39}$$

We are interested in estimating the true state $\mathbf{x}_T = [1.5, 4, 1.5]' \in \mathbb{R}^3$ for given erroneous measurements $\mathbf{z} = F(\mathbf{x}) + \mathbf{w}$ and a constant error vector $\mathbf{w} = 0.5[1, 1, 1, 1] \in \mathbb{R}^4$.

Since the nonlinear LSM relies on linearization, we also need to define a state $\mathbf{x}_0$ close enough to the true state $\mathbf{x}_T$. We set $\mathbf{x}_0 = \mathbf{x}_T + 6.5[0.1, 0.1, 0.1]'$ for instance.

We implement the nonlinear LSM with the following Python code:

```
def nlLeastSquares(z,x0,h_x0,dh_x0,Sgm):
# INPUT:
# z: measurements/data
# h: nonlinear measurement map
# dh_x0: measurement tensor (Jacobian of the linear measurement
#                  map h(x)) where x represents the parameters/systems
#                  state of interest
# Sgm: weighting matrix
#
# OUTPUT:
# xEst: (least squares) estimate of system parameters/state

iSig = np.linalg.inv(Sgm)
i_dhT_iSig_dh =
     np.linalg.inv(np.matmul(dh_x0.T,np.matmul(iSig,dh_x0)))
dhT_iSig_z = np.matmul(dh_x0.T,np.matmul(iSig,z))
dhT_iSig_linH =
     np.matmul(dh_x0.T,(np.matmul(iSig,np.matmul(dh_x0,x0))))
#
dhT_iSig_h = np.matmul(dh_x0.T,np.matmul(iSig,h_x0))
#

return   np.matmul(i_dhT_iSig_dh,
     ( dhT_iSig_z - dhT_iSig_h +dhT_iSig_linH))
```

which we import with the command `import estimationTools as et`. After execution, we get the following results:

```
xEst = et.nlLeastSquares(z,x0,F(x0,a,b,c,d),dF(x0,a,b,c,d),Sgm)
print("xEst=",xEst)
print("xTrue=",xTrue)
print("|xEst-xTrue|=",np.linalg.norm(xEst-xTrue))
```

```
xEst= [1.59825581 4. 1.59825581]
xTrue= [1.5 4. 1.5]
|xEst-xTrue|= 0.1389547046750318
```

Let us briefly look at the influence of choosing a different fixed linearization state, $\mathbf{x}_0 = \mathbf{x}_T + 11.5[0.1, 0.1, 0.1]'$. As a result, we obtain the following:

```
# Choosing a x0, which is hopefully close to
#             xTrue (in practice, not known):
x0=xTrue+11.5*np.array([0.1,0.1,0.1]) # linearization point (Taylor)
xEst = et.nlLeastSquares(z,x0,F(x0,a,b,c,d),dF(x0,a,b,c,d),Sgm)
print("xEst=",xEst)
print("xTrue=",xTrue)
print("|xEst-xTrue|=",np.linalg.norm(xEst-xTrue))
```

```
xEst= [1.7495283 4. 1.7495283]
xTrue= [1.5 4. 1.5]
|xEst-xTrue|= 0.35288630872423027
```

Naturally, we have silently ignored the lines of code defining the nonlinear function $F(\mathbf{x})$ and its Jacobian $dF(\mathbf{x})$ in the code blocks above. ◇

**Remark 4.2.7** (Linear versus nonlinear LSM). The previous Example 4.2.6 is referred to as *nonlinear* since it nonlinearly maps its wanted state into the measured data. However, we can have a nonlinear measurement map that, however, depends linearly in the wanted parameter, e. g., fitting a polynomial of degree $d > 1$ to the measured data. In this case the LSM is called *linear* as we will see in the next example. ◇

**Example 4.2.8** (Linear LSM despite nonlinear map). Consider a polynomial of degree $n > 0$, i. e.,

$$z = h_x(\boldsymbol{\theta}) = \sum_{i=0}^{n} \theta_i x^i , \tag{4.40}$$

where $\boldsymbol{\theta} := [\theta_0, \theta_1, \dots, \theta_n]'$. Clearly, $h(\boldsymbol{\theta})$ is linear in $\theta_i$, $0 \le i \le n$. Naturally, the least squares problem associated with the mapping $h$ consists in finding the optimal parameters $\boldsymbol{\theta} \in \mathbb{R}^{n+1}$ such that the following quadratic error is minimized,

$$E(\boldsymbol{\theta}) = \frac{1}{2} \sum_{k=1}^{K} (z_k - h_{x_k}(\boldsymbol{\theta}))^2 . \tag{4.41}$$

◇

Recall that the map $h_{x_k}(\boldsymbol{\theta})$ is linear in $\boldsymbol{\theta}$,

$$\mathbf{z} = h(\boldsymbol{\theta}) = \hat{\mathrm{H}}\boldsymbol{\theta} := \begin{bmatrix} 1 & x_1 & x_1^2 & \dots & x_1^n \\ 1 & x_2 & x_2^2 & \dots & x_2^n \\ \dots & & & & \\ 1 & x_K & x_K^2 & \dots & x_K^n \end{bmatrix} \begin{bmatrix} \theta_0 \\ \theta_1 \\ \dots \\ \theta_n \end{bmatrix} , \tag{4.42}$$

where $\mathbf{z} := [z_1, z_2, \dots, z_K]'$, and $\hat{\mathrm{H}} \in \mathbb{R}^{K \times n}$ is called the *Vandermonde matrix*.

**Solution to the polynomial LSM:** A look at the above polynomial LSM problem shows that after identifying $\hat{H} = \nabla_{\theta} h_{x_k}(\boldsymbol{\theta})$, we immediately can apply the well-known *normal equations* resulting from minimizing the quadratic error $E(\boldsymbol{\theta})$,

$$\boldsymbol{\theta} = (\hat{H}^T \hat{H})^{-1} \hat{H}^T \mathbf{z}. \tag{4.43}$$

◇

We can easily solve this problem in *Python*. First, let us construct the *measurement data* with the help of the exact/known polynomial parameters.

```
import numpy as np

# Measured data z = h(x):
x = np.array([1.5,2.1,3.2,1.0,3.2,1.4,9.0]) # size K=4, input
n = 4 # degree of approximating polynomial
H_x = np.vander(x,n+1,increasing=True)
theta = np.array([1.0,1.5,2.5,1.25,3.3]) #
z = np.matmul(H_x,theta)
print("z=",z)
zReal = z+0.1
print("zReal=",zReal)
```

```
       z= [2.9800000e+01 9.0929980e+01 4.1839008e+02 9.5500000e+00
4.1839008e+02 2.4107280e+01 2.2779550e+04]
       zReal= [2.9900000e+01 9.1029980e+01 4.1849008e+02 9.6500000e+00
4.1849008e+02 2.4207280e+01 2.2779650e+04]
```

With these measurements, we are now able to compute the parameters $\theta_i$, $i = 0, 1, \ldots, n$, with the help of (4.43):

```
# Least squares batch estimate (using the standard pseudoinverse):
thetaEst = np.matmul(np.linalg.inv(np.matmul(H_x.T,H_x)),…
   np.matmul(H_x.T,zReal))
thetaEst
```

```
array([1.10000491, 1.49999809, 2.49999952, 1.25000006, 3.3])
```

◇

In the last example, we motivate the versatility of the LSM with an application for passive radar systems as provided by Malanowski [103]. For the interested reader, Malanowski's book provides a general introduction into passive radar systems.

**Example 4.2.9** (Euclidean localization with LSM). Let us look at a scenario with $K > 0$ *transmitters* located at $\mathbf{x}_t^{(k)} \in \mathbb{R}^3, 1 \leq k \leq K, L > 0$, *receivers* located at $\mathbf{x}_r^{(l)} \in \mathbb{R}^3, 1 \leq l \leq L$, and $I > 0$ *targets* located at $1 \leq n \leq N \in \mathbb{N}$ time steps at position $\mathbf{x}^{n,(i)} \in \mathbb{R}^3$. The velocity of target $i$, $1 \leq i \leq I$, for the number of targets $I > 0$ at time step $n$ is $\mathbf{v}^{n,(i)}$. Herewith, we can identify various basic distances such as the *base line/transmitter-receiver range* $R_{\text{tr}}^{k,l} := \|\mathbf{x}_t^{(k)} - \mathbf{x}_r^{(l)}\|$, *target-transmitter range* $R_{\text{ft}}^{(i,k)}(n) := \|\mathbf{x}^{n,(i)} - \mathbf{x}_t^{(k)}\|$, and the *target-receiver range* $R_{\text{fr}}^{(i,l)}(n) := \|\mathbf{x}^{n,(i)} - \mathbf{x}_r^{(l)}\|$. Generally, passive radar systems measure the following *specific range distance,*

$$R_i^{(k,l)}(n) := R_{\text{ft}}^{(i,k)}(n) + R_{\text{fr}}^{(i,l)}(n) - R_{\text{tr}}^{(k,l)}, \tag{4.44}$$

as well as the time derivative of the *bistatic range*, i. e., the following bistatic velocity,

$$V_i^{(k,l)}(n) := \frac{\langle(\mathbf{x}^{n,(i)} - \mathbf{x}_t^{(k)}), \mathbf{v}^{n,(i)}\rangle}{\|\mathbf{x}^{n,(i)} - \mathbf{x}_t^{(k)}\|} + \frac{\langle(\mathbf{x}^{n,(i)} - \mathbf{x}_r^{(l)}), \mathbf{x}^{n,(i)}\rangle}{\|\mathbf{x}^{n,(i)} - \mathbf{x}_r^{(l)}\|}. \tag{4.45}$$

In fact, passive radar systems measure the time difference $\tau_i^{(k,l)}$ of arrival between the direct and the reflected signal such that $R_i^{(k,l)}(n) = c\tau^{(k,l)}$.

However, to derive a feasible computational algorithm for target localization with the help of such measurements, the following *range sum* is very convenient,

$$R_s^{(i,k,l)}(n) := R_i^{(k,l)}(n) + R_{\text{tr}}^{(k,l)}. \tag{4.46}$$

Let us restrict ourselves to the case of a *single receiver* $L = 1$. This motivates us to employ the following vector notation,

$$\mathbf{R}_s^{(i)}(n) := \mathbf{R}_s(\mathbf{x}^{n,(i)}) := \mathbf{R}_i(n) + \mathbf{R}_{\text{tr}}, \tag{4.47}$$

where $\mathbf{R}_i(n) := [R_i^{1,1}, R_i^{2,1}, \ldots, R_i^{K,1}]'$, and $\mathbf{R}_{\text{tr}} := [R_{\text{tr}}^{1,1}, R_{\text{tr}}^{2,1}, \ldots, R_{\text{tr}}^{K,1}]'$.

For the subsequent derivation of the *Spherical Intersection (SIS)* algorithm, we denote locations of target $i$ by $\tilde{\mathbf{x}}^{n,(i)}$ and the measured bistatic range sum by $\tilde{\mathbf{R}}_s^{(i)}(n)$, where the components of the vector are related to the $K$-transmitters.

These preliminary and notational considerations allow us to immediately write the *target localization problem* of finding the $i$th target location $\mathbf{x}^{n,(i)}$ as the following least squares formulation depending on $K$ bistatic measurements $\tilde{\mathbf{R}}_s^{(i)}$ for each target $i$, i. e.,

$$\mathbf{x}^{n,(i)} = \underset{\overline{\mathbf{x}}^{n,(i)}}{\operatorname{argmin}} \|\mathbf{R}_s(\overline{\mathbf{x}}^{n,(i)}) - \tilde{\mathbf{R}}_s^{(i)}(n)\|^2. \tag{4.48}$$

Our goal is to obtain a formula directly computing the target location without the need of an iterative solver/minimizer. To this end, we rewrite the equation for the range sum

$R_s^{(i,k)} = R_s^{(i,k,1)}$ for a single receiver $l = 1$ $\mathbf{x}_r^{(1)} := [0,0,0]'$ as follows,

$$R_s^{(i,k)}(n) - \|\mathbf{x}^{n,(i)}\| = \|\mathbf{x}^{n,(i)} - \mathbf{x}_t^{(k)}\| . \tag{4.49}$$

Next, we can square each side,

$$(R_s^{(i,k)}(n))^2 - 2R_s^{(i,k)}(n)\|\mathbf{x}^{n,(i)}\| + \|\mathbf{x}^{n,(i)}\|^2 = \|\mathbf{x}^{n,(i)} - \mathbf{x}_t^{(k)}\|^2 , \tag{4.50}$$

which can be further restated as,

$$\sum_{j=1}^{3} x_j^{n,(i)} x_{t,j}^{(k)} - R_s^{(i,k)}(n)\|\mathbf{x}^{n,(i)}\| = \frac{1}{2}\left(\|\mathbf{x}_t^{(k)}\|^2 - (R_s^{(i,k)}(n))^2\right) . \tag{4.51}$$

We set $R_f := \|\mathbf{x}^{n,(i)}\| = R_{\text{fr}}^{(i,1)}$ for our receiver location $\mathbf{x}_r^{(1)} = [0,0,0]'$. Employing the notation $\mathbf{x}^{n,(i)} := [x_1^{n,(i)}, x_2^{n,(i)}, x_3^{n,(i)}]'$,

$$\hat{\mathrm{A}} := \begin{bmatrix} x_{t,1}^{(1)} & x_{t,2}^{(1)} & x_{t,3}^{(1)} \\ x_{t,1}^{(2)} & x_{t,2}^{(2)} & x_{t,3}^{(2)} \\ \cdots & \cdots & \cdots \\ x_{t,1}^{(K)} & x_{t,2}^{(K)} & x_{t,3}^{(K)} \end{bmatrix} , \quad \text{and} \quad \mathbf{z} := \frac{1}{2}\begin{bmatrix} \|\mathbf{x}_t^{(1)}\|^2 - (R_s^{(i,1)}(n))^2 \\ \|\mathbf{x}_t^{(2)}\|^2 - (R_s^{(i,2)}(n))^2 \\ \cdots \\ \|\mathbf{x}_t^{(K)}\|^2 - (R_s^{(i,K)}(n))^2 \end{bmatrix} \tag{4.52}$$

allows us to rewrite (4.51) in the following simple way,

$$\hat{\mathrm{A}}\mathbf{x}^{n,(i)} = \mathbf{z} + \mathbf{R}_s^{(i)}(n) R_f . \tag{4.53}$$

Applying the so-called pseudoinverse $\hat{\mathrm{A}}^{\#} := (\hat{\mathrm{A}}'\hat{\mathrm{A}})^{-1}\hat{\mathrm{A}}'$, which represents also the general least squares solutions, and setting $\boldsymbol{\alpha} := \hat{\mathrm{A}}^{\#}\mathbf{z}$ and $\boldsymbol{\beta} := \hat{\mathrm{A}}^{\#}\mathbf{R}_s^{(i)}(n)$, we obtain an expression for the target location,

$$\mathbf{x}^{n,(i)} = \boldsymbol{\alpha} + \boldsymbol{\beta} R_f , \tag{4.54}$$

in which the right-hand side still depends on the target location itself. This immediately leads to a quadratic equation for $R_f$ after exploiting the fact that $R_f^2 = \|\mathbf{x}^{n,(i)}\|^2$,

$$0 = \boldsymbol{\alpha}'\boldsymbol{\alpha} + 2\boldsymbol{\alpha}'\boldsymbol{\beta} R_f + (\boldsymbol{\beta}'\boldsymbol{\beta} - 1)R_f^2 . \tag{4.55}$$

Therefore, we end up with,

$$R_f = \frac{-2\boldsymbol{\alpha}'\boldsymbol{\beta} \pm \sqrt{4(\boldsymbol{\alpha}'\boldsymbol{\beta})^2 - 4(\boldsymbol{\beta}'\boldsymbol{\beta} - 1)\boldsymbol{\alpha}'\boldsymbol{\alpha}}}{2(\boldsymbol{\beta}'\boldsymbol{\beta} - 1)} . \tag{4.56}$$

Herewith, we have completely derived the following:

---

**Spherical Inter-Section Algorithm (SISA)** For a single receiver location ($L = 1$), $K$ transmitter locations stored row-by-row in a matrix $\hat{A} \in \mathbb{R}^{K\times 3}$, and available measurements in the form of range sums $R_s^{(i,k)}(n) = R_s^{(i,k,1)}$ for $1 \le k \le K$ (see (4.46)), we can compute for $\boldsymbol{\alpha} := \hat{A}^{\#}\mathbf{z}$ and $\boldsymbol{\beta} := \hat{A}^{\#}\mathbf{R}_s^{(i)}(n)$ the location of target $i$, $1 \le i \le I$, with the help of the following equations,

$$\begin{cases} R_f = \frac{-2\boldsymbol{\alpha}'\boldsymbol{\beta} \pm \sqrt{4(\boldsymbol{\alpha}'\boldsymbol{\beta})^2 - 4(\boldsymbol{\beta}'\boldsymbol{\beta}-1)\boldsymbol{\alpha}'\boldsymbol{\alpha}}}{2(\boldsymbol{\beta}'\boldsymbol{\beta}-1)}, \\ \mathbf{x}^{n,(i)} = \boldsymbol{\alpha} + \boldsymbol{\beta} R_f, \end{cases} \tag{4.57}$$

where the matrix $\hat{A}^{\#}$ and vector $\mathbf{z}$ are as defined in (4.52).

---

We can realize this target localization algorithm in *Python*. First, we load the necessary package for numerical vector and matrix computations:

```
import numpy as np
```

Next, we define the methods implementing essential components of the algorithm derived above:

```
# Create the z-vector of the Spherical Intersection Method:
def createZvec(x_t,R_s):
# INPUT:

# x_t: array containing in each row transmitter locations
# R_s: array of range sums to the different transmitters
#        (for a single receiver)
#
# OUTPUT:
# zVec: z-vector of the Spherical Intersection Method

K = np.shape(x_t)[0]
zVec = np.zeros((K,1))

for i in range(K):
zVec[i] = 0.5*(np.linalg.norm(x_t[i])**2-R_s[i]**2)

return   zVec
```

```
# Create the B-matrix of the Spherical Intersection Method:
def createBmat(x_t):
# INPUT:
# x_t: matrix of transmitters with locations in each row
#
# OUTPUT:
# B: auxiliary matrix, B=(x_t'x_t)^-1 x_t

B = np.zeros(np.shape(x_t))
B = np.matmul(np.linalg.inv(np.matmul(np.transpose(x_t),(x_t))),
np.transpose(x_t))

return  B
```

The main function implementing the SIS localization method reads as follows:

```
 # SIS algorithm
def algoSIS(x_t,R_s):
# INPUT:
# z: auxiliary vector of SIS method
# B: auxiliary matrix of SIS method
# R_s: vector of range sums to a single target (elements related
#         to different transmitters)
#
# OUTPUT:
# x: target location

z = createZvec(x_t,R_s)
B = createBmat(x_t)
alpha, beta, R_f_plus, R_f_minus = compSISparam(z,B,R_s)
x = np.zeros(np.shape(alpha))
x_plus = alpha+np.matmul(beta,R_f_plus)
x_minus = alpha+np.matmul(beta,R_f_minus)
if  :x_plus[2] >= x_minus[2]:
  x=x_plus
  R_f = R_f_plus
else:
  x=x_minus
  R_f = R_f_minus
return  x, x_minus, x_plus
```

To test this algorithm, we need to define a test scenario with given transmitters $\mathbf{x}_t^{(k)}$, $1 \leq k \leq K$, which are collected row by row in the matrix $\mathrm{x_t}$. Here, we choose $K = 4$ transmitters. Moreover, we place a single receiver at the origin, i. e., $\mathbf{x}_r^{(1)} := [0, 0, 0]$.

```
# INPUT DATA:
# Given geometric data:
x_t = np.array([[20.0,0.0,0.0],[-30.0,5.0,0.15],[-10.0,-15.0,0.1],
[10.0,-25.0,0.05]]) # transmitter locations matrix
x_r = np.array([0.0,0.0,0.0]) # receiver location
```

To simulate/compute measurements and being able to validate our algorithm with an error measure, we also need to know the exact target location $\mathbf{x}^{0,(1)}$, where we set the time step $n = 0$ and the targets identification $i = 1$. That is, we restrict ourselves to a single target:

```
# EXACT TARGET LOCATION: (for test purposes, e. g.
by computing/simulating of measurements)
x = np.array([-110.0,244.0,-153.0]) # target location
```

In the absence of real bistatic range measurements $R_s^{(1)}$ associated with target $\mathbf{x}^{(0,(1)} := [-110, 244, -153]'$, we directly compute this range sum $R_s^{(1)}$ via (4.47) as follows,

```
      R_s= [[624.2762856 ]
 [603.21049721]
 [625.34354527]
 [640.23540666]]

      R_tr= [[20.        ]
 [30.41418255]
 [18.02803373]
 [26.92587046]]
```

Since the SISA only requires the transmitter locations `x_t` and the range sum `R_s`, we have now all the required data to perform the computation by running the following Python code:

```
# LOCALIZATION: Spherical Intersection Algorithm
xSol, x_minus, x_plus = algoSIS(x_t,R_s) # NOTE: R_s = R + R_tr
   for given bistatic measurements R and baseline R_tr
print("Computed location: xSol=",xSol)
print("Exact location: x=",x)
print("Error: ||xSol-x||=", np.linalg.norm(xSol.T-x))
```

```
Computed location: xSol= [[-109.99999999  244.         153.00000437]]
Exact location: x= [-110.  244.  153.]
Error: ||xSol-x||= 4.368855726410558e-06
```

The error `||xSol - x||` is the $L^2$-error between the exact location `x` and the computed location `xSol`. ◇

### 4.2.2 Maximum Likelihood Estimator (MLE)

The likelihood function $L_\ell(\boldsymbol{\theta})$ introduced in (4.30) provides the quantity of interest for the following basic estimation strategy.

---

**Maximum Likelihood Estimator (MLE)**

**Method 1.** *An estimator $\boldsymbol{\theta}_m$ that maximizes the likelihood function $L_\ell(\boldsymbol{\theta})$, that is,*

$$\boldsymbol{\theta}_m := \operatorname*{argmax}_{\boldsymbol{\theta}} L_\ell(\boldsymbol{\theta}) := \operatorname*{argmax}_{\boldsymbol{\theta}} p\left(\boldsymbol{Z}^\ell \mid \boldsymbol{\theta}\right), \tag{4.58}$$

*represents a* Maximum Likelihood Estimator (MLE). ◇

---

As motivated by basic calculus, the usual step to compute an extremum, i. e., a minimum or maximum of $L_\ell(\boldsymbol{\theta})$, is looking for the parameter $\boldsymbol{\theta}$ that satisfies the equation,

$$\frac{\partial L_\ell(\boldsymbol{\theta})}{\partial \boldsymbol{\theta}} = 0\,, \tag{4.59}$$

which is called the *likelihood equation.*

**Remark 4.2.10.** We note that randomness in the observations $\boldsymbol{Z}^\ell$ implies that the MLE $\boldsymbol{\theta}_m$ is a random variable. ◇

Let us look at a very simple and fundamental application, which generally allows us to understand the main mechanisms of estimating a specific quantity from measurements.

**Example 4.2.11** (Purely noisy measurements). Possibly, the simplest form of a measurement is one that allows us to directly measure the quantity of interest, e. g., the weight of an object. Let us denote the exact weight of different persons by $\overline{\theta}$. From experience, we know that measuring some quantity of interest generally involves errors, e. g., thermal fluctuations in sensors or state-dependent errors such as the degree of hydration. Therefore, we can describe such a simple measurement by adding a noise/random error term $w \sim \mathcal{N}(0,\sigma^2)$,[2] i. e.,

$$z = \overline{\theta} + w\,, \tag{4.60}$$

2 Note that a Gaussian uncertainty is generally motivated by the central limit theorem (CLT).

where $\sigma^2$ is the variance, and $\mu = 0$ is the mean. This immediately leads to the following conditional probability density for the measurement variable,

$$p(z|\theta) = \frac{1}{\sqrt{2\pi}\sigma}\exp\left(\frac{-(z-\theta)^2}{2\sigma^2}\right), \tag{4.61}$$

since $z - \theta \sim \mathcal{N}(0, \sigma^2)$ for unknown/to be estimated $\theta$. Finally, the form of the probability density $p$ allows us to immediately identify the value, where $\theta$ becomes maximal, that is, $\theta_m = z$. ◇

A rigorous and quantitative result about the quality of an estimator is provided by the following:

**Cramer–Rao lower bound**

**Lemma 4.2.12.** *Consider a random parameter $\theta$ that we would like to estimate with the help of an unbiased and non-Bayesian/Fisher estimator, that is,*

$$\tilde{\theta} = \overline{\Theta}(z). \tag{4.62}$$

*The mean square error between the exact parameter value $\bar{\theta}$ and the estimate $\tilde{\theta}$ satisfies the following inequality,*

$$\mathbb{E}\left[(\tilde{\theta} - \bar{\theta})^2\right] \geq a_{\mathrm{CRL}}, \tag{4.63}$$

*where the lower bound $a_{\mathrm{CRL}}$ is called the* Fisher information, *which satisfies for the likelihood function $L_z(\theta)$ the equation,*

$$\begin{aligned} a_{\mathrm{CRL}} &:= -\mathbb{E}\left[\frac{\partial^2 \ln L_z(\theta)}{\partial \theta^2}\right]\Bigg|_{\theta=\bar{\theta}} \\ &= \mathbb{E}\left[\left(\frac{\partial \ln L_z(\theta)}{\partial \theta}\right)^2\right]\Bigg|_{\theta=\bar{\theta}}. \end{aligned} \tag{4.64}$$

For practical aspects, it is important to note that the Cramer–Rao lower bound destroys any hope for the search of an algorithm that achieves a mean-square error below this bound $a_{\mathrm{CRL}}$.

**Remark 4.2.13** (Cramer–Rao lower bound and machine learning). Due to widely communicated and advertised use and developments of *machine learning (ML)* and *artificial intelligence (AI)*, it is more and more important to note its limits and to comment its actual state and level of reliability for sensitive applications. In particular, the wording AI is generally applied in news and media even if the actual meaning is rather ML. The origin of this confusion relies on the fact that there are multiple definitions of AI introduced in science and industry.[3] Fortunately, the existence of an AI has not been officially

3 A widely accepted (industrial) definition of AI is motivated by *autonomous system* applications such as self-driving cars or unmanned aerial vehicles (UAVs) and reads in short: [AI] = [physical descriptions

verified until now, that is, a human-built machine showing consciousness primarily by indicating in its responses/feedback either a form of creativity going beyond its training data and by showing an awareness, which implies an understanding/reasoning following Descartes' famous phrase "cogito, ergo sum", which means "I am thinking, hence I am/exist". Since ML and AI depend on training data, playing the role of measurements as introduced in the context of estimation theory here, we emphasize that the Cramer–Rao lower bound implies the existence of such a bound in applications of ML and AI and hence limits the achievable precision possible in ML and AI applications.

Therefore, rigorous mathematical results, such as the above lower bound (Lemma 4.2.12) on the precision, motivates the practical importance of estimation theory beyond classical autonomous system approaches relying on systematic optimal control frameworks to rigorously demonstrate limits in precision of ML-realized control and decision algorithms, in particular, in safety critical contexts. Therefore, the application of ML and AI allows us to quickly gain first computational solutions (for the price of limited precision) for complex real-world problems such as CHeSs, where a highly interdisciplinary understanding is required to systematically extract exact answers based on an intrinsic understanding of the underlying processes and principles governing a specific CHeS. To conclude, ML requires awareness of its users to carefully assess the reliability and the ethical applicability of the results hereby produced. ◇

Before we prove the latter result, we introduce a generally valid term in the subsequent definition, which captures an optimal class of estimators, see [88].

**Efficient estimator**

**Definition 4.2.14.** Consider the subset $\mathcal{U}$ of unbiased and non-Bayesian/Fisher estimators. An ideal subset of $\mathcal{U}$, which is defined by the set of estimators that satisfy the equality in (4.63), represents the class of *efficient estimators*.

**Remark 4.2.15** (Efficient versus optimal). Note that the wording "efficient", a label generally found in the literature, seems to imply some kind of optimality in the sense of being "more efficient". However, the form of the inequality with the lower bound does not say anything about the computational demand and speed. Hence, let us promote the term "optimal estimator", which more closely captures the nature of the lower bound (4.63). ◇

*Proof.* [4] We first recall that we understand an unbiased estimator $\overline{\Theta}$ as that one that satisfies,

$$\mathbb{E}[(\overline{\theta} - \overline{\Theta}(z))] = 0 \,, \tag{4.65}$$

of a system] + [descriptions and parameters defining the system evolution are selected and extracted, respectively, by trained ML models from live (/sensor) and used training data].

4 Proof of Lemma 4.2.12 (Cramer–Rao lower bound).

which by the definition of expectation reads,

$$\int_{-\infty}^{\infty} (\bar{\theta} - \bar{\Theta}(z))p(z|\bar{\theta})\, dz\,. \tag{4.66}$$

Assuming enough regularity of the integrand in (4.66), we can differentiate (4.66) with respect to $\bar{\theta}$ and take the derivative inside the integral:

$$\begin{aligned}\frac{d}{d\bar{\theta}} \int_{-\infty}^{\infty} (\bar{\theta} - \bar{\Theta}(z))p(z|\bar{\theta})\, dz &= \int_{-\infty}^{\infty} p(z|\bar{\theta})\, dz - \int_{-\infty}^{\infty} (\bar{\theta} - \bar{\Theta}(z))\frac{\partial p(z|\bar{\theta})}{\partial \bar{\theta}}\, dz \\ &= 0\,. \end{aligned}\tag{4.67}$$

Then, the lemma follows after using the properties $\frac{\partial p(z|s)}{\partial s} = \frac{\partial \ln p(z|s)}{\partial s} p(z|s)$ and $p(z|s) = \sqrt{p(z|s)}\sqrt{p(z|s)}$ together with a subsequent application of Schwarz's inequality. □

## 4.3 State prediction of CHeSs by measurements: Kalman filter

To connect modeling, i. e., the theoretical understanding of the dynamics of a CHeS/(thermodynamic) system $\mathcal{S}$, with measurements (e. g., real data or experiments), the prediction/estimation theory (filtering of essential information from measurements) provides a well-established and systematic approach. Moreover, this theory builds a fundamental basis for developing an optimal control framework for CHeSs, e. g., using techniques such as *dynamic programming (DP)*. Hence, let us look at a widely applied class of filters, the so-called *Kalman filter*. In applications, two main areas indicate a wide range of beneficial use: (i) *stochastic processes* and (ii) *signal processing*.

**(i) The filtering problem from a stochastic point of view.** Here, the main goal is to obtain a “best estimate” for the evolution of a system state under uncertainty/randomness. The problem of establishing a “best estimate” is translated into a mathematical *minimal variance principle*. Kalman filters belong to this class of problems.

**(ii) Filters motivated by signal processing.** Low- or high-pass filters represent classical signal processing applications to removing/suppressing undesired frequencies. However, filters are not restricted to the frequency domain and hence find application in image processing, for instance.

R. E. Kalman proposed in 1960 a recursive solution to a linear filtering and prediction problem, see [82]. To introduce the meaning and difference of filtering and prediction, we first provide relevant notational aspects of the theory. Let $t_0 > 0$ be a time when the measurements/observations start, and let $t \geq t_0 > 0$ be the present time. Moreover, let $t^i > 0$ be an arbitrary time point of interest. We distinguish the following scenarios,

$$\begin{cases} t^i < t\,, & \text{referred to as } \textbf{smoothing}, \\ t^i = t\,, & \text{called } \textbf{filtering}, \\ t^i > t\,, & \text{representing a } \textbf{prediction}. \end{cases} \tag{4.68}$$

Let the random state (i. e., random variable) of a system $\mathcal{S}$ at present time be $\mathbf{s} = \mathbf{s}(t)$, whereas in a time-discrete setting, we write,

$$\mathbf{s}^k := \mathbf{s}(t_k)\,, \tag{4.69}$$

where $t_k = k\Delta t$, $k \in \mathbb{N}$, is a discrete time step for the case of a uniform time step size $\Delta t > 0$. Note that $\mathbf{s} = [s_1, s_2, \ldots, z_{n_d}]'$ is a state vector of dimension $n_d \in \mathbb{N}$, which contains the minimal number of necessary and independent *state variables* $s_i$, $1 \le i \le n_d$. The subsequently described formalism additionally depends on a random variable $\mathbf{z} = \mathbf{z}(t) = [z_1(t), z_2(t), \ldots, z_{m_d}(t)]$ that represents actual/real measurements taken at time $t \ge t_0$, i. e., the present time. In fact, we are interested in being able to provide an optimal estimate $\tilde{\mathbf{s}}(t)$ of the system state $\mathbf{s}(t)$ with the help of measurements $\mathbf{z}^k = \mathbf{z}(t_k)$, where $t_0 \le t_k \le t$, $1 \le k \le K$, and $t_K = t$. This leads to the following conditional probability distribution for the system state $\mathbf{s}$,

$$\mathbb{P}[\mathbf{s} \le \mathbf{s}^\nu \,|\, \mathbf{z}^0, \mathbf{z}^1, \ldots, \mathbf{z}^K] =: F_{\mathbf{z}}(\mathbf{s}\nu)\,, \tag{4.70}$$

where $F_{\mathbf{z}}$ is the associated conditional distribution function, and $\mathbf{s}^\nu$ denotes a vector of explicit state values and hence has the same dimension as $\mathbf{s}$. Note that the $\le$-operation on fields/vectors has the canonical meaning of applying the $\le$-operation component-by-component.

### 4.3.1 The Kalman filter problem formulation

For simplicity and with a focus on applications, we look at the following simple discrete-time evolution equations describing the state $\mathbf{s}^k$ (e. g., location) and measurement $z^k$ of a CHeS of interest,

$$\begin{cases} \mathbf{s}^{k+1} = \hat{\mathrm{A}}\mathbf{s}^k + \mathbf{w}^k\,, \\ \mathbf{z}^{k+1} = \hat{\mathrm{H}}\mathbf{s}^k + \mathbf{v}^k\,, \end{cases} \tag{4.71}$$

where the random variables $\mathbf{w}^k$ and $\mathbf{v}^k$ denote state and measurement noises, respectively. This system leads to a linear estimation problem/Kalman filter and update process obtained by minimizing the mean square error (variance), see Figure 4.3. We note that a straightforward generalization of (4.71) would also account for a *control variable* in the state evolution equation $(4.71)_1$ and possibly also in the measurement equation $(4.71)_2$.

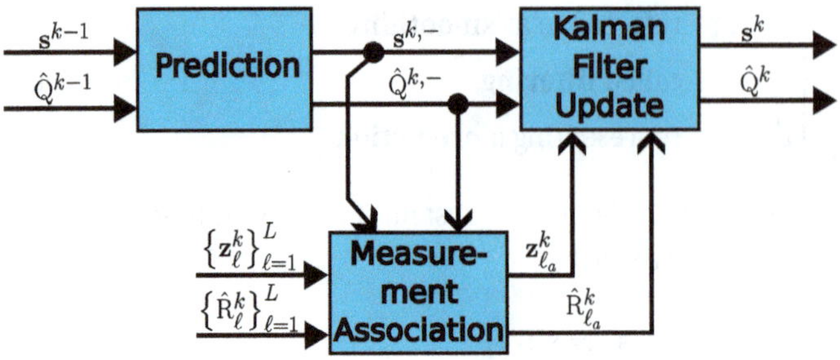


Figure 4.3: Kalman filter steps: Prediction, measurement association, and update.

**Example 4.3.1** (Classical velocity-driven motion). A very basic and simple example belonging to formulation (4.71) is the following Newtonian movement described by the state vector $\mathbf{s} = [x_0, x_1, x_2, v_0, v_1, v_2]^T \in \mathbb{R}^6$,

$$\begin{cases} \mathbf{s}^{k+1} = \hat{\mathrm{A}}\mathbf{s}^k + \mathbf{w}^k , \\ \mathbf{x}^0 = \mathbf{x}_0 , \\ \mathbf{z}^k = \hat{\mathrm{H}}\mathbf{s}^k + \mathbf{v}^k , \end{cases} \tag{4.72}$$

where $\mathbf{v}^k \in \mathbb{R}^6$ and $\mathbf{w}^k \in \mathbb{R}^6$ are zero-mean Gaussian noise variables with *Measurement Correlation Matrix* $\hat{\mathrm{R}} \in \mathbb{R}^{3\times3}$ and *State Correlation Matrix* $\hat{\mathrm{Q}} \in \mathbb{R}^{6\times6}$, respectively, such that $\mathbf{v}^k \sim \mathcal{N}(\mathbf{0}, \hat{\mathrm{R}})$ and $\mathbf{w}^k \sim \mathcal{N}(\mathbf{0}, \hat{\mathrm{Q}})$. Moreover, the Newtonian movement is reflected in the following definition of the state transition matrix $\hat{\mathrm{A}} \in \mathbb{R}^{6\times6}$:

$$\hat{\mathrm{A}} := \begin{bmatrix} 1 & 0 & 0 & k & 0 & 0 \\ 0 & 1 & 0 & 0 & k & 0 \\ 0 & 0 & 1 & 0 & 0 & k \\ 0 & 0 & 0 & 1 & 0 & 0 \\ 0 & 0 & 0 & 0 & 1 & 0 \\ 0 & 0 & 0 & 0 & 0 & 1 \end{bmatrix} , \tag{4.73}$$

where $k = t_{n+1} - t_n$, $0 \le n \le N$, denotes a constant time step size. As we can see from the dimensions of the Measurement Correlation Matrix, the measurement vector $\mathbf{z} \in \mathbb{R}^3$ is lower dimensional than the state vector $\mathbf{s} \in \mathbb{R}^6$. As a consequence, the measurement map $\hat{\mathrm{H}} \in \mathbb{R}^{3\times6}$ is defined for solely spatial measurements by,

$$\hat{\mathrm{H}} := \begin{bmatrix} 1 & 0 & 0 & 0 & 0 & 0 \\ 0 & 1 & 0 & 0 & 0 & 0 \\ 0 & 0 & 1 & 0 & 0 & 0 \end{bmatrix} . \tag{4.74}$$

◇

Before we can introduce the main result and derivation of the Kalman filter for CHeSs applicable to (4.71), we need to precisely characterize under which circumstances it can be applied. To this end, we restrict ourselves to situations where the state and measurement spaces are both one dimensional, i. e., at time step $k > 0$, we denote by $s^k$ and $z^k$ the state and measurement variables, respectively.

**Assumptions/prerequisites** (Noise and initial conditions). We suppose that the CHeS of interest has the following properties:

**(A1)** *Let $w^k$ and $v^k$ be state and measurement noises, that is, these are white*[5] *and have the following statistical properties:*

$$\begin{aligned}
&\textbf{(N1)} \quad \mathbb{E}\left[w^k\right] = \mathbb{E}\left[v^k\right] = 0 \quad \text{for all } k\,,\\
&\textbf{(N2)} \quad \mathbb{E}\left[(w^k)^T w^l\right] = \begin{cases} \hat{Q}^k & \text{for } k = l\,,\\ 0 & \text{else}\,,\end{cases}\\
&\textbf{(N3)} \quad \mathbb{E}\left[(v^k)^T v^l\right] = \begin{cases} \hat{R}^k & \text{k=l}\,,\\ 0 & \text{else}\,,\end{cases}\\
&\textbf{(N4)} \quad \mathbb{E}\left[(w^k)^T v^l\right] = 0 \quad \text{for all } k, l\,.
\end{aligned} \tag{4.75}$$

**(A2)** *The random initial state $s_0$ is uncorrelated with state and measurement noises $w^k$ and with $v^k$ for all k.*

**(A3)** *The mean and covariance of the initial state is known, i. e.,*

$$\begin{cases} \bar{s}^0 = \mathbb{E}\left[s^0\right]\,,\\ \hat{P}^0 = \mathbb{E}\left[(s^0 - \bar{s}^0)^T (s^0 - \bar{s}^0)\right]\,. \end{cases} \tag{4.76}$$

### 4.3.2 The discrete-time Kalman filter algorithm

Based on **Assumptions (A1)–(A3)**, we can immediately state the following Kalman filter algorithm, see also Figure 4.3 again.

---

**Recursive Kalman Filter (RKF) Algorithm**

**Algorithm 1.**

**Step 1.** *(Project)* For $k \geq 1$, compute the a priori estimates $\tilde{s}^{k+1,-}$ and $\tilde{\hat{P}}^{k+1,-}$ for level $k+1$, i. e.,

$$\begin{cases} \tilde{s}^{k+1,-} = \hat{A}\tilde{s}^k\,,\\ \tilde{\hat{P}}^{k+1,-} = \hat{A}\hat{P}^k\hat{A}^T + \hat{Q}^k\,, \end{cases} \tag{4.77}$$

and at $k = 0$, we use (4.76).

---

5 We say that a stationary random process with zero autocorrelation represents white noise. Examples are Gaussian White Noise (GWN) or Uniform White Noise (UWN), which represent independent and identically distributed Gaussian and Uniform Random Processes.

**Step 2.** *(Kalman gain)* Compute,

$$\hat{K}^k = \frac{\hat{P}^{k,-}\hat{H}^T}{\hat{H}\hat{P}^{k,-}\hat{H}^T + \hat{R}^k} . \tag{4.78}$$

**Step 3.** *(Update estimate & covariance)* Compute the a posteriori estimate and error covariance by,

$$\begin{cases} \tilde{s}^k = \tilde{s}^{k,-} + \hat{K}^k(z^k - H\tilde{s}^{k,-}) , \\ \hat{P}^k = (\hat{I} - \hat{K}^k\hat{H})\hat{P}^{k,-} . \end{cases} \tag{4.79}$$

◇

---

**Remark 4.3.2. 1.** We note that equation $(4.79)_1$ represents an intuitive, linear update strategy for an unknown Kalman gain $\hat{K}^k$. The difficulty in applying the Kalman filter strategy to general systems relies on finding the right equation for the *Kalman gain.* **2.** The denominator $\hat{S}^k = \hat{H}\hat{P}^{k,-}\hat{H}^T + \hat{R}^k$ in (4.78) is called the *measurement prediction covariance.*

### 4.3.3 Derivation of the discrete-time Kalman filter

As motivated in Remark 4.3.2.1, the crucial ingredient in the discrete-time Kalman filter is the definition of the Kalman gain, which we derive in this section together with the other equations.

**Derivation of (4.78).** We start with rewriting the a posteriori estimate error covariance $\hat{P}^k$ by using $(4.71)_2$ in (4.79),

$$\tilde{s}^k = \tilde{s}^{k,-} + \hat{K}^k(\hat{H}s^k + v_k - \hat{H}\tilde{s}^{k,-}) , \tag{4.80}$$

which after insertion into the definition of $\hat{P}^k$ gives,

$$\begin{aligned} \hat{P}^k &= \mathbb{E}\left[(s^k - \tilde{s}^k)(s^k - \tilde{s}^k)^T\right] \\ &= \mathbb{E}\left[(\hat{I}(s^k - \tilde{s}^{k,-}) - \hat{K}^k(\hat{H}s^k + v^k - \hat{H}\tilde{s}^{k,-}))\right. \\ &\quad \left.(\hat{I}(s^k - \tilde{s}^{k,-}) - \hat{K}^k(\hat{H}s^k + v^k - \hat{H}\tilde{s}^{k,-}))^T\right] \\ &= \mathbb{E}\left[((\hat{I} - \hat{K}^k\hat{H})(s^k - \tilde{s}^{k,-}) - \hat{K}^k v^k)((\hat{I} - \hat{K}^k\hat{H})(s^k - \tilde{s}^{k,-}) - \hat{K}^k v^k)^T\right] . \end{aligned} \tag{4.81}$$

Due to assumptions **(A1)**, **(N3)**, and **(A3)**, we can rewrite (4.81) as follows,

$$\begin{aligned} \hat{P}^k &= (\hat{I} - \hat{K}^k\hat{H})\mathbb{E}\left[(s^k - \hat{s}^{k,-})(s^k - \hat{s}^{k,-})^T\right](\hat{I} - \hat{K}^k\hat{H})^T + \hat{K}^k\mathbb{E}\left[v^k(v^k)^T\right](\hat{K}^k)^T \\ &= (\hat{I} - \hat{K}^k\hat{H})\hat{P}^{k,-}(\hat{I} - \hat{K}^k\hat{H})^T + \hat{K}^k\hat{R}^k(\hat{K}^k)^T . \end{aligned} \tag{4.82}$$

Minimizing the a *posteriori estimate error covariance* with respect to the Kalman gain $\hat{K}^k$, we have to differentiate $\hat{P}^k$ with respect to $\hat{K}^k$ and then to set the result to zero. To this end, we first expand (4.82) as follows,

$$\hat{P}^k = \hat{P}^{k,-} - \hat{K}^k\hat{H}\hat{P}^{k,-} - \hat{P}^{k,-}(\hat{K}^k\hat{H})^T + \hat{K}^k(\hat{H}\hat{P}^{k,-}\hat{H}^T + \hat{R}^k)(\hat{K}^k)^T , \tag{4.83}$$

and since $\mathrm{tr}[\hat{P}^k] = \mathrm{tr}[(\hat{P}^k)^T]$, (4.83) becomes,

$$\mathrm{tr}\left[\hat{P}^k\right] = \mathrm{tr}\left[\hat{P}^{k,-}\right] - 2\mathrm{tr}\left[\hat{K}^k\hat{H}\hat{P}^{k,-}\right] + \mathrm{tr}\left[\hat{K}^k(\hat{H}\hat{P}^{k,-}\hat{H}^T + \hat{R}^k)(\hat{K}^k)^T\right] . \tag{4.84}$$

Finally, differentiating (4.84) with respect to $\hat{K}^k$ gives,

$$\frac{d\mathrm{tr}[\hat{P}^k]}{d\hat{K}^k} = -2(\hat{H}\hat{P}^{k,-})^T + 2\hat{K}^k(\hat{H}\hat{P}^{k,-}\hat{H}^T + \hat{R}^k), \tag{4.85}$$

and after setting it to zero, we obtain the so-called Kalman gain/blending factor (4.78).

**Derivation of (4.79)$_2$.** Inserting (4.78) into (4.83) leads to,

$$\begin{aligned}
\hat{P}^k &= \hat{P}^{k,-} - \hat{P}^{k,-}\hat{H}^T(\hat{H}\hat{P}^{k,-}\hat{H}^T + \hat{R}^k)^{-1}\hat{H}\hat{P}^{k,-} \\
&\quad - \hat{P}^{k,-}\hat{H}^T(\hat{P}^{k,-}\hat{H}^T(\hat{H}\hat{P}^{k,-}\hat{H}^T + \hat{R}^k)^{-1})^T \\
&\quad + \hat{P}^{k,-}\hat{H}^T(\hat{H}\hat{P}^{k,-}\hat{H}^T + \hat{R}^k)^{-1}(\hat{H}\hat{P}^{k,-}\hat{H}^T + \hat{R}^k)(\hat{P}^{k,-}\hat{H}^T(\hat{H}\hat{P}^{k,-}\hat{H}^T + \hat{R}^k)^{-1})^T \\
&= \hat{P}^{k,-} - \hat{K}^k\hat{H}\hat{P}^{k,-} \\
&= (\hat{I} - \hat{K}^k\hat{H})\hat{P}^{k,-} .
\end{aligned} \tag{4.86}$$

**Derivation of (4.77)$_2$.** First, we note that,

$$e^{k+1,-} = s^{k+1} - \bar{s}^{k+1,-} = (\hat{A}s^k + w^k) - \hat{A}\bar{s}^k = \hat{A}e^k + w^k , \tag{4.87}$$

so that $\hat{P}^{k,-}$ reads,

$$\begin{aligned}
\hat{P}^{k+1,-} &= \mathbb{E}\left[e^{k+1,-}(e^{k+1,-})^T\right] = \mathbb{E}\left[\hat{A}e^k(\hat{A}e^k)^T\right] + \mathbb{E}\left[w^k(w^k)^T\right] \\
&= \hat{A}\hat{P}^k\hat{A}^T + \hat{Q}^k .
\end{aligned} \tag{4.88}$$

Since the a priori state estimate is implemented in a straightforward manner by (4.77)$_1$, this completes the derivation.

# 5 Reliability and degradation in CHeSs

## 5.1 Introduction

The scope of *Reliability* is very broad in practical contexts and generally ranges from *Availability* over *Maintainability* to *Safety* and which all can be collected in the acronym RAMS, as in [20]. Let us first clarify what exactly we mean by reliability.[1]

**Measured Reliability**

**Definition 5.1.1.** Consider $N$ identical and independent systems[a] that are operational since $t = 0$ up to a specified mission time $t = T$. We identify the ratio $p(n|N) := n/N$ as the *measured reliability* of the $N$ systems, where $n$ denotes the number of systems completing the mission without failure and $p(n|N)$ represents a realization of the random variable approximating the true reliability defined as the limit $p_\infty := \lim_{N\to\infty} p(n|N)$. ◇

*a* We motivate the term system by its meaning in thermodynamics, i. e., an entity $\mathcal{S}$ that interacts (e. g., exchanges information, energy/heat, work, etc.) with its surroundings $\mathcal{N}$. Both system $\mathcal{S}$ and surroundings $\mathcal{N}$ together form a so-called universe $\mathcal{U} := \mathcal{S} \cup \mathcal{N}$.

**Remark 5.1.2** (Acceptance test). In the context of an *acceptance test*,[2] we aim to give a binary answer "YES" or "NO" to the question whether a system operates failure-free. However, sometimes we can only provide a *probability*, which is generally called *reliability* and denoted by $R$. Naturally, this is closely related to the duration/depth of tests/validations, i. e., the smaller the depth (and hence the shorter the duration), the more likely the test turns out to be less meaningful for statements about reliability. ◇

**Statements about reliability.** To phrase appropriate/precise enough reliability statements, it is necessary to provide detailed assumptions on the system $\mathcal{S}$ and surroundings $\mathcal{N}$, i. e., operational conditions, such as

- **required function (of the system),**[3] which is required before any reliability analysis,
- **environmental, operational, and maintenance conditions,**[4]
- **mission time/duration** $T$,[5]
- **state of the system at mission beginning** $t = 0$, e. g., generally assumed to be a new system.

1 The concept of reliability does not only apply to systems; it can also be applied to processes and services.

2 Factory Acceptance Test (FAT) and Site Acceptance Test (SAT) are such examples.

3 which thereby also defines failures.

4 For example, for semiconductors, a large enough increase in temperature can double the failure rate.

5 We are generally interested in the reliability $R(T)$, i. e., the probability of no failure during the mission time $(0, T]$ such that explicit initial conditions (state of system at $t = 0$) have to be provided.

https://doi.org/10.1515/9783110579543-005

**Remark 5.1.3** (System performance parameters). The aspects of RAMS represent performance parameters, which we need to introduce during the *design and development phase* of a product/system. Moreover, these parameters have to be updated/maintained during the production and operation of the system of interest. ◇

**Reliability analysis.** The following two main tasks are part of a reliability analysis:
- **Prediction.** Under the consideration of failure and repair rates as well as the specific design of the system, we calculate the reliability of systems such as CHeSs.
- **Estimation.** The two tasks, performing reliability tests and the subsequent statistical evaluation, belong to the estimation process during the reliability investigation.

**Redundancy.** We also note that a general strategy to increase redundancy (e. g., parallel/horizontal extension of "the number of systems") of a system relies on smart designs and architectures of a system.[6] In general, we distinguish between the following three types of redundancies [20]:

- **Hot/active/parallel redundancy.** The redundant system is from the beginning (at $t = 0$) exposed to the same load as the officially working system. Hence, the failure rate for the redundant system is exactly the same as for the working system.
- **Warm/low-load redundancy.** This redundancy scheme has the advantage that it clearly has a lower failure rate due to the lower load of the redundant system compared to the working system, which obviously has a higher failure rate.
- **Cold/stand-by/no-load redundancy.** In this context, the redundant system does not experience any load, and hence we generally assume a zero failure rate for the system in the no-load state.

## 5.2 Fundamental concepts

Let us first introduce assumptions and notation for subsequent discussions.

**Assumption 1. (A1)** *The likelihood of observing a failure of system $\mathcal{S}_i$, $0 \leq i \leq N$ ($i \in \mathbb{N}$), during the interval $(0, T]$ for a specific mission time $T$ will follow the following distribution,*

$$F_i(T) := P[B_i \leq T] = 1 - \exp(-\lambda_i T)\,, \tag{5.1}$$

*where $\lambda_i$ is a constant failure rate, and $B_i$ is a random variable representing the appearance of a failure at a specific time.*

6 In fact, the increasing interest in hosting IT services in a central and possibly location-redundant data center (DC) is exactly due to this horizontal scalability, which primarily is used to deal with heavy load scenarios but also allows for highly redundant architectures/designs.

**(A2)** *All involved systems* $0 \le i \le N$ *and failures are independent.*

**(A3)** *Our interest in reliability is for a specific mission time* $T > 0$. *With* **(A1)**, *we can express the reliability* $R_i(T)$ *of system* $i$ *for the mission time* $T$ *as follows:*

$$R_i(T) := 1 - P[B_i \le T] = P[B_i > T] = \exp(-\lambda_i T)\,. \tag{5.2}$$

◇

## 5.3 Serial arrangement: *N* independent systems

Let us first look at a system-of-systems design that does not account for any redundancy by simply *connecting identical systems serially*. Note that if the connected systems $\mathcal{S}_i$, $0 \le i \le N$, do not only transfer/process inputs in one direction but even in both directions, then the failure rate cannot be safely assumed to be time-independent. Figure 5.1 clearly shows that the group of $N$ systems fails, if one of the single subsystems $[\mathcal{S}]$ fails. This immediately allows us to compute the *reliability of the group* as,

$$R_g(T) = R^N = \left(\exp(-\lambda T)\right)^N, \tag{5.3}$$

where $\exp(-\lambda T) = R = R_i = \exp(-\lambda_i T)$ is the same reliability for all subsystems $\mathcal{S} = \mathcal{S}_i$ with integers $i$ such that $0 \le i \le N$. Therefore, the *failure rate of the group* $\lambda_g$ follows from the failure rate $\lambda = \lambda_i$ of the identical subsystems $0 \le i \le N$ by,

$$\lambda_g := N\lambda\,. \tag{5.4}$$

This immediately gives the *Mean Time To Failure (MTTF)*,

$$\mu_g = 1/\lambda_g\,. \tag{5.5}$$

$$\text{input} - [\mathcal{S}] - [\mathcal{S}] - \cdots - [\mathcal{S}] - > \text{output}$$

Figure 5.1: Serial arrangement of *N* identical and independent systems. This amounts to the so-called worst-case scenario, where the failure of a single system $\mathcal{S}$ leads to a failure of the whole group of *N* systems.

In case that the *systems are not identical*, i. e., each system $\mathcal{S}_i$ has a different reliability $R_i$, we can compute the availability of such a group in the serial setup as follows,

$$\begin{aligned} R_g(T) &= R_1 R_2 \cdots R_N \\ &= \exp(-\lambda_1 T)\exp(-\lambda_2 T)\cdots\exp(-\lambda_N T) \\ &= \exp\left(-T\sum_{i=1}^{N}\lambda_i\right). \end{aligned} \tag{5.6}$$

This immediately gives the *failure rate of the group of N different subsystems* as follows,

$$\lambda_g = \sum_{i=1}^{N} \lambda_i \,. \tag{5.7}$$

Accordingly, the *MTTF for a serial arrangement of different subsystems* is obtained with $\lambda_g$ from (5.7) by,

$$\mu_g = 1/\lambda_g = \frac{1}{\sum_{i=1}^{N} \lambda_i} \,. \tag{5.8}$$

## 5.4 Parallel arrangement: *N* independent systems

Let us first look at the case of *identical systems arranged in a parallel configuration*, see Figure 5.2. Since the goal of a parallel architecture is to increase redundancy, we expect higher reliabilities in this case. A straightforward way of obtaining a formula giving higher reliability for multiple and identical connected subsystems $\mathcal{S}_i$ is the following *reliability of the group*,

$$\begin{aligned} R_g(T) &= 1 - (1 - R)^N \\ &= 1 - (1 - \exp(-\lambda T))^N \,, \end{aligned} \tag{5.9}$$

where $R = R_i$ is the identical reliability of all subsystems $\mathcal{S}_i$ for $0 \le i \le N$. We can compute the *MTTF of the group of parallel subsystems* based on the group reliability $R_g$ as follows,

$$\mu_g = \int_0^\infty R_g(T)\, dT \,. \tag{5.10}$$

$$\begin{array}{l} \quad\quad\quad - -[\mathcal{S}] - - > \text{output} \\ \quad\quad\quad\ | \\ \text{input} - + -[\mathcal{S}] - - > \text{output} \\ \quad\quad\quad\ | \\ \quad\quad\quad\ \vdots \\ \quad\quad\quad\ | \\ \quad\quad\quad - -[\mathcal{S}] - - > \text{output} \end{array}$$

Figure 5.2: Parallel arrangement of *N* identical and independent systems. This represents a system with redundancy where the failure of a single system $\mathcal{S}$ does not imply the failure of the whole group.

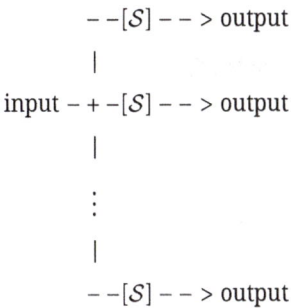

Similarly, we get the *failure rate of the group of parallel subsystems* based on the group reliability $R_g$ by,

$$\lambda_g = \frac{-\frac{dR_g(T)}{dt}}{R_g(T)} . \tag{5.11}$$

If we have nonidentical subsystems $\mathcal{S}_i$, then we only have to adjust the equation for the reliability of the group $R_g(T)$, equation (5.9), as follows,

$$R_g(T) = 1 - \prod_{i=1}^{N}(1 - R_i(T)) , \tag{5.12}$$

whereas equations (5.10) for the *MTTF* $\mu_g$ *of the group* and (5.11) for the *failure rate* $\lambda_g$ *of the group* remain the same.

A special formulation of redundancy using the parallel architecture is the case, where a mission/system operation is considered to be successful, if at least $n$ subsystems of the total of $N$ subsystems configured in a parallel manner are still working/operational at the end of mission time $T$. However, for simplicity, let us assume that all the subsystems $\mathcal{S}$ are identical. This gives for the *reliability of n-out-of-N for N identical subsystems* the following equation,

$$R_g(T) = \sum_{j=n}^{N} \binom{N}{j} p^j (1-p)^{N-j} = 1 - \sum_{j=0}^{n-1} \binom{N}{j} p^j (1-p)^{N-j} . \tag{5.13}$$

## 5.5 Reliability in CHeSs: serial and parallel combinations

General systems represent usually a combination of both, parallel and serial combinations of subsystems. Let us denote parallel subsystems by $P_1, P_2, \ldots, P_M$, where $M \in \mathbb{N}$ is the number of parallel systems. Accordingly, we denote by $S_1, S_2, \ldots, S_N$, $N \in \mathbb{N}$, specific serial subsystems. With the help of Sections 5.3 (serial) and 5.4 (parallel), we can determine the reliabilities of the corresponding subsystems as subsequently described. Let $R_{S_i}$ and $R_{P_j}$ denote the reliabilities determined for serial subsystems, $1 \leq i \leq N$, and parallel subsystems, $1 \leq j \leq M$, respectively.

These considerations then allow us to analyze and compute the reliabilities of *system of systems* designs, see Figure 5.3 and Example 5.5.1, for instance.

**Example 5.5.1** (System of systems analysis). Consider Figure 5.3 as a simple setup of a system of systems. Let us first compute the reliabilities of the single serial and parallel components based on the known reliabilities $R_{S_1}$, $R_{P_1}$, and $R_{P_3}$ of the serial subsystem

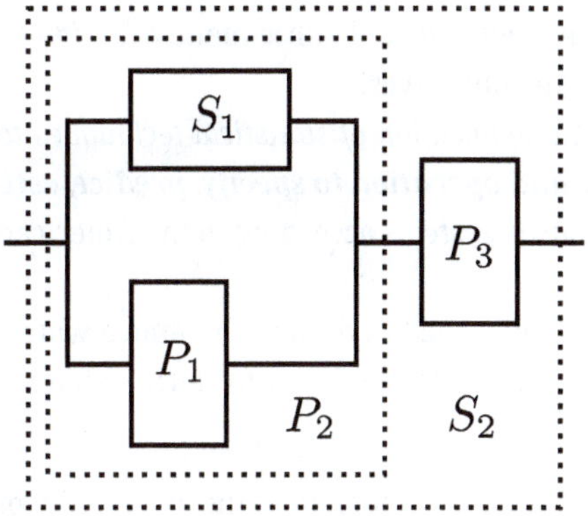


Figure 5.3: System of Systems.

$S_1$ and the parallel subsystems $P_1$ and $P_3$, respectively. We can calculate the reliability of the parallel subsystem $P_2$ composed of $S_1$ and $P_1$,

$$R_{P_2} = 1 - (1 - R_{S_1})(1 - R_{P_1}) . \tag{5.14}$$

This then gives the following total reliability for this specific system of systems design,

$$R_{S_2} = R_{P_2} R_{P_3} . \tag{5.15}$$

◇

### 5.5.1 Software reliability

Before we discuss the details of *software reliability*, let us motivate that a computer (i. e., a hardware system) running a specific software (i. e., a system consisting of a set of logical instructions/algorithms) represents a CHeS. Therefore, we consider software and hence also software reliability to be an integral part of a general framework/theory trying to grasp the essential features, characteristics, and patterns underlying various types/classes of CHeSs.

The following considerations rely on the fundamental approach by Schneidewind [148, 149] to *Software Reliability Engineering (SRE)*. Two essential goals can be identified:

**(G1)** develop standard software reliability processes/practices applicable in Systems Support Organizations (SSOs);

**(G2)** build (software) tools for a simplified and systematic way to apply a Software Reliability Model suitable for a particular organization of interest.

From a general management point of view, goal **(G2)** supports the application of Software Reliability measures thanks to intuitive and easy-to-use software tools. However, the basis for such software tools supporting SRE, the scope of the software reliability program must be developed as part of goal **(G1)**. In fact, the activities in **(G1)** are twofold.

First, so-called *reliability objectives need to be identified*, and subsequently the *implemented software has to be tested* how it fulfills the objectives.

As noted in [148], SRE is understood as *the application of statistical techniques to* ***data collected during system development and operation to specify, predict, estimate, and assess*** *the reliability of software-based systems* according to the American Institute of Aeronautics & Astronautics.

Since we are talking about *Software*, which over the last decades became a widely and especially also colloquially used term, it is therefore necessary to clarify what we consider as *Software* in the context of SRE.

**Definition 5.5.2** (Software). Under the term software we understand the availability of the following parts:
(S1) *source code*,
(S2) *supporting documentation*, and
(S3) *test results*. ◇

A key challenge in SRE is the fact that it is difficult to *validate*, *measure*, and to *predict* the *reliability* of software. To obtain a systematic framework for a qualitative and quantitative investigation of the reliability of software, Schneidewind [148] suggests the following procedure as a starting point for developing an SRE method in a specific organization of interest.

**Basics steps of SRE:**
**Step 1.** State the reliability requirements.
**Step 2.** Establish a measurement framework.
**Step 3.** Collect data.
**Step 4.** Establish problem severity levels.
**Step 5.** Estimate reliability model parameters.
**Step 6.** Select optimal set of failure data.
**Step 7.** Identify the operational profile.
**Step 8.** Make reliability predictions.
**Step 9.** Validate the model.
**Step 10.** Make reliability decisions.
**Step 11.** Use software reliability tools.

For a detailed account on these eleven steps, we refer the interested reader to [148], where a possible approach is well explained. Nevertheless, we also refer the reader to recently emerging Commercial off-the-Shelf (COTS) software products such as *SonarQube* [61], which can be applied during a software development process and has built in automated functionalities as suggested by [148]. Clearly, companies or public organizations requiring the highest security and reliability standards might not want to solely rely on a general COTS product but rather use it as a complementary analysis/validation platform next to their own designed, built, and maintained special purpose SRE tool.

Finally, Shooman [154] clearly represents the basic elements, which build up the reliability of a (complex) system, that is,

$$R_s = R_h R_{sw} R_o \,, \tag{5.16}$$

where $R_i$ for $i \in \{s, h, sw, o\}$ represent reliabilities of hardware $i = h$, software $i = sw$, operators/users $i = o$, and the entire system $i = s$. At the same time, we can well imagine more complex situations, where the different reliabilities $R_h$, $R_{sw}$, and $R_o$ all are interdependent.

# 6 Multiscale CHeSs: upscaling, effective properties, and macroscopic descriptions

## 6.1 Introduction

Let us look at CHeSs consisting of subsystems showing a much smaller length scale $\ell$ than the macroscopic system of interest having the length $L$. The resulting dimensionless parameter $\epsilon := \frac{\ell}{L}$ is called the heterogeneity of the CHeS. We are interested in a systematic and reliable framework for deriving effective macroscopic equations describing the CHeS starting from a detailed microcscopic formulation. Such upscaling procedures rely on a systematic averaging method, which allows us to retain essential and characteristic microscopic features on the macroscale.

A very basic and simple multiscale problem represents the conductivity of a composite material $\mathcal{D}$, which contains two material phases with conductivity $0 < \sigma_1 \ll \sigma_2$. In fact, in this case, we even say that the material shows a *high contrast* (in conductivity) due to $\sigma_1 \ll \sigma_2$. The subsequent techniques and methods are not specifically refined to account for such high-contrast situations. Nevertheless, techniques to deal with such kind of situations are presented in [136]. Now, the material $\mathcal{D}$ shows a highly heterogeneous/oscillating conductivity $\sigma^\epsilon(\mathbf{x}) := \sigma(\mathbf{x}/\epsilon) = \sigma_1\chi_{Y_1}(\mathbf{x}/\epsilon) + \sigma_2\chi_{Y_2}(\mathbf{x}/\epsilon)$, where $Y := Y_1 \cup Y_2$ denotes the microscopic subsystem/*reference cell*. This cell captures the microscopic features such as the conductivity here of the macroscopic material $\mathcal{D}$. This finally leads to the following microscopic conductivity formulation,

$$\begin{cases} -\operatorname{div}(\sigma^\epsilon(\mathbf{x})\nabla u^\epsilon(\mathbf{x})) = 0\,, & \text{in } \mathcal{D}\,, \\ u^\epsilon(\mathbf{x}) = V\,, & \text{on } \Gamma_l\,, \\ u^\epsilon(\mathbf{x}) = 0\,, & \text{on } \Gamma_r\,, \end{cases} \tag{6.1}$$

where $u^\epsilon(\mathbf{x})$ is the microscopic electric potential in $\mathcal{D}$. Before we look at this kind of continuous upscaling ($\epsilon \to 0$), we present an intuitive discrete upscaling in the next section.

## 6.2 Network/graph-based upscaling: flow/circuit laws

Consider a regular resistor network $\mathcal{N} = \mathbb{Z}^d \cap \mathcal{D}$, where $\mathcal{D}$ is the macroscopic domain of the conductor of interest, see Figure 6.1. Let us first recall the following basic laws of electrical circuits [92]:

**Ohm's law**

*The current $I_{i-1/2,j}$ through a 2D-material $\mathcal{D}$ with left end point $(i-1,j) \in \mathbb{Z}^2 \cap \mathcal{D}$ and right end point $(i,j) \in \mathbb{Z}^2 \cap \mathcal{D}$ is directly proportional to the voltage difference $u_{i,j} - u_{i-1,j}$ applied across it.*

https://doi.org/10.1515/9783110579543-006

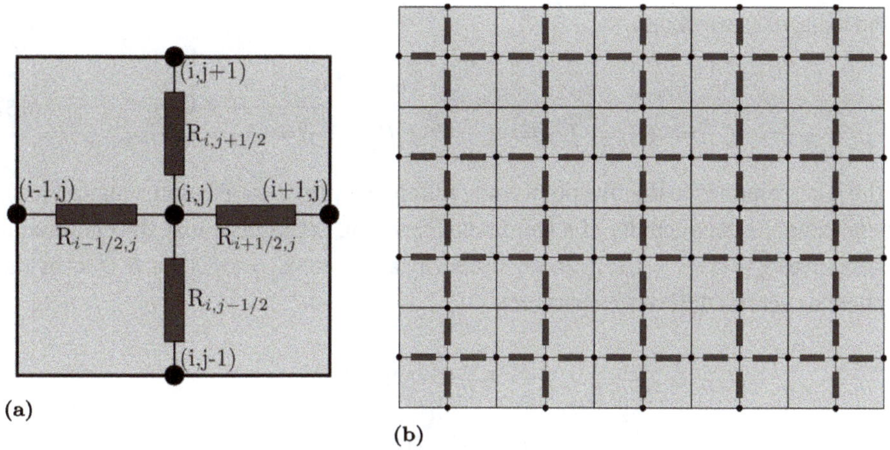


Figure 6.1: (a) Reference cell of characteristic length $\ell$. (b) Resistor network $\mathcal{N}$ of macroscopic length $L$ with heterogeneity $\epsilon = \frac{\ell}{L}$.

**Kirchhoff's law**
*The algebraic sum of currents in a network $\mathcal{N}$ of conductors meeting at a point $(i,j) \in \mathbb{Z} \cap \mathcal{D}$ is zero.*

Next, we apply these laws to a network $\mathcal{N}$ for which a reference element in the form of a five-point star with center $(i,j)$ is depicted in Figure 6.1. The index $i \in \mathbb{Z}$ increases from left to the right along the horizontal axis, and accordingly the index $j \in \mathbb{Z}$ increases along the vertical axis from the bottom to the top. Hence, we can measure voltages $u_{ij}$ at the nodes of the five-point star. The edge connecting the nodes $(i,j)$ and $(i+1,j)$, $(i,j) \in \mathbb{Z}^2$, represents the resistance $R_{i+1/2,j}$, and the edge connecting nodes $(i,j)$ and $(i,j+1)$, $(i,j) \in \mathbb{Z}^2$, represents the resistance $R_{i,j+1/2}$. Thanks to Ohm's law, we can compute the following currents,

$$I_{i+1/2,j} = \frac{u_{i+1,j} - u_{i,j}}{R_{i+1/2,j}} \quad \text{and} \quad I_{i,j+1/2} = \frac{u_{i,j+1} - u_{i,j}}{R_{i+1/2,j}} \quad \text{for } (i,j) \in \mathbb{Z}^2. \tag{6.2}$$

Additionally, we apply the following sign convention at the node $(i,j) \in \mathcal{N} = \mathcal{D} \cap \mathbb{Z}^2$,

$$\begin{cases} -I_{i+1/2,j}\,, & \text{current out of the node}\,, \\ +I_{i-1/2,j}\,, & \text{current into the node}\,, \end{cases} \tag{6.3}$$

where we used the fact that the current flows from the nodes with higher voltage/potential toward the node with lower voltage/potential (Dirichlet boundary conditions, see $(6.1)_2$ and $(6.1))_3$. After applying Kirchhoff's law to the central node $(i,j)$,

$$I_{i-1/2,j} - I_{i+1/2,j} + I_{i,j-1/2} - I_{i,j+1/2} = 0\,, \tag{6.4}$$

and using (6.2), we obtain,

$$\frac{u_{i,j}-u_{i-1,j}}{R_{i-1/2,j}} - \frac{u_{i+1,j}-u_{i,j}}{R_{i+1/2,j}} + \frac{u_{i,j}-u_{i,j-1}}{R_{i,j-1/2}} - \frac{u_{i,j+1}-u_{i,j}}{R_{i,j+1/2}} = 0\,, \tag{6.5}$$

which corresponds to the five-point finite difference star of the Poisson equation $(6.1)_1$ describing the conductivity of a continuous medium. Let us discretize the continuous derivatives $\frac{\partial}{\partial x^k} f(x)$, $k = 1, 2$, in 2D for simplicity at a point $x_{i,j} = (x_i^1, x_j^2) \in \mathcal{D} \subset \mathbb{R}^2$ by the following central difference operators,

$$\begin{aligned}
\delta^1_{i,\frac{h}{2}} f(x_{i,j}) &= \frac{f(x_{i+\frac{1}{2},j}) - f(x_{i-\frac{1}{2},j})}{h}\,,\\
\delta^1_{j,\frac{h}{2}} f(x_{i,j}) &= \frac{f(x_{i,j+\frac{1}{2}}) - f(x_{i,j-\frac{1}{2}})}{h}\,,\\
\delta^2_{i,\frac{h}{2}} f(x_{i,j}) &= \delta^1_{i,\frac{h}{2}}\left(\frac{f(x_{i+\frac{1}{2},j}) - f(x_{i-\frac{1}{2},j})}{h}\right),\\
\delta^2_{j,\frac{h}{2}} f(x_{i,j}) &= \delta^1_{j,\frac{h}{2}}\left(\frac{f(x_{i,j+\frac{1}{2}}) - f(x_{i,j-\frac{1}{2}})}{h}\right),
\end{aligned} \tag{6.6}$$

where we apply the same grid spacing $h$ in the $x^1$- and $x^2$-directions. Then, the second central difference discretization of the Laplace equation (6.1) gives, for $u_{i,j} = u(x_{i,j})$,

$$\begin{aligned}
0 &= -\delta^1_{i,\frac{h}{2}}\big(\sigma(x_{i,j}/\epsilon)\delta^1_{i,\frac{h}{2}} u_{i,j}\big) - \delta^1_{j,\frac{h}{2}}\big(\sigma(x_{i,j}/\epsilon)\delta^1_{j,\frac{h}{2}} u_{i,j}\big)\\
&= -\delta^1_{i,\frac{h}{2}}\left(\sigma(x_{i,j}/\epsilon)\frac{u_{i+1/2,j}-u_{i-1/2,j}}{h}\right) - \delta^1_{j,\frac{h}{2}}\left(\sigma(x_{i,j}/\epsilon)\frac{u_{i,j+1/2}-u_{i,j-1/2}}{h}\right)\\
&= -\left(\sigma(x_{i+1/2,j}/\epsilon)\frac{u_{i+1,j}-u_{i,j}}{h^2} - \sigma(x_{i-1/2,j}/\epsilon)\frac{u_{i,j}-u_{i-1,j}}{h^2}\right)\\
&\quad - \left(\sigma(x_{i,j+1/2}/\epsilon)\frac{u_{i,j+1}-u_{i,j}}{h^2} - \sigma(x_{i,j-1/2}/\epsilon)\frac{u_{i,j}-u_{i,j-1}}{h^2}\right),
\end{aligned} \tag{6.7}$$

which leads to (6.5) after the identifying $\sigma(x_{i,j}/\epsilon) = 1/R_{i,j}$ for $\epsilon = \frac{\ell}{L} \in \mathbb{R}$ and after multiplying with $h^2$.

Hence, the upscaling limit consists here in passing $h \to 0$, which is equivalent to the convergence of discrete solutions of the finite difference scheme (6.7) toward its continuous limit represented as the solution of the Laplace equation (6.1).

**Remark 6.2.1** (Continuous limit, modeling, and numerics). These considerations demonstrate that the natural character of real/practical aspects show a finite scale, and, as a consequence, we motivate that discrete models have an equally strong validity from a physical point of view as continuous formulations. The latter are generally preferred in mathematical analysis (of PDEs). Hence, the scope of numerical analysis grows under this view to also include approximation studies of discrete numerical algorithms to capture discrete models of systems, in particular, CHeSs. In fact, such numerical approaches very likely are methods to reduce the dimensional complexity at hand such as for CHeSs, which generally lead to high-dimensional multiscale problems. ◇

## 6.3 Upscaling based on homogenization theory: asymptotic two-scale expansion

Let us look again at the conductivity problem (6.1) which is a classical two-scale problem defined by a material specific microscale $Y$ and an empirically measurable macroscale $\mathcal{D}$. We denote the Hilbert space of periodic $H^1(Y)$-functions by $H^1_\sharp(Y)$. Moreover, we rely on the following definition.

**Strongly Elliptic Operator**

**Definition 6.3.1.** We call a tensor $\hat{\Sigma} = \{\sigma_{ij}\}_{i,j=1}^{d,d}$ to be *strongly elliptic*, if for $\mathbf{r} \in \mathbb{R}^d$ it holds that,

$$\text{(Strong Ellipticity)}\quad \begin{cases} \sigma_{ij}(\mathbf{x})r_i r_j \geq a_e |\mathbf{r}|^2\,, & \text{a. e. for } a_e \in \mathbb{R} \text{ with } a_e > 0\,,\\ |\sigma_{ij}(\mathbf{x})r_j| \leq A_e|\mathbf{r}|\,, & \text{a. e. in } \mathcal{D} \text{ for } A_e \in \mathbb{R} \text{ with } A_e > a_e\,. \end{cases} \tag{6.8}$$

In this section, with the help of continuous methods, we systematically derive the following:

**Upscaled Macroscopic Conductivity Equation**

**Homogenization result:** *The microscopic conductivity problem (6.1) governing the electric potential $u^\epsilon$ inside a composite material $\mathcal{D}$, which is characterized by a material specific conductivity tensor $\hat{\Sigma}^\epsilon = \{\sigma^\epsilon_{ij}\}_{i,j=1}^{d,d}$, admits the following effective macroscopic conductivity equation in the limit as $\epsilon \to 0$, that is,*

$$\begin{cases} -\mathrm{div}_{\mathbf{x}}\left(\hat{\Sigma}^0 \nabla_{\mathbf{x}} u^0(\mathbf{x})\right) = 0\,, & \text{in } \mathcal{D}\,,\\ u^\epsilon = V\,, & \text{on } \Gamma_l\,,\\ u^\epsilon = 0\,, & \text{on } \Gamma_r\,, \end{cases} \tag{6.9}$$

*where the effective conductivity tensor $\hat{\Sigma}^0 = \{\sigma^0_{ij}\}_{i,j=1}^{d,d}$ is defined by (6.22),*

$$\sigma^0_{ij} = \frac{1}{|Y|}\int_Y \left(\sigma_{ik}(\mathbf{y})\frac{\partial}{\partial y_k}\xi^j(\mathbf{y}) + \sigma_{kj}\right) d\mathbf{y}\,, \tag{6.10}$$

*and the correctors $\xi^j \in H^1_\sharp(Y)$, $j = 1, 2, \dots, d$, solve the following reference cell problem,*

$$\begin{cases} -\frac{\partial}{\partial y_i}\left(\sigma(\mathbf{y})\delta_{ij}\left(\frac{\partial}{\partial y_j}\xi^k(\mathbf{y}) + \delta_{jk}\right)\right) = 0\,, & \text{in } Y,\\ \xi^k(\mathbf{y}) \text{ is } Y\text{-periodic, and } M_Y(\xi^k) := \frac{1}{|Y|}\int_Y \xi^k\, d\mathbf{y} = 0\,. \end{cases} \tag{6.11}$$

The derivation is based on the following basic steps:

**Asymptotic Two-Scale Expansion (TE)**

**(TE1)** *Make the ansatz of the following (asymptotic) two-scale expansion,*

$$u^\epsilon(\mathbf{x}) = u^0(\mathbf{x}, \mathbf{x}/\epsilon) + \epsilon u^1(\mathbf{x}, \mathbf{x}/\epsilon) + \epsilon^2 u^2(\mathbf{x}, \mathbf{x}/\epsilon) + \mathcal{O}(\epsilon^3), \tag{6.12}$$

*where the microscale* $\mathbf{y} = \mathbf{x}/\epsilon$ *is* $Y$*-periodic.*

**(TE2)** *Collect equal terms of equal order in* $\epsilon$*. This leads to a sequence of equations at the orders* $\mathcal{O}(\epsilon^{-2})$, $\mathcal{O}(\epsilon^{-1})$, *and* $\mathcal{O}(\epsilon^0)$.

**(TE3)** *The equation at order* $\mathcal{O}(\epsilon^{-1})$ *is called the cell problem, and its solution is called (first-)order corrector. The upscaled solution* $u^0(\mathbf{x}, \mathbf{x}/\epsilon)$*, which is generally independent of the microscale* $\mathbf{y} = \frac{\mathbf{x}}{\epsilon} \in Y$*, is the solvability constraint of the equation at order* $\mathcal{O}(\epsilon^0)$.

Subsequently, we will work out this method in detail under the assumption that all the functions are smooth. This is obviously not realistic as we can already see on the conductivity problem of a composite material, where the conductivity jumps from one material to the other. However, the sequence of problems as outlined in **(TE2)** will subsequently be discussed rigorously.

**Step 1. (TE1)** and **(TE2)**: *Insert the two-scale expansion* (6.12) *into* (6.1) We obtain the following sequence of problems having equal order in $\epsilon$:

$$\mathcal{O}(\epsilon^{-2}): \quad \begin{cases} -\mathrm{div}_{\mathbf{y}}(\sigma(\mathbf{y})\nabla_{\mathbf{y}} u^0(\mathbf{x},\mathbf{y})) = 0 \quad \text{in } Y, \\ u^0 \quad Y\text{-periodic in } \mathbf{y}, \end{cases} \tag{6.13}$$

$$\mathcal{O}(\epsilon^{-1}): \quad \begin{cases} -\mathrm{div}_{\mathbf{y}}(\sigma(\mathbf{y})\nabla_{\mathbf{y}} u^1(\mathbf{x},\mathbf{y})) = \mathrm{div}_{\mathbf{x}}(\sigma(\mathbf{y})\nabla_{\mathbf{y}} u^0(\mathbf{x},\mathbf{y})) \\ \qquad\qquad \mathrm{div}_{\mathbf{y}}(\sigma(\mathbf{y})\nabla_{\mathbf{x}} u^0(\mathbf{x},\mathbf{y})) \quad \text{in } Y, \\ u^1 \quad Y\text{-periodic in } \mathbf{y}, \end{cases} \tag{6.14}$$

$$\mathcal{O}(\epsilon^{0}): \quad \begin{cases} -\mathrm{div}_{\mathbf{y}}(\sigma(\mathbf{y})\nabla_{\mathbf{y}} u^2(\mathbf{x},\mathbf{y})) = \mathrm{div}_{\mathbf{x}}(\sigma(\mathbf{y})\nabla_{\mathbf{y}} u^1(\mathbf{x},\mathbf{y})) \\ \qquad\qquad + \mathrm{div}_{\mathbf{y}}(\sigma(\mathbf{y})\nabla_{\mathbf{x}} u^1(\mathbf{x},\mathbf{y})) \\ \qquad\qquad + \mathrm{div}_{\mathbf{x}}(\sigma(\mathbf{y})\nabla_{\mathbf{x}} u^0(\mathbf{x},\mathbf{y})) \quad \text{in } Y, \\ u^2 \quad Y\text{-periodic in } \mathbf{y}. \end{cases} \tag{6.15}$$

In the subsequent steps, we work out **(TE3)**.

**Step 2.** ($\mathcal{O}(\epsilon^{-2})$*-problem*) Due to periodicity, this problem implies that the leading-order solution $u^0(\mathbf{x}, \mathbf{y})$ is independent of the microscale $\mathbf{y}$.[1] We note that this is not necessarily true for other type of problems in homogenization theory, where a dependence of the upscaled variable on the reference cell $Y$ remains part of the homogenized solution.

1 Testing equation (6.13) with $u^0(\mathbf{x}, \mathbf{y}) \in H^1_\sharp(Y)$ immediately implies that $u^0(\mathbf{x}, \mathbf{y})$ is invariant with respect to $\mathbf{y} \in Y$.

**Step 3.** ($\mathcal{O}(\epsilon^{-1})$*-problem*) This problem is often called the cell problem whose solution provides the leading-order corrector that takes the geometric or material (e. g., conductivity) features of porous media/heterogeneous systems into account. The cell problem is obtained after making the ansatz $u^1(\mathbf{x},\mathbf{y}) = -\sum_{j=1}^{d} \xi^j(\mathbf{y})\frac{\partial u^0(\mathbf{x},\mathbf{y})}{\partial x_j}$ and using the fact the leading-order solution $u^0$ is independent of $\mathbf{y} \in Y$,

$$\begin{aligned} 0 &= -\mathrm{div}_{\mathbf{y}}(\sigma(\mathbf{y})(\nabla_{\mathbf{y}} u^1(\mathbf{x},\mathbf{y}) - \nabla_{\mathbf{x}} u^0(\mathbf{x},\mathbf{y}))) \\ &= \frac{\partial}{\partial y_i}\left(\sigma(\mathbf{y})\delta_{ij}\left(\frac{\partial}{\partial y_j}\xi^k(\mathbf{y})\frac{\partial u^0(\mathbf{x},\mathbf{y})}{\partial x_k} + \frac{\partial u^0(\mathbf{x},\mathbf{y})}{\partial x_k}\frac{\partial x_k}{\partial x_j}\right)\right) \\ &= \frac{\partial}{\partial y_i}\left(\sigma(\mathbf{y})\delta_{ij}\left(\frac{\partial}{\partial y_j}\xi^k(\mathbf{y}) + \delta_{jk}\right)\right) \quad \text{in } Y, \end{aligned} \tag{6.16}$$

where we used Einstein's summation convention, and $\delta_{jk}$ denotes the Kronecker delta. Equation (6.16) is called the *(reference) cell problem* and plays a crucial role in the homogenization theory as its solution $\xi^k(\mathbf{y})$, called the corrector, accounts for material/geometric information available on the microscale $Y$. The corrector $\xi^k(\mathbf{y}) \in H^1_\sharp(Y)$ is a unique solution of,

$$\text{(Cell Problem)} \qquad \begin{cases} -\frac{\partial}{\partial y_i}\left(\sigma(\mathbf{y})\delta_{ij}\left(\frac{\partial}{\partial y_j}\xi^k(\mathbf{y}) + \delta_{jk}\right)\right) = 0 & \text{in } Y, \\ \xi^k(\mathbf{y}) \text{ is } Y\text{-periodic and } M_Y(\xi^k) = 0, \end{cases} \tag{6.17}$$

which is an immediate result of the Lax–Milgram theorem [27, p. 140, Corollary 5.8]. The operator $M_Y(f(\mathbf{y}))$ denotes the mean value of a $Y$-integrable function $f$, that is,

$$M_Y(f(\mathbf{y})) := \frac{1}{|Y|}\int_Y f(\mathbf{y})\,d\mathbf{y}. \tag{6.18}$$

**Step 4.** ($\mathcal{O}(\epsilon^{0})$*-problem*) To analyze equation (6.15), we make use of the following solvability result:

**Solvability of elliptic equations**

**Theorem 6.3.2.** *Let $\hat{D}(\mathbf{y})$ be a strongly elliptic and $Y$-periodic tensor. Then, for $f \in (W^1_\sharp(Y))'$, the equation,*

$$(\hat{D}\nabla u, \nabla v) = (f, v), \quad \textit{for all } v \in W^1_\sharp(Y), \tag{6.19}$$

*has a unique solution $u \in W^1_\sharp(Y)$.*

*Proof.* This result is a direct consequence of the Lax–Milgram theorem. □

**Remark 6.3.3.** We recall that $(W^1_\sharp(Y))' = \{F \in (H^1_\sharp(Y))' \mid F(c) = 0, \forall c \in \mathbb{R}\}$ is the dual space of $W^1_\sharp(Y) := H^1_\sharp(Y)/\mathbb{R}$. ◇

The right-hand side of problem (6.15) reads for all $v \in W^1_\sharp(Y)$ in the weak sense as follows,

$$0 = (F, v) := (\operatorname{div}_{\mathbf{x}}(\sigma(\mathbf{y})\nabla_{\mathbf{y}} u^1(\mathbf{x}, \mathbf{y})), v) - (\sigma(\mathbf{y})\nabla_{\mathbf{x}} u^1(\mathbf{x}, \mathbf{y}), \nabla_{\mathbf{y}} v) + (\operatorname{div}_{\mathbf{x}}(\sigma(\mathbf{y})\nabla_{\mathbf{x}} u^0(\mathbf{x}, \mathbf{y})), v)\,, \tag{6.20}$$

since $v(\mathbf{y})$ only depends on $Y$. Hence, according to Theorem 6.3.2 and Remark 6.3.3, a necessary and sufficient condition for the solvability of equation (6.15) is that (6.20) holds for $v = 1$. Based on this and following **(TE3)**, we first rewrite the remaining terms as follows,

$$- \operatorname{div}_{\mathbf{x}}\left( \int_Y \sigma(\mathbf{y})(\nabla_{\mathbf{y}} u^1(\mathbf{x}, \mathbf{y}) + \nabla_{\mathbf{x}} u^0(\mathbf{x}))\, d\mathbf{y} \right) = -\frac{\partial}{\partial x_i}\left( \left( \int_Y \sigma(\mathbf{y})\delta_{ij}\left( \frac{\partial}{\partial y_j}\xi^k(\mathbf{y}) + \delta_{jk} \right) d\mathbf{y} \frac{\partial}{\partial x_k} u^0(\mathbf{x}) \right) \right) = 0\,, \tag{6.21}$$

and for $\sigma^0_{ij} := \sigma^0\delta_{ij}$ and $\sigma_{ij} := \sigma\delta_{ij}$, we can define the so-called *effective conductivity tensor* $\hat{\Sigma}^0 := \{\sigma^0_{ij}\}_{i,j=1}^d$ by,

$$\textbf{(Effective Conductivity)} \quad \left\{ \sigma^0_{ij} := \frac{1}{|Y|} \int_Y \left( \sigma_{ik}(\mathbf{y}) \frac{\partial}{\partial y_k} \xi^j(\mathbf{y}) + \sigma_{kj} \right) d\mathbf{y}\,. \right. \tag{6.22}$$

This finally leads to,

$$\textbf{(Homogenized Equation)} \quad \left\{ -\operatorname{div}_{\mathbf{x}}(\hat{\Sigma}^0 \nabla_{\mathbf{x}} u^0(\mathbf{x})) = 0 \quad \text{in } \mathcal{D}\,, \right. \tag{6.23}$$

supplemented by the same boundary conditions as in (6.1). We note that the same procedure is also valid for a nonisotropic conductivity tensor $\{\sigma_{ij}(\mathbf{y})\}_{i,j=1}^d$.

## 6.4 Properties of the effective conductivity tensor (6.22)

In this section, we state and derive fundamental properties of the upscaled material or transport coefficients/tensors such as symmetry and positive definiteness. Moreover, since these coefficients/tensors, referred to as correction tensors, generally can only be obtained numerically, except for special cases such as laminates, we are highly interested in upper and lower bounds on these effective macroscopic tensors.

### 6.4.1 (A) Symmetry and positive definiteness

Our goal for this subsection is to rigorously derive in two steps that the homogenized conductivity tensor $\hat{\Sigma}^0$ is symmetric positive definite.

**Correction tensor is symmetric**

**Lemma 6.4.1** (Symmetry). *Assume that the microscopic conductivity tensor $\sigma_{ij}$ is symmetric in each of the material phases $Y^1$ and $Y^2$. Then, the effective homogenized conductivity tensor $\hat{\Sigma}^0$ defined in equation* (6.22) *via the cell problem* (6.17) *is symmetric.*

*Proof.* First, we bring the cell problem (6.17) into weak form by testing it with a smooth enough $Y$-periodic function $\psi \in H^1_\sharp(Y)$, which is continuous across the interface of the different materials, that is, we multiply (6.17) by $\psi$ and subsequently integrate over $Y$. By the divergence theorem and periodicity we obtain,

$$\left(\sigma_{ij}(\mathbf{y})\frac{\partial(\xi^k + y_k)}{\partial y_j}, \frac{\partial \psi}{\partial y_i}\right) = \left(\sigma_{ij}(\mathbf{y})\frac{\partial \Xi^k}{\partial y_j}, \frac{\partial \psi}{\partial y_i}\right) = 0\,, \tag{6.24}$$

where we set $\Xi^k := \xi^k + y_k$. Using (6.22), we can write the effective conductivity as follows,

$$|Y|\sigma^0_{ij} = \left(\sigma_{ik}(\mathbf{y})\frac{\partial \Xi^j}{\partial y_k}, 1\right) = \left(\sigma_{lk}(\mathbf{y})\frac{\partial \Xi^j}{\partial y_k}\frac{\partial y_i}{\partial y_l}, 1\right) = \left(\sigma_{lk}(\mathbf{y})\frac{\partial \Xi^j}{\partial y_k}, \frac{\partial \Xi^i - \xi^i}{\partial y_l}\right). \tag{6.25}$$

Taking into account the $Y$-periodicity of $\xi^i$ together with equation (6.24), which holds for $Y$-periodic $\psi$ and hence, in particular for $\psi = \xi^i$, we end up with the following expression for the effective conductivity,

$$\sigma^0_{ij} = \frac{1}{|Y|}\left(\sigma_{lk}(\mathbf{y})\frac{\partial \Xi^j}{\partial y_k}, \frac{\partial \Xi^i}{\partial y_l}\right), \tag{6.26}$$

which is symmetric if the microscopic conductivity tensor $\sigma_{ij}$ is symmetric in both material phases $Y^1$ and $Y^2$. □

Similarly, we can also conclude the next result.

**Correction tensor is positive definite**

**Lemma 6.4.2** (Positive definiteness). *Assume that the microscopic conductivity tensor $\hat{\Sigma} = \{\sigma_{ij}\}_{i,j=1}^{d,d}$ is strongly elliptic according to Definition* 6.3.1. *Then, the effective, homogenized conductivity tensor $\hat{\Sigma}^0$, defined in equation* (6.22) *with the solution of the cell problem* (6.17), *is positive definite.*[a]

[a] Recall that $(6.8)_1$ denotes the classical *uniform ellipticity.*

*Proof.* Thanks to (6.26) and (6.8), we have that,

$$\sigma^0_{ij} r_i r_j = \left(\sigma_{lk}(\mathbf{y})\frac{\partial(r_j \Xi^j)}{\partial y_k}, \frac{\partial(r_i \Xi^i)}{\partial y_k}\right) \geq a_e\left(\nabla_{\mathbf{y}}(r_i \Xi^i), \nabla_{\mathbf{y}}(r_i \Xi^i)\right) \geq 0\,, \tag{6.27}$$

after identifying $v = r_i \Xi^i = \sum_{i=1}^d r_i \Xi^i$ by Einstein's summation convention. This gives the positive definiteness of $\hat{\Sigma}^0$ under (6.8).

A $d \times d$-tensor $\hat{\mathrm{A}} \in \mathbb{R}^{d\times d}$ is positive definite, if $\mathbf{r}^{\mathrm{T}}\hat{\mathrm{A}}\mathbf{r} > 0$ for every nonzero column vector $\mathbf{r} \in \mathbb{R}^d$. Hence, we have to prove that $\sigma^0_{ij}r_i r_j = 0$ only holds, if $r_i = 0$. Hence, equality in (6.27) leads to the requirement,

$$\frac{\partial(r_i \Xi^i)}{\partial y_k} = r_i\left(\frac{\partial \xi^i}{\partial y_i} + \delta_{ik}\right) = 0\,, \tag{6.28}$$

which induces after integrating over $Y$ the required results $0 = r_k = r_i\delta_{ik}$ due to the $Y$-periodicity of $\xi^i$. □

### 6.4.2 (B) Simple upper and lower bounds

We can derive upper and lower bounds on the effective conductivity based on variational principles. The first principle that we present gives means to derive upper bounds.

**Variational Principle I (VPI)**

**Proposition 6.4.3.** *For $u \in H^1_\sharp(Y)$ and for any $\mathbf{r} = [r_1, r_2, \ldots, r_d] \in \mathbb{R}^d$, we consider the functional,*

$$F(u) = \frac{1}{|Y|}\int_Y \sigma_{ij}(\mathbf{y})\left(\frac{\partial u}{\partial y_i} + r_i\right)\left(\frac{\partial u}{\partial y_j} + r_j\right) d\mathbf{y}\,. \tag{6.29}$$

*For any $\mathbf{r} \in \mathbb{R}^d$, we have,*

$$\sigma^0_{ij}r_i r_j = \min_{u\in H^1_\sharp} F(u)\,, \tag{6.30}$$

*where $\sigma^0_{ij}$ is the effective conductivity defined in (6.22).*

**Remark 6.4.4** (Upper bounds based on VPI). The VPI can now be used to obtain upper bounds on the effective conductivity tensor, since any function $u \in H^1_\sharp(Y)$ that is not a minimizer of $F$ is greater than the left-hand side in (6.30) and hence an upper bound. ◇

*Proof of Proposition* 6.4.3. The proof is a based on deriving the inequalities,

$$\min_{u\in H^1_\sharp(Y)} F(u) \le \sigma^0_{ij}r_i r_j \le \min_{u\in H^1_\sharp(Y)} F(u)\,.$$

This then proves VPI.

*Step 1.* ($\min F \le \cdot$) Identity (6.26) allows us to rewrite (6.30) by $\sigma^0_{ij}r_i r_j = F(r_i\xi^i)$, where $\xi^i$ is the corrector solving (6.17) in the weak sense, and hence, we obtain,

$$\sigma^0_{ij}r_i r_j \ge \min_{u\in H^1_\sharp(Y)} F(u)\,. \tag{6.31}$$

*Step 2.* ($\cdot \leq \min F$) This step is a bit more involved. Let $v = \xi^i r_i - u$ be another $Y$-periodic function that allows us to substitute $u = \boldsymbol{\xi} \cdot \mathbf{r} - v$ into (6.29), i. e.,

$$\begin{aligned}\left(\sigma_{ij}\left(\frac{\partial u}{\partial y_j}+r_j\right),\left(\frac{\partial u}{\partial y_i}+r_i\right)\right) &= \left(\sigma_{ij}\left(\frac{\partial(\boldsymbol{\xi}\cdot\mathbf{r}-v)}{\partial y_j}+r_j\right),\left(\frac{\partial(\boldsymbol{\xi}\cdot\mathbf{r}-v)}{\partial y_i}+r_i\right)\right)\\ &= \left(\sigma_{ij}\left(\frac{\partial(\boldsymbol{\xi}\cdot\mathbf{r})}{\partial y_j}+r_j\right),\left(\frac{\partial(\boldsymbol{\xi}\cdot\mathbf{r})}{\partial y_i}+r_i\right)\right)+\left(\sigma_{ij}\frac{\partial v}{\partial y_j},\frac{\partial v}{\partial y_i}\right)\\ &\quad -2\left(\sigma_{ij}\frac{\partial v}{\partial y_i},\left(\frac{\partial(\boldsymbol{\xi}\cdot\mathbf{r})}{\partial y_j}+r_j\right)\right),\end{aligned} \tag{6.32}$$

due to symmetry of $\sigma_{ij}$. With identity (6.26), after rearranging terms in (6.32), we end up with,

$$\begin{aligned}|Y|\sigma^0_{ij}r_ir_j &= \left(\sigma_{ij}\left(\frac{\partial u}{\partial y_j}+r_j\right),\left(\frac{\partial u}{\partial y_i}+r_i\right)\right)+2\left(\sigma_{ij}\frac{\partial v}{\partial y_i},\left(\frac{\partial(\boldsymbol{\xi}\cdot\mathbf{r})}{\partial y_j}+\xi_j\right)\right)\\ &\quad -\left(\sigma_{ij}\frac{\partial v}{\partial y_j},\frac{\partial v}{\partial y_i}\right).\end{aligned} \tag{6.33}$$

The second term in (6.33) vanishes after integrating by parts using Gauss' theorem and due to $Y$-periodicity. Moreover, the last term in (6.33) is negative, since the microscopic conductivity $\sigma_{ij}$ is symmetric positive definite. Hence, for all $u \in H^1_\sharp(Y)$, we have,

$$|Y|\sigma^0_{ij}r_ir_j \leq \left(\sigma_{ij}\left(\frac{\partial u}{\partial y_j}+r_j\right),\left(\frac{\partial u}{\partial y_i}+r_i\right)\right), \tag{6.34}$$

and therefore,

$$\sigma^0_{ij}r_ir_j \leq \min_{u\in H^1_\sharp(Y)} F(u), \tag{6.35}$$

as required. □

**Simple upper bound with VPI**

**Example 6.4.5.** 1. The simplest $Y$-periodic functions to which the VP1 applies are the constant functions. For $u$ = const., we immediately obtain,

$$\sigma^0_{ij}r_ir_j \leq \frac{1}{|Y|}\int_Y \sigma_{ij}(\mathbf{y})\,d\mathbf{y}r_ir_j = M_Y(\sigma_{ij})r_ir_j, \tag{6.36}$$

where $M_Y(\sigma_{ij})$ denotes the volume average (6.18) of the microscopic conductivity tensor $\sigma_{ij}(\mathbf{y})$.

2. A further simplification is the case of an isotropic microscopic conductivity tensor, that is,

$$\sigma_{ij}(\mathbf{y}) = \sigma_1\delta_{ij}\chi_{Y^1}(\mathbf{y}) + \sigma_2\delta_{ij}\chi_{Y^2}(\mathbf{y}).$$

If we denote the volume fractions of the phases $Y^1$ and $Y^2$ by $p_1$ and $p_2$, respectively, then we obtain for $M_Y(\sigma_{ij})$ the expression,

$$M_Y(\sigma_{ij}) = (p_1\sigma_1 + p_2\sigma_2)\delta_{ij}\,. \tag{6.37}$$

Since the effective conductivity tensor $\hat{\Sigma}^0 = \{\sigma_{ij}^0\}_{i,j=1}^d$ is symmetric, we can compute its eigenvalues $\lambda_n$ by,

$$\sigma_{ij}^0 v_i = \lambda\delta_{ij}v_i\,, \tag{6.38}$$

where, in general, $\lambda_1 \neq \lambda_2 \neq \cdots \neq \lambda_d$, i. e., the effective conductivity tensor is anisotropic. We also note that the additional property of positive definiteness guarantees that all the eigenvalues are positive. Herewith, we can conclude that the right-hand side of (6.37) is an upper bound of the largest principle conductivity on the macroscale, that is,

$$\lambda_n \leq p_1\sigma_1 + p_2\sigma_2 \quad \text{for n} = 1, 2, \ldots, d\,.$$

The next variational principle can be used to derive lower bounds and relies on the invertibility of the microscopic conductivity tensor.

**Variational Principle II (VPII)**

**Proposition 6.4.6.** *For* $\mathbf{j} \in H^1_\sharp(Y, \text{div})$, *where the space,*

$$H^1_\sharp(Y, \text{div}) := \left\{\mathbf{j} \in H^1_\sharp\left(Y, \mathbb{R}^d\right) \,\middle|\, \text{div}\,\mathbf{j} = 0 \textit{ in } Y \textit{ and } M_Y(\mathrm{j}_i) =: \bar{\mathrm{j}}_i \in \mathbb{R} \textit{ for } 1 \leq i \leq d\right\},$$

*contains periodic and solenoidal $H^1$-functions on $Y$, we consider the functional,*

$$L(\mathbf{j}) = \frac{1}{|Y|}\int_Y \sigma_{ij}^{-1}(\mathbf{y})\mathrm{j}_i\mathrm{j}_j\, d\mathbf{y}\,. \tag{6.39}$$

*Then we have the following variational principle:*

$$\left(\sigma_{ij}^0\right)^{-1}\bar{\mathrm{j}}_i\bar{\mathrm{j}}_j = \inf_{\mathbf{j}\in H^1_\sharp(Y,\text{div})} L(\mathbf{j})\,, \tag{6.40}$$

*where $\sigma_{ij}^0$ is the effective conductivity defined in* (6.22).

The proof of this dual variational principle is based on the following inequality.

**Tensor inequality**

**Lemma 6.4.7.** *Let $\hat{A} = \{a_{ij}\}_{i,j=1}^d$ be a symmetric and positive definite matrix. Then, for vectors $\mathbf{r} \in \mathbb{R}^d$ and vector functions $\mathbf{j} \in H^1_\sharp(Y, \text{div})$, we have the following inequality,*

$$\frac{1}{2}a_{ij}r_ir_j \geq r_i\mathrm{j}_i - \frac{1}{2}a_{ij}^{-1}\mathrm{j}_i\mathrm{j}_j \tag{6.41}$$

*with equality if and only if $\mathrm{j}_j = a_{ij}r_i$.*

*Proof of Lemma* 6.4.7. First, we identify $r_i = v_i + a_{ij}^{-1}\mathrm{j}_j$ for an appropriate vector $\mathbf{v}$ and then compute,

$$\frac{1}{2}a_{ij}r_ir_j = \frac{1}{2}a_{ij}(v_i + a_{il}^{-1}\mathrm{j}_l)(v_j + a_{jk}^{-1}\mathrm{j}_k) = \frac{1}{2}a_{ij}(a_{il}^{-1}\mathrm{j}_l a_{jk}^{-1}\mathrm{j}_k + 2v_i a_{jk}^{-1}\mathrm{j}_k + v_iv_j)$$

$$
\begin{aligned}
&= \frac{1}{2}\mathrm{j}_l a_{lk}\mathrm{j}_k + \frac{1}{2}r_j\mathrm{j}_j - \frac{1}{2}a_{ji}^{-1}\mathrm{j}_i\mathrm{j}_j + \frac{1}{2}r_i\mathrm{j}_i - \frac{1}{2}a_{ij}^{-1}\mathrm{j}_j\mathrm{j}_i + \frac{1}{2}a_{ij}v_iv_j \\
&= \mathrm{j}_i r_i - \frac{1}{2}a_{ij}^{-1}\mathrm{j}_j\mathrm{j}_i + \frac{1}{2}a_{ij}v_iv_j\,, \qquad (6.42)
\end{aligned}
$$

where we used the identities $\frac{1}{2}\mathrm{j}_l a_{li}^{-T} a_{ij}(a_{jk}^{-1}\mathrm{j}_k) = \frac{1}{2}\mathrm{j}_l a_{lk}\mathrm{j}_k$, $\frac{1}{2}a_{ij}v_i a_{jk}^{-1}\mathrm{j}_k = \frac{1}{2}v_i\mathrm{j}_i$, $\frac{1}{2}a_{ij}a_{kl}^{-1}\mathrm{j}_l v_j = \frac{1}{2}v_j\mathrm{j}_j$, $\frac{1}{2}v_j\mathrm{j}_j = \frac{1}{2}r_j\mathrm{j}_j - \frac{1}{2}a_{ji}^{-1}\mathrm{j}_i\mathrm{j}_j$, and $\frac{1}{2}v_i\mathrm{j}_i = \frac{1}{2}r_i\mathrm{j}_i - \frac{1}{2}a_{ij}^{-1}\mathrm{j}_j\mathrm{j}_i$. Inequality (6.41) now follows immediately, since the last term in (6.42) is positive. Additionally, we have equality in (6.41), if and only if $v_i = 0$, which leads to $r_i = a_{ij}^{-1}\mathrm{j}_j$. Herewith, we proved the lemma. □

*Proof of Proposition* 6.4.6. For the effective conductivity tensor $\hat{\Sigma}^0 = \{\sigma_{ij}^0\}_{i,j=1}^d$, we have derived identity (6.27), which we can estimate with the help of Lemma 6.4.7 as follows,

$$
\begin{aligned}
\frac{1}{2}\sigma_{ij}^0 r_i r_j = \frac{1}{2}\left(\sigma_{ij}\frac{\partial(r_k\Xi^k)}{\partial y_i}, \frac{\partial(r_k\Xi^k)}{\partial y_j}\right) &\geq \left(\frac{\partial(r_k\Xi^k)}{\partial y_i}, \mathrm{j}_i\right) - \frac{1}{2}(\sigma_{ij}^{-1}\mathrm{j}_i, \mathrm{j}_j) \\
&= \left(\frac{\partial(r_k\xi^k)}{\partial y_i}, \mathrm{j}_i\right) + r_i\bar{\mathrm{j}}_i - \frac{1}{2}(\sigma_{ij}^{-1}\mathrm{j}_i, \mathrm{j}_j) \\
&= r_i\bar{\mathrm{j}}_i - \frac{1}{2}(\sigma_{ij}^{-1}\mathrm{j}_i, \mathrm{j}_j)\,, \qquad (6.43)
\end{aligned}
$$

where we used the fact that the first term on the right-hand side of the second line in (6.43) is zero after integrating by parts and using Gauss' theorem under the periodicity of $\xi^k$ as well as $\mathbf{j} \in H_\sharp^1(Y, \mathrm{div})$. Hence, according to Lemma 6.4.7, we have equality if and only if,

$$
\mathrm{j}_i = \sigma_{ij}\frac{\partial(r_k\Xi^k)}{\partial y_j}\,, \qquad (6.44)
$$

and therefore, the maximum is attained for $\mathbf{j} \in H_\sharp^1(Y, \mathrm{div})$ satisfying,

$$
\bar{\mathrm{j}}_i = \left(\sigma_{ij}, \frac{\partial(r_k\Xi^k)}{\partial y_j}\right) = \left(\sigma_{ij}, \frac{\partial\Xi^k}{\partial y_j}\right) r_k = \sigma_{ik}^0 r_k\,, \qquad (6.45)
$$

where we used the definition of the effective conductivity tensor (6.22). Herewith, inequality (6.43) reads,

$$
\frac{1}{2}\sigma_{ij}^0 r_i r_j = \max_{\mathbf{j}\in H_\sharp^1(Y,\mathrm{div})}\left\{r_i\bar{\mathrm{j}}_i - \frac{1}{2}\left((\sigma_{ij}^0)^{-1}\mathrm{j}_i\mathrm{j}_j\right)\right\}, \qquad (6.46)
$$

and turns with (6.45) into,

$$
\sigma_{ij}^0 r_i r_j = \min_{\mathbf{j}\in H_\sharp^1(Y,\mathrm{div})}(\sigma_{kl}^{-1}\mathrm{j}_k\mathrm{j}_l) = \min_{\mathbf{j}\in H_\sharp^1(Y,\mathrm{div})} L(\mathbf{j})\,. \qquad (6.47)
$$

We can again use (6.45) to substitute $r_i = \sigma_{ij}^{-1}\bar{\mathrm{j}}_j$ into (6.47), which finally leads to the VPII (6.40). □

**Simple lower bound with VPII**

**Example 6.4.8** (Simple lower bound). A straightforward application of VPII consists of choosing the constant vector $\bar{\mathbf{j}} = M_Y(\mathbf{j})$ in (6.40), i. e.,

$$\left(\sigma_{ij}^{0}\right)^{-1}\bar{\mathrm{j}}_i\bar{\mathrm{j}}_j \leq \frac{1}{|Y|}\int_Y \sigma_{ij}^{-1}\,d\mathbf{y}\,\bar{\mathrm{j}}_i\bar{\mathrm{j}}_j\,, \tag{6.48}$$

and therefore,

$$\sigma_{ij}^{0}\bar{\mathrm{j}}_i\bar{\mathrm{j}}_j \geq \left(\frac{1}{|Y|}\int_Y \sigma_{ij}^{-1}\,d\mathbf{y}\right)^{-1}\bar{\mathrm{j}}_i\bar{\mathrm{j}}_j\,. \tag{6.49}$$

As in Example 6.4.5, for an isotropic microscopic conductivity,

$$\sigma_{ij}(\mathbf{y}) = \sigma_1\delta_{ij}\chi_{Y^1}(\mathbf{y}) + \sigma_2\delta_{ij}\chi_{Y^2}(\mathbf{y})\,,$$

we can write equation (6.49) as follows,

$$\sigma_{ij}^{0}\bar{\mathrm{j}}_i\bar{\mathrm{j}}_j \geq \left(\frac{p_1}{\sigma_1} + \frac{p_2}{\sigma_2}\right)^{-1}\bar{\mathrm{j}}_i\bar{\mathrm{j}}_j\,. \tag{6.50}$$

Finally, if we let $\bar{\mathbf{j}}$ be an arbitrary eigenvector of $\hat{\Sigma}^0$, then equation (6.38) allows us to rewrite (6.50) as,

$$\lambda_n \geq \left(\frac{p_1}{\sigma_1} + \frac{p_2}{\sigma_2}\right)^{-1} \quad \text{for } n = 1, 2, \ldots, d\,, \tag{6.51}$$

where $p_1 := \frac{|Y_1|}{|Y|}$ and accordingly $p_2 := \frac{|Y_2|}{|Y|}$.

Examples 6.4.5 and 6.4.8 show that the arithmetic/volume average and the harmonic average are upper and lower bounds, respectively, of the eigenvalues of the homogenized/upscaled conductivity tensor. Let us summarise these immediate results as follows:

**Bounds based on the variational principles VPI and VPII**

*For any $\mathbf{j} \in H_\#^1(Y;\mathrm{div})$, we can bound the effective macroscopic conductivity tensor $\hat{\Sigma}^0 = \{\sigma_{ij}^0\}_{i,j=1}^d$ above and below by,*

$$\left(\frac{1}{|Y|}\int_Y \sigma_{ij}^{-1}\,d\mathbf{y}\right)^{-1}\bar{\mathrm{j}}_i\bar{\mathrm{j}}_j \leq \sigma_{ij}^{0}\bar{\mathrm{j}}_i\bar{\mathrm{j}}_j \leq M_Y(\sigma_{ij})\bar{\mathrm{j}}_i\bar{\mathrm{j}}_j\,, \tag{6.52}$$

*where $\bar{\mathrm{j}}_i = M_Y(\mathrm{j}_i)$. In the case of isotropic microscopic conductivities, i. e.,*

$$\sigma_{ij}(\mathbf{y}) = \sigma_1\delta_{ij}\chi_{Y^1}(\mathbf{y}) + \sigma_2\delta_{ij}\chi_{Y^2}(\mathbf{y})\,,$$

*the bounds (6.52) can be written more intuitively in terms of principle eigenvalues $\{\lambda_n\}_{n=1}^d$ of the effective macroscopic conductivity tensor $\hat{\Sigma}^0 = \{\sigma_{ij}^0\}_{i,j=1}^d$,*

$$\left(\frac{p_1}{\sigma_1} + \frac{p_2}{\sigma_2}\right)^{-1} \leq \lambda_n \leq p_1\sigma_1 + p_2\sigma_2\,, \tag{6.53}$$

*where $p_1 := \frac{|Y_1|}{|Y|}$ and accordingly $p_2 := \frac{|Y_2|}{|Y|}$.*

However, note that these variational principles do not account for imperfect conduction through the interfaces. This can occur, for instance, by also taking into account a jump across the interface, e. g., as considered in [109]. In the subsequent remark, we also comment on improved estimates based on our VPI and VPII.

**Remark 6.4.9** (Hashin–Shtrikman bounds). We note that the above variational principles for upper and lower bounds are rather simple and there are more refined principles available [109]. The optimal bounds available without imposing additional characteristic properties on volume fractions of the composite material are the so-called Hashin–Shtrikman bounds [71]. ◇

## 6.5 Upscaling for slowly varying microstructure

The purpose of this section is o discuss possibilities of generalizing the periodic homogenization strategy towards nonperiodic systems. To this end, we first consider systems with slowly varying microstructure (SVM) over the (slow) macroscale $\mathbf{x}$. An example of a perturbed periodic pore-structure $\mathcal{D}_\epsilon^1$ is depicted in Figure 6.2. As a consequence, the single characteristic reference cell $Y$ from the periodic case will turn into a spatially depending reference cell $Y_{[\mathbf{x}/\epsilon]}$, where $[x]$ represents the nearest smaller integer part of $x \in \mathbb{R}$. Naturally, $[\cdot]$ operates componentwise for vectorial arguments. The generalized microstructures as depicted in Figure 6.2 (right), where the porosity $p(\mathbf{x})$ depends on the macroscale $\mathbf{x} \in \mathcal{D}$, require the theory of spatial averaging (volume averages), which has been systematically initiated by Slattery [158], Gray and Lee [64], and Whitaker [174]. Here, we recall the averaging result derived by Mei [108, Appendix 3A].

**Spatial Averaging Result**
*Let $\mathfrak{h}(\mathbf{x}, \mathbf{x}/\epsilon) \in C^1(\mathcal{D}, C^1(Y_\mathbf{x}, \mathbb{R}))$ be a parameterization of the porous medium such that,*

$$\begin{aligned} Y_\mathbf{x}^1 &:= \{\mathbf{y} \in Y_\mathbf{x} \mid \mathfrak{h}(\mathbf{x}, \mathbf{y}) > 0\}, \quad Y_\mathbf{x}^2 := \{\mathbf{y} \in Y_\mathbf{x} \mid \mathfrak{h}(\mathbf{x}, \mathbf{y}) < 0\}, \\ I_\mathbf{x} &:= \{\mathbf{y} \in Y_\mathbf{x} \mid \mathfrak{h}(\mathbf{x}, \mathbf{y}) = 0\}. \end{aligned} \tag{6.54}$$

*Then for any $\mathbf{u}(\mathbf{x}, \mathbf{x}/\epsilon) = [u_1(\mathbf{x}, \mathbf{x}/\epsilon), u_2(\mathbf{x}, \mathbf{x}/\epsilon), \dots, u_d(\mathbf{x}, \mathbf{x}/\epsilon)]' \in C^1(\mathcal{D}, C^1(Y_\mathbf{x}, \mathbb{R}^d))$, we have the following volume averaging property,*

$$\frac{\partial}{\partial x_i} \int_{Y_\mathbf{x}^1} u_i \, d\mathbf{y} = \int_{Y_\mathbf{x}^1} \frac{\partial u_i}{\partial x_i} \, d\mathbf{y} + \int_{I_\mathbf{x}} u_i \frac{\partial \mathfrak{h}}{\partial x_i} \frac{1}{|\nabla \mathfrak{h}|} \, do(\mathbf{y}), \tag{6.55}$$

*which is often written with averaging symbols $\langle \cdot \rangle$ denoting the volume average over the reference cell $Y_\mathbf{x}$ depending on the macroscale, as well as,*

$$\frac{\partial \langle u_i \rangle}{\partial x_i} = \left\langle \frac{\partial u_i}{\partial x_i} \right\rangle + \frac{1}{|Y_\mathbf{x}|} \int_{I_\mathbf{x}} u_i \frac{\partial \mathfrak{h}}{\partial x_i} \frac{1}{|\nabla \mathfrak{h}|} \, do(\mathbf{y}), \tag{6.56}$$

*where $do(\mathbf{y})$ represents the associated surface element.*

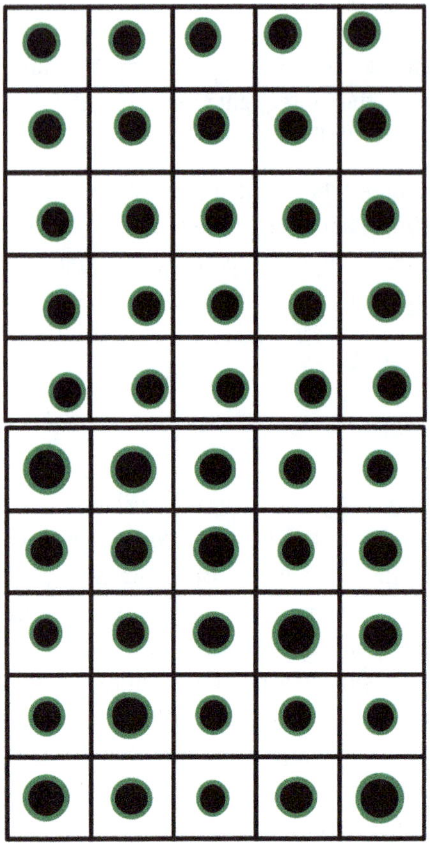

Figure 6.2: *Above:* Slowly varying pore-structure $\mathcal{D}_\epsilon^1$ over the macroscopic length scale composed of non-$Y$-periodic reference cells. Only the microscopic barycenter $\mathbf{y}_\mathbf{x} \in Y_{[\mathbf{x}/\epsilon]}$ of a circular obstacle $B_R([\mathbf{x}/\epsilon] + \mathbf{y}_{[\mathbf{x}/\epsilon]})$ is slightly shifted from cell to cell. *Below:* Spatially slowly varying porosity $p(\mathbf{x}) = \frac{|Y_\mathbf{x}^1|}{|Y_\mathbf{x}|}$.

The derivation relies upon integrating over a $\delta$-neighborhood around the interface $I_\mathbf{x}$, change of variables, and finally passing to the limit as $\delta \to 0$. For more detail, we refer the interested reader to [108, Appendix 3A].

**Remark 6.5.1.** 1. The above spatial averaging result (SAR) shows a formal character, since frequently in applications, we do not have such high regularities of functions that are to be averaged. However, the same situation appears if we apply the formal asymptotic two-scale expansion without discussing the existence and uniqueness for the arising sequence of problems. Even if we rigorously investigate the well-posedness of this sequence of problems, then we still do not have a rigorous proof of the homogenization limit as $\epsilon \to 0$. A convenient and constructive tool, which provides the form and rigorous justification of the upscaled/homogenized equations, is the so-called 2-scale convergence methods, see Chapter 10.

2. We note that the gradient $\nabla\mathfrak{h}$ is proportional to the unit normal $\mathbf{n}$ pointing outward of domain $Y_\mathbf{x}^1$. ◇

## 6.6 Hydraulic conductivity: Darcy's law

We consider the flow of an incompressible fluid through a porous medium under low Reynolds numbers. We assume that the flow already reached equilibrium such that we can neglect temporal effects. As in every upscaling/homogenization procedure, we first identify the physical problem on the microscale. There are two essential scenarios that we aim to investigate subsequently: (i) deterministic pore structure and (ii) random pore structure.

### 6.6.1 Stokes flow through deterministic pore spaces: formal derivation

The subsequent formal homogenization study is very similar to the conductivity problem, after suitable scalings of the involved physical quantities are identified.

#### 6.6.1.1 Physical and mathematical problem on the microscale

As in the case of conductivity (Section 6.3), we denote by $\mathcal{D}$ the macroscopic material that consists of the pore space $\mathcal{D}_\epsilon^1 := \mathcal{D} \cap \bigcup_{\mathbf{z}\in\mathbb{Z}^d} \epsilon(\mathbf{z} + Y^1)$ and the solid (obstacle) phase $\mathcal{D}_\epsilon^2 := \mathcal{D} \cap \bigcup_{\mathbf{z}\in\mathbb{Z}^d} \epsilon(\mathbf{z} + Y^2)$, where the sets $Y^1$ and $Y^2$ are the pore and solid phases, respectively, identified with respect to a characteristic periodic reference cell $Y = Y^1 \cup Y^2 = [0,1]^d$. Moreover, we denote the interface between the pore and solid phases by $I_\epsilon := \partial\mathcal{D}_\epsilon^1 \cap \partial\mathcal{D}_\epsilon^2$.

With the help of this notation, we can introduce the following microscopic problem,

$$\begin{cases} -\epsilon^2 \mu \Delta \mathbf{u}^\epsilon + \nabla p^\epsilon = \mathbf{0}\,, & \text{in } \mathcal{D}_\epsilon^1\,, \\ \operatorname{div} \mathbf{u}^\epsilon = 0\,, & \text{in } \mathcal{D}_\epsilon^1\,, \\ \mathbf{u}^\epsilon = \mathbf{0}\,, & \text{on } I_\epsilon\,, \end{cases} \tag{6.57}$$

where $\mu$ is the viscosity of the fluid. We note that the scaling of the viscosity by $\epsilon^2$ implies that the fluid velocity scales with an $\mathcal{O}(\epsilon^2)$ higher than the pressure. Hence this scaling allows us to follow the usual asymptotic two-scale expansion method (6.12) based on the three steps (TE1)–(TE3) for the fluid velocity and pressure.

#### 6.6.1.2 Effective macroscopic equations and derivation

Next, we derive the following widely used and fundamental result in porous media research.

**Upscaling result**

*The microscopic formulation based on the incompressible Stokes equations* (6.57), *where solutions* $(\mathbf{u}^\epsilon, p^\epsilon)$ *are sought in the fully resolved microscopic pore space* $\mathcal{D}_\epsilon^1$, *admits the following (homogenized) porous media approximation,*

$$\text{(Darcy's law)} \qquad \begin{cases} \mathbf{v} = -\frac{\hat{k}}{\mu}\nabla p^0(\mathbf{x}), & \text{in } \mathcal{D}, \\ \operatorname{div}\mathbf{v} = 0, & \text{in } \mathcal{D}, \end{cases} \tag{6.58}$$

*for the upscaled variables* $(\mathbf{v}, p^0)$ *defined on the whole macroscopic domain* $\mathcal{D}$. *The effective hydraulic permeability/conductivity tensor* $\hat{k} := \{\kappa_{ij}\}_{i,j=1}^d$ *is defined by,*

$$\kappa_{ij} := \frac{1}{|Y|}\int_Y w_i^j(\mathbf{y})\,d\mathbf{y}, \tag{6.59}$$

*where the microscopic flow fields* $\mathbf{w}^j$, $j = 1, 2, \ldots, d$, *are the solutions of the following reference cell problem,*

$$\begin{cases} \Delta_{\mathbf{y}}\mathbf{w}^j(\mathbf{y}) + \nabla_{\mathbf{y}}\pi^j(\mathbf{y}) = \mathbf{e}_j, & \text{in } Y^1, \\ \operatorname{div}_{\mathbf{y}}\mathbf{w}^j(\mathbf{y}) = 0, & \text{in } Y^1, \\ \mathbf{w}^j(\mathbf{y}) = 0, & \text{on } I_0, \end{cases} \tag{6.60}$$

*where* $\mathbf{e}_j$ *is the jth canonical basis element, and hence d is the spatial dimension.*

**Derivation of (6.58). Step 1: (TE1)** We make the ansatz,

$$\begin{aligned} \mathbf{u}^\epsilon(\mathbf{x}) &= \mathbf{u}^0(\mathbf{x},\mathbf{y}) + \epsilon\mathbf{u}^1(\mathbf{x},\mathbf{y}) + \epsilon^2\mathbf{u}^2(\mathbf{x},\mathbf{y}) + \cdots, \\ p^\epsilon(\mathbf{x}) &= p^0(\mathbf{x},\mathbf{y}) + \epsilon p^1(\mathbf{x},\mathbf{y}) + \epsilon^2 p^2(\mathbf{x},\mathbf{y}) + \cdots. \end{aligned} \tag{6.61}$$

The microscale $\mathbf{y} = \mathbf{x}/\epsilon \in Y$ is $Y$-periodic as in Section 6.3, where we introduced the formal homogenization by asymptotic two-scale expansion (6.12) for the conductivity (Laplace) equation.

After inserting (6.61) into (6.57), we end up with the following problem, where only the leading-order terms are stated,

$$\begin{aligned} &-\epsilon^{-1}\nabla_{\mathbf{y}}p^0(\mathbf{x},\mathbf{y}) \\ &\quad + \epsilon^0\{\mu\Delta_{\mathbf{y}}\mathbf{u}^0(\mathbf{x},\mathbf{y}) - (\nabla_{\mathbf{y}}p^1(\mathbf{x},\mathbf{y}) + \nabla_{\mathbf{x}}p^0(\mathbf{x},\mathbf{y}))\} \\ &\quad + \epsilon^1(\ldots) + \text{h.o.t.} = \mathbf{0}, \quad \text{in } Y^1, \end{aligned} \tag{6.62}$$

$$\begin{aligned} &\epsilon^{-1}\operatorname{div}_{\mathbf{y}}\mathbf{u}^0(\mathbf{x},\mathbf{y}) \\ &\quad + \epsilon^0(\operatorname{div}_{\mathbf{y}}\mathbf{u}^1(\mathbf{x},\mathbf{y}) + \operatorname{div}_{\mathbf{x}}\mathbf{u}^0(\mathbf{x},\mathbf{y})) \\ &\quad + \epsilon^1(\ldots) + \text{h.o.t.} = 0, \quad \text{in } Y^1, \end{aligned} \tag{6.63}$$

$$\begin{aligned} &\epsilon^0\mathbf{u}^0(\mathbf{x},\mathbf{y}) \\ &\quad + \epsilon^1\mathbf{u}^1(\mathbf{x},\mathbf{y}) + \text{h.o.t.} = \mathbf{0}, \quad \text{on } I_0, \end{aligned} \tag{6.64}$$

where $I_0 := \partial Y^1 \cap \partial Y^2$ is the solid-pore interface on the level of the reference cell $Y$, and h. o. t. denotes higher-order terms in $\epsilon$.

**Step 2: (TE2)** The lowest order problem in Step 1 (TE1) is obtained by comparing the coefficients of the terms of order $\epsilon^{-1}$, i. e., we obtain the equations,

$$\begin{aligned} \nabla_{\mathbf{y}} p^0(\mathbf{x}, \mathbf{y}) &= \mathbf{0}, \quad \text{in } Y^1, \\ \operatorname{div}_{\mathbf{y}} \mathbf{u}^0(\mathbf{x}, \mathbf{y}) &= 0, \quad \text{in } Y^1. \end{aligned} \tag{6.65}$$

This allows us only to conclude that the leading order pressure component $p^0$ is independent of the microscale $\mathbf{y}$, but not so for the leading-order fluid velocity $\mathbf{u}^0$.

At the next order $\mathcal{O}(\epsilon^0)$, we obtain,

$$\begin{aligned} -\mu\Delta_{\mathbf{y}}\mathbf{u}^0(\mathbf{x}, \mathbf{y}) + (\nabla_{\mathbf{y}} p^1(\mathbf{x}, \mathbf{y}) + \nabla_{\mathbf{x}} p^0(\mathbf{x}, \mathbf{y})) &= 0, \quad \text{in } Y^1, \\ \operatorname{div}_{\mathbf{y}} \mathbf{u}^1(\mathbf{x}, \mathbf{y}) + \operatorname{div}_{\mathbf{x}} \mathbf{u}^0(\mathbf{x}, \mathbf{y}) &= 0, \quad \text{in } Y^1. \end{aligned} \tag{6.66}$$

**Step 3: (TE3)** This step is not entirely standard and requires some physical intuition. If we make now the usual ansatz for the pressure $p^1$ in the asymptotic expansion and treat the fluid velocity $\mathbf{u}^0$ instead of $\mathbf{u}^1$ in the same way, which is motivated by its dependence on the microscale $\mathbf{y}$, i. e.,

$$\begin{aligned} p^1(\mathbf{x}, \mathbf{y}) &= -\pi^j(\mathbf{y})\nabla_{x_j} p^0(\mathbf{x}), \\ \mathbf{u}^0(\mathbf{x}, \mathbf{y}) &= -\frac{1}{\mu}\mathbf{w}^j(\mathbf{y})\nabla_{x_j} p^0(\mathbf{x}), \end{aligned} \tag{6.67}$$

then problem $(6.66)_1$ turns into,

$$\text{(Cell Problem)} \quad \begin{cases} -\Delta_{\mathbf{y}}\mathbf{w}^j(\mathbf{y}) - \nabla_{\mathbf{y}}\pi^j(\mathbf{y}) = -\mathbf{e}_j. & \text{in } Y^1, \\ \operatorname{div}_{\mathbf{y}} \mathbf{w}^j(\mathbf{y}) = 0. & \text{in } Y^1, \\ \mathbf{w}^j(\mathbf{y}) = 0. & \text{on } I_0. \end{cases} \tag{6.68}$$

Herewith, ansatz $(6.67)_2$ finally leads us to Darcy's law after integrating it over the pore space $Y^1$,

$$\mathbf{v}(\mathbf{x}) = \frac{|Y^1|}{|Y|} M_{Y^1}(\mathbf{u}^0(\mathbf{x}, \mathbf{y})) = -\frac{1}{\mu}\hat{\kappa}\nabla p^0(\mathbf{x}), \tag{6.69}$$

where the hydraulic conductivity/permeability tensor $\hat{\kappa} := \{\kappa_{ij}\}$ is defined by the volume averages,

$$\kappa_{ij} := M_{Y^1}(w_i^j(\mathbf{y})) := \frac{1}{|Y|}\int_{Y^1} w_i^j(\mathbf{y})\, d\mathbf{y}. \tag{6.70}$$

It remains to show that the effective Darcy velocity $\mathbf{v}$ is divergence free, that is, it satisfies $(6.58)_2$. To this end, we can make use of the still untouched equation $(6.66)_2$, which leads after integration over the pore space $Y^1$ and with the help of Gauss' theorem to,

$$\operatorname{div}_{\mathbf{x}} \mathbf{v}(\mathbf{x}) = \int_{\partial Y} \mathbf{n}\cdot\mathbf{u}^1(\mathbf{x}, \mathbf{y})\, do(\mathbf{y}) - \int_{I_0} \mathbf{n}\cdot\mathbf{u}^1(\mathbf{x}, \mathbf{y})\, do(\mathbf{y}) = 0, \tag{6.71}$$

where we used the $Y$-periodicity and the homogeneous boundary condition (6.64) at order $\mathcal{O}(\epsilon)$.

## 6.7 Convection diffusion problems

Let us consider reactive fluid flow in a periodic porous medium $\mathcal{D} = \mathcal{D}^1_\epsilon \cup \mathcal{D}^2_\epsilon$, where $\mathcal{D}^1_\epsilon$ is again the pore space, and $\mathcal{D}^2_\epsilon$ is the solid phase. We will again investigate transport of diffusing species in a low Reynolds number fluid described by the Stokes equations on the microscale, where the species of density $c^\epsilon(\mathbf{x}, t)$ undergo a *linear reaction* $R(c^\epsilon)$ in the pore space $\mathcal{D}^1_\epsilon$, i. e.,

$$\textbf{(Microscopic problem)}\quad \begin{cases} -\epsilon^2\mu\Delta\mathbf{u}^\epsilon + \nabla p^\epsilon = \mathbf{0}\,, & \text{in } \mathcal{D}^1_\epsilon\,,\\ \operatorname{div}\mathbf{u}^\epsilon = 0\,, & \text{in } \mathcal{D}^1_\epsilon\,,\\ \mathbf{u}^\epsilon = \mathbf{0}\,, & \text{on } I_\epsilon\,,\\ c^\epsilon_t - \operatorname{div}(\hat{\mathrm{D}}^\epsilon(\mathbf{x})\nabla c^\epsilon - \mathbf{u}^\epsilon c^\epsilon) = R(c^\epsilon)\,, & \text{in } \mathcal{D}^1_\epsilon\,,\\ \nabla_{\mathbf{n}} c^\epsilon = 0\,, & \text{on } I_\epsilon\,. \end{cases} \tag{6.72}$$

The microscopic reactive convection diffusion problem (6.72) is equipped with the no-slip boundary condition ($\mathbf{u}^\epsilon = \mathbf{0}$) and a no-penetration boundary condition ($\nabla_{\mathbf{n}} c^\epsilon = 0$, since the convective term in the flux vanishes due to the no-slip boundary condition).

Making the usual ansatz of the asymptotic two-scale expansion method for the solutions of (6.72), that is,

$$\begin{aligned} \mathbf{u}^\epsilon(\mathbf{x}) &= \mathbf{u}^0(\mathbf{x},\mathbf{y}) + \epsilon\mathbf{u}^1(\mathbf{x},\mathbf{y}) + \epsilon^2\mathbf{u}^2(\mathbf{x},\mathbf{y}) + \cdots\,,\\ c^\epsilon(\mathbf{x},t) &= c^0(\mathbf{x},\mathbf{y},t) + \epsilon c^1(\mathbf{x},\mathbf{y},t) + \epsilon^2 c^2(\mathbf{x},\mathbf{y},t) + \cdots\,, \end{aligned} \tag{6.73}$$

leads to the following:

**Upscaling Result**

*The solutions* $(\mathbf{u}^\epsilon, c^\epsilon)$ *of the microscopic porous media reaction–diffusion–convection problem* (6.72) *turns after upscaling/homogenization into the effective macroscopic solutions* $(\mathbf{v}, c^0)$ *of the following* upscaled/homogenized system,

$$\begin{cases} \mathbf{v} = -\frac{\hat{k}}{\mu}\nabla p^0\,, & \text{in } \mathcal{D}\,,\\ \operatorname{div}\mathbf{v} = 0\,, & \text{in } \mathcal{D}\,,\\ p_1 c^0_t - \operatorname{div}\big(\hat{\mathrm{D}}^0(\mathbf{x})\nabla c^0 + \mathbf{v}c^0\big) = p_1 R(c^0)\,, & \text{in } \mathcal{D}\,, \end{cases} \tag{6.74}$$

*where* $\hat{k} := -\frac{1}{\mu|Y|}\int_{Y^1} w^j_i(\mathbf{y})\,d\mathbf{y}$ *is defined as in* (6.59), *and the effective diffusion tensor* $\hat{\mathrm{D}}^0 = \{d^0_{ij}\}^D_{i,j=1}$ *is defined by,*

$$d^0_{ij} = \frac{1}{|Y|}\int_{Y^1}\left(d^1_{ik}\frac{\partial\xi^j(\mathbf{y})}{\partial y_k} + d^1_{kj}\right)d\mathbf{y}\,, \tag{6.75}$$

*for correctors $\xi^j \in H^1_\sharp(Y), j = 1, 2, \ldots, d$, solving the classical reference cell problem (6.11), i. e.,*

$$\begin{cases} -\frac{\partial}{\partial y_i}\left(d_{ij}(\mathbf{y})\left(\frac{\partial}{\partial y_j}\xi^k(\mathbf{y}) + \delta_{jk}\right)\right) = 0\,, & \text{in } Y, \\ \xi^k(\mathbf{y}) \text{ is } Y\text{-periodic, and } M_Y(\xi^k) = 0\,. \end{cases} \tag{6.76}$$

*As previously, in (6.75) (or in (6.11)), $p_1 = \frac{|Y^1|}{|Y|}$ denotes the porosity.*

**Derivation of (6.74).** The derivation of Darcy's law $(6.74)_1$–$(6.74)_2$ is the same as in Section 6.6.1. Hence, we only demonstrate here the asymptotic two-scale expansion for the convective reaction–diffusion equation $(6.72)_4$–$(6.72)_5$.

**Step 1. (TE1)** We insert the ansatz $(6.73)_2$ into $(6.72)_4$–$(6.72)_5$.

**Step 2. (TE2)** Collecting terms of equal order in $\epsilon$ gives,

$$\mathcal{O}(\epsilon^{-2}): \quad \begin{cases} -\mathrm{div}_{\mathbf{y}}(\hat{\mathrm{D}}(\mathbf{y})\nabla_{\mathbf{y}}c^0(\mathbf{x},\mathbf{y},t)) = 0\,, \quad \text{in } Y^1\,, \\ c^0 \quad Y\text{-periodic in } \mathbf{y}\,, \end{cases} \tag{6.77}$$

$$\mathcal{O}(\epsilon^{-1}): \quad \begin{cases} -\mathrm{div}_{\mathbf{y}}(\hat{\mathrm{D}}(\mathbf{y})\nabla_{\mathbf{y}}c^1(\mathbf{x},\mathbf{y},t)) = \mathrm{div}_{\mathbf{x}}(\hat{\mathrm{D}}(\mathbf{y})\nabla_{\mathbf{y}}c^0(\mathbf{x},\mathbf{y},t)) \\ \qquad\qquad + \mathrm{div}_{\mathbf{y}}(\hat{\mathrm{D}}(\mathbf{y})\nabla_{\mathbf{x}}c^0(\mathbf{x},\mathbf{y},t)) \\ \qquad\qquad + \mathrm{div}_{\mathbf{y}}(\mathbf{u}^0(\mathbf{x},\mathbf{y})c^0(\mathbf{x},\mathbf{y},t))\,, \quad \text{in } Y^1\,, \\ c^1 \quad Y\text{-periodic in } \mathbf{y}\,, \end{cases} \tag{6.78}$$

$$\mathcal{O}(\epsilon^{0}): \quad \begin{cases} -\mathrm{div}_{\mathbf{y}}(\hat{\mathrm{D}}(\mathbf{y})\nabla_{\mathbf{y}}c^2(\mathbf{x},\mathbf{y},t)) = R(c^0(\mathbf{x},t)) + \mathrm{div}_{\mathbf{x}}(\hat{\mathrm{D}}(\mathbf{y})\nabla_{\mathbf{y}}c^1(\mathbf{x},\mathbf{y},t)) \\ \qquad\qquad + \mathrm{div}_{\mathbf{y}}(\hat{\mathrm{D}}(\mathbf{y})\nabla_{\mathbf{x}}c^1(\mathbf{x},\mathbf{y},t)) \\ \qquad\qquad + \mathrm{div}_{\mathbf{x}}(\mathbf{u}^0c^0(\mathbf{x},\mathbf{y},t)) \\ \qquad\qquad + \mathrm{div}_{\mathbf{x}}(\hat{\mathrm{D}}(\mathbf{y})\nabla_{\mathbf{x}}c^0(\mathbf{x},\mathbf{y},t)) + c^0_t(\mathbf{x},\mathbf{y},t)\,, \quad \text{in } Y^1 \\ c^2 \quad Y\text{-periodic in } \mathbf{y}\,. \end{cases} \tag{6.79}$$

**Step 3. (TE3)** Clearly, at leading order (6.77), we can conclude that $c^0(\mathbf{x}, t)$ is independent of the microscale $\mathbf{y} \in Y^1$.

For the next order (6.78), we recall that thanks to $(6.65)_2$, $\mathrm{div}_{\mathbf{y}}\mathbf{u}^0(\mathbf{x},\mathbf{y}) = 0$ in $Y^1$. With Gauss' theorem, the no-slip boundary condition ($\mathbf{u}^\epsilon = \mathbf{0}$ on $I_\epsilon$, i. e., $\mathbf{u}^0 = \mathbf{0}$ on $I_0$), and the $Y$-periodicity of $\mathbf{u}^\epsilon$ and $c^\epsilon$, we can deduce that,

$$\mathrm{div}_{\mathbf{y}}(\mathbf{u}^0(\mathbf{x},\mathbf{y})c^0(\mathbf{x},\mathbf{y},t)) = 0\,. \tag{6.80}$$

Hence, the $\mathcal{O}(\epsilon^{-1})$ is the same as that for the conductivity problem (6.14), and therefore the same reference cell problem (6.11) holds here in the convective reaction–diffusion case (6.72). Finally, the homogenized/upscaled convection–diffusion equation is obtained by applying the solvability constraint (see Theorem 6.3.2) to equation (6.79). That is, using the test function $\mathbf{v} = 1$ in,

$$
\begin{aligned}
(R(c^0(\mathbf{x},t)), v(\mathbf{y}))_{Y^1} = {} & (\mathrm{div}_{\mathbf{x}}(\hat{\mathrm{D}}(\mathbf{y})(\nabla_{\mathbf{y}} c^1(\mathbf{x},\mathbf{y},t) + \nabla_{\mathbf{x}} c^0(\mathbf{x},\mathbf{y},t))), v(\mathbf{y}))_{Y^1} \\
& + (\mathrm{div}_{\mathbf{x}}(\mathbf{u}^0(\mathbf{x},\mathbf{y}) c^0(\mathbf{x},t)), v(\mathbf{y}))_{Y^1} \\
& + (\hat{\mathrm{D}}(\mathbf{y})\nabla_{\mathbf{x}} c^1(\mathbf{x},\mathbf{y},t), \nabla_{\mathbf{y}} v(\mathbf{y}))_{Y^1} \\
& + (c_t^0(\mathbf{x},t), v(\mathbf{y}))_{Y^1}\,, 
\end{aligned}
\tag{6.81}
$$

leads to the homogenized/upscaled reaction–diffusion equation $(6.74)_3$.

## 6.8 CHeSs showing fractal designs: theory and examples

The natural emergence of fractals is exemplified in a wide range of different contexts such as trees, plants, family trees of rabbits, animals' pelts, shells, and rocks, see Figure 6.3, for instance. These examples motivate our subsequent considerations of fundamental concepts and principles governing fractals as a naturally occurring subset of CHeSs. This wide spread presence of fractal structures in nature (see [6, 104, 105, 55, 150, 176]) seems unlikely to be a coincidence. In fact, we can observe, for instance, sub- and upper-surface fractal structures in the growth of plants and trees. This immediately leads to the central question why do plants or trees develop such almost symmetric subsurface fractals called roots and upper-surface fractals called branches and leaves?

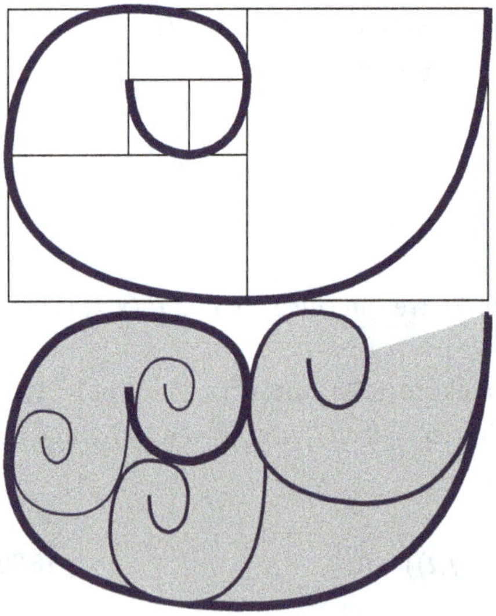

Figure 6.3: *Top:* Design of a *spiral* using Fibonacci numbers, i. e., the numbers recursively defined by $f_{n+1} = f_n + f_{n-1}$ with $f_0 = 0$ and $f_1 = 1$. In fact, the sequence $\{f_n\}$ is used to define the edge length of squares whose corners represent points of an outward going spiral where the origin is just a point corresponding to $f_0 = 0$. *Bottom: Fractal shell* built with scaled "Fibonacci spirals". The scaling is here governed by the growth of the Fibonacci numbers.

To answer this question, we follow a logical deduction based on physical, biological, and economic aspects: from a biological (genetic) point of view, fractals emerge as self-similar replicating structures. Hence, the relevant evolutionary optimized genetic subcode, which defines the essential unit pattern of a fractal, can be repeatedly used, and therefore, this represents also a first economic motivation. The physical contribution to our answer is that the time-scale for optimizing shapes to certain outer/environmental conditions are very long and hence not accessible to the lifetime of a particular plant/tree. Hence, it is easier to grow self-similar (structures) to increase energy and nutrient uptake for further growth. Finally, a main economic contribution to our motivation relies on the assumption that a plant/tree aims to optimize its nutrient/energy uptake (e.g., sun, water, and minerals) to grow as much as possible (to achieve as high reproduction rates as possible). These biological examples demonstrate nicely *fractal growth* as graphically visualized by the Sierpiński triangle on the bottom of Figure 6.4.

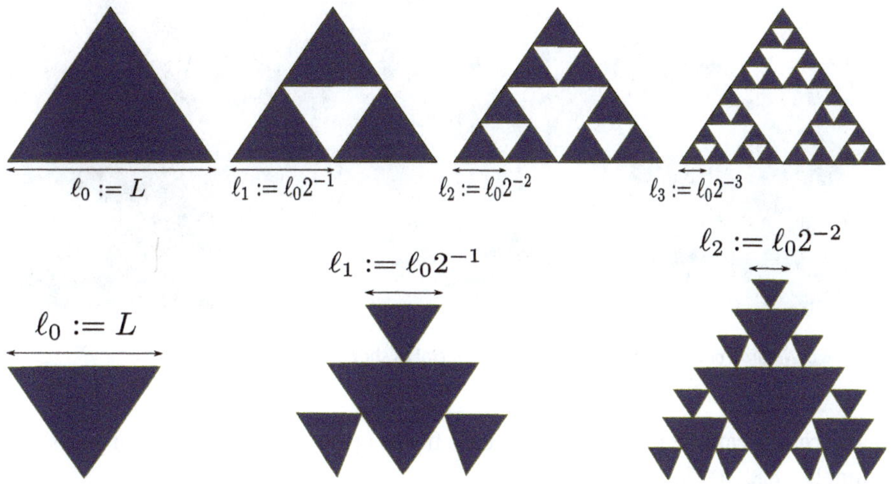


Figure 6.4: *Top:* Fractal decay depicted by an area-shrinking Sierpiński triangle. *Bottom:* Fractal growth exemplified by an area-growing Sierpiński triangle.

Conversely, a naturally formed *decaying fractal* can be discovered in rocks, which can be identified as a *natural porous medium* on various scales, see [89] and Figure 6.5. Of course, decay in rocks can happen for various reasons such as climate change with its abrupt temperature changes and its local shifts in humidity levels, for instance. Note that we generally distinguish three different types of rocks such as *sedimentary*, *igneous*, and *metamorphic rocks*. The sedimentary rocks are either from pieces of existing rock or organic material, which leads to coal (which is classified as rock). The *igneous rocks* are generally the product of cooling and solidifying molten hot material, e.g., volcanic rocks. Finally, *metamorphic rocks* have been transformed by very high pressure or heat, e.g., the transformation of *igneous granite* into *gneiss*.

Fractals can also appear in the form of graphs/networks, see [131], for instance. Of course, these graphs and networks can emerge naturally. However, they can also be the result of an analytical investigation such as the estimation of the likelihood of the ap-

Figure 6.5: *Natural rock:* An example of a decaying fractal as shown in [89].

pearance of a catastrophe in a certain time frame (generally, on the scale of years), see Example 6.8.8.

The above physical, biological, and economic motivations for fractal growth allow us to conclude that the fractal shapes of plants and trees represent catalytic structures as economic building blocks for a sustained, continued, and increased *catalytic activity*. In fact, the purpose of subsequent considerations and investigations is to better consider the fractal characteristics and their effect on *catalytic activities*.

### 6.8.1 Characterizations of fractals and generalizations to CHeSs

Almost every discussion of fractals includes a characterization such as the *fractal dimension*. Examples of such dimensions are the *Minkowski* or *Hausdorff dimension*. Let us first look at a more practical and easily accessible dimension.

**Box-counting dimension**

**Definition 6.8.1.** Consider an arbitrary bounded nonempty set $B \subset \mathbb{R}^n$. Count the smallest number of sets $U_r \subset \mathbb{R}^n$ with diameter $r = \operatorname{diam} U_r := \sup\{|x - y| \mid x, y \in U_r\}$ that fully cover $B$. Let us denote this number by $n_r(B)$.

This allows us to introduce two kinds of box-counting dimensions, the *lower box-counting dimension*

$$\underline{\dim}_b B := \liminf_{r\to 0} \frac{\log n_r(B)}{-\log r} \tag{6.82}$$

and the *upper box-counting dimension*,

$$\overline{\dim}_b B := \limsup_{r\to 0} \frac{\log n_r(B)}{-\log r}\,. \tag{6.83}$$

If $\underline{\dim}_b B = \overline{\dim}_b B$, then this common value is called the *box-counting dimension* of *B* and denoted by $\dim_b B$.

**Remark 6.8.2** (Set $B$ and CHeSs). For some CHeSs, the fractal set $B \subset G$ can represent a scaled self-similar subgraph of the complex system/graph $G$ or immediately the entire set $G$. If $G$ appears as a continuous set on the macroscale without immediately showing a graph-like characteristic or a clearly visible structure, then it might be necessary that we first need to zoom in, to discover the design of the system $G$. In this case, $G$ hides its underlying *microscopic* structure and principles on the *macroscale*. Naturally, such structures allow us to distinguish the different elements in $G$, which become vertices/nodes of the graph, and the associated links are interactions or exchange of material, information, or energy for instance. ◇

Let us first apply this box-counting definition to the examples presented in the introductory Section 6.8.

**Example 6.8.3** (Fractal dimension of growing and decaying Sierpiński triangles). Let us compute the box-counting dimension of the decaying and growing Sierpiński triangles shown in Figure 6.4. For the decaying Sierpinski triangle $S_{\text{dec}}$, we can identify the number of boxes $\eta_n$ and the box-diameter $\ell_n$ at fractal level $n$ by,

$$\begin{aligned} \eta_n &= 3^n\,, \\ \ell_n &= \ell_0 2^{-n}\,. \end{aligned} \tag{6.84}$$

Setting $\ell_0 = 1$ for simplicity, this immediately leads to the following well-known box-counting dimension (e. g., [55]),

$$\dim_b S_{\text{dec}} = \lim_{n\to\infty} \frac{\log(3^n)}{-\log(2^{-n})} = \frac{\log(3)}{\log(2)}\,. \tag{6.85}$$

Similarly, the growing Sierpiński triangle $S_{\text{grw}}$ shows the following box-counter $\eta_n$ and diameter $\ell_n$,

$$\begin{aligned} \eta_n &= \begin{cases} 1\,, & \text{if } n = 0\,, \\ 1 + 3^n\,, & \text{if } n \geq 1\,, \end{cases} \\ \ell_n &= \ell_0 2^{-n}\,. \end{aligned} \tag{6.86}$$

The associated fractal growth scenario requires us to consider the following lower and upper bounds,

$$a := \liminf_{n\to\infty} \frac{\log(3^n)}{-\log(2^{-n})} \leq \dim_b S_{\text{grw}} := \lim_{n\to\infty} \frac{\log(1+3^n)}{-\log(2^{-n})} \leq \limsup_{n\to\infty} \frac{\log(3^{n+1})}{-\log(2^{-n})} =: b\,. \quad (6.87)$$

For the lower and upper bounds, we immediately obtain the following results,

$$\begin{aligned} a &= \liminf_{n\to\infty} \frac{\log(3^n)}{-\log(2^{-n})} = \frac{\log(3)}{\log(2)}\,, \\ b &= \limsup_{n\to\infty} \frac{\log(3^{n+1})}{-\log(2^{-n})} = \limsup_{n\to\infty}\left(1+\frac{1}{n}\right)\frac{\log(3)}{\log(2)} = \frac{\log(3)}{\log(2)}\,, \end{aligned} \quad (6.88)$$

so that we finally obtain the fractal box-counting dimension $\dim_b S_{\text{grw}} = \frac{\log(3)}{\log(2)}$.

These results for the growing and decaying Sierpiński triangles allow us to conclude that the concept of *fractal dimension* does not account for the underlying constructive and destructive nature that builds the fractal. The reason is that by definition the fractal dimension is only obtained in its ideal, complete design, that is, in the limit of infinite fractal depth/level by passing to the limit as $n \to \infty$ and not for a finite $n \in \mathbb{N}$ being part of the construction/destruction process. ◇

Unfortunately, the above intuitive box-counting dimension relies on boxes showing the same size/diameter. A generalization represents the *Hausdorff dimension*, which has been developed already before the box-counting dimension. In fact, the definition of the Hausdorff dimension depends on the following measure.

**δ-dimensional Hausdorff measure**

**Definition 6.8.4.** Let $U$ be an arbitrary set. For the parameters $\delta > 0$ and $r > 0$, we call the measure defined by,

$$\mathcal{H}_r^\delta(U) := \inf\left\{ \sum_i \left(\operatorname{diam}(V_i)\right)^\delta \,\middle|\, \{V_i\}_i \text{ such that } \operatorname{diam}(V_i) \leq r \text{ and } U = \bigcup_i V_i \right\}, \quad (6.89)$$

the *r-approximate δ-dimensional Hausdorff measure* of $U$. Its limit as $r \to 0$, that is,

$$\mathcal{H}^\delta(U) := \lim_{r\to 0} \mathcal{H}_r^\delta(U), \quad (6.90)$$

is called the *δ-dimensional Hausdorff measure*.

To get an understanding of the subsequent definition of *Hausdorff dimension*, let us first look at the following classical and simple example, see [56, 131], for instance.

**Example 6.8.5** (Hausdorff measure and dimension). Consider the unit square $U := [0,1]\times[0,1]$. Recall that $\lfloor x \rfloor$ denotes the lower integer part of $x$, i. e., $\lfloor x \rfloor \leq x$. In the sense of the $r$-approximate $\delta$-dimensional Hausdorff measure $\mathcal{H}_r^\delta(U)$, we introduce coverings $\{V_i\}_i$ with $V_i := x_i + [\rho_i, \rho_i]$ for suitable $x_i \in \mathbb{R}^2$ and $0 \leq \rho_i \leq r$. Assuming for simplicity that the uniform cover $V_i = V$ with $\mathrm{diam}(V) = r$ represents the necessary infimum in (6.89), we can immediately establish the following limit property,

$$\mathcal{H}_r^\delta(U) = \frac{1}{r^2}(\mathrm{diam}(V_i))^\delta \leq \frac{1}{r^2}(\sqrt{2}r)^\delta = \sqrt{2}r^{\delta-2} \overset{(r\to 0)}{\longrightarrow} \begin{cases} \infty, & \delta < 2, \\ 1, & \delta = 2, \\ 0, & \delta > 2. \end{cases} \tag{6.91}$$

Despite the simplification of our assumption, this formally motivates that a meaningful Hausdorff dimension $\delta$ can be identified as the smallest (not necessarily admitted) $\delta$, for which the Hausdorff measure is zero. ◇

As a consequence, we can state the following:

**Hausdorff dimension**

**Definition 6.8.6.** The Hausdorff dimension $\delta$ of a set $U$ is defined by,

$$\delta := \inf\{d \in \mathbb{R} \,|\, \mathcal{H}^d(U) = 0\}. \tag{6.92}$$

**Remark 6.8.7** (Box-counting versus Hausdorff dimension). Note that in practice, the box-counting dimension is generally used due to its simplicity compared to the Hausdorff dimension, which requires to take the often rather complicated infimum of all possible $r$-coverings $\{V_i\}_i$, that is, coverings $\{V_i\}_i$ such that $0 \leq \mathrm{diam}(V_i) \leq r$. Historically, the Hausdorff dimension has been introduced before the box-counting dimension despite its definition being more involved. ◇

Next, we consider ways of finding meaningful expressions for the *fractal dimension of graphs/networks*. Recall that the purpose of fractal dimension is to capture the fraction of the area of a fractal from its enclosing domain extended up to its outer boundary, which itself can be identified as the smallest possible path enclosing the entire fractal. As a consequence, from this kind of dimensional concept it immediately becomes apparent that we can introduce various types of fractal dimensions by extending the classical geometric realization of a classical fractal by general physical, chemical, biological, etc. realizations showing self-similar and scaled repetitions of elementary patterns, see [131] for possible fractal dimensions of networks/graphs, for instance. Before recalling some elementary concepts from [131], we discuss immediate attempts in the subsequent example of a growing binary risk graph (i. e., fractal growth).

**Example 6.8.8** (Fractal growth of risk and extinction dimension). Suppose that a catastrophic event is expected to happen during a year with probability $p \in (0, 1)$. This immediately leads to the central question of how many years $n \in \mathbb{N}$ need to pass such that the probability estimate $p_{\mathrm{est}}(n)$ for the catastrophe to happen satisfies,

$$p_{\mathrm{est}}(n) > 0.5. \tag{6.93}$$

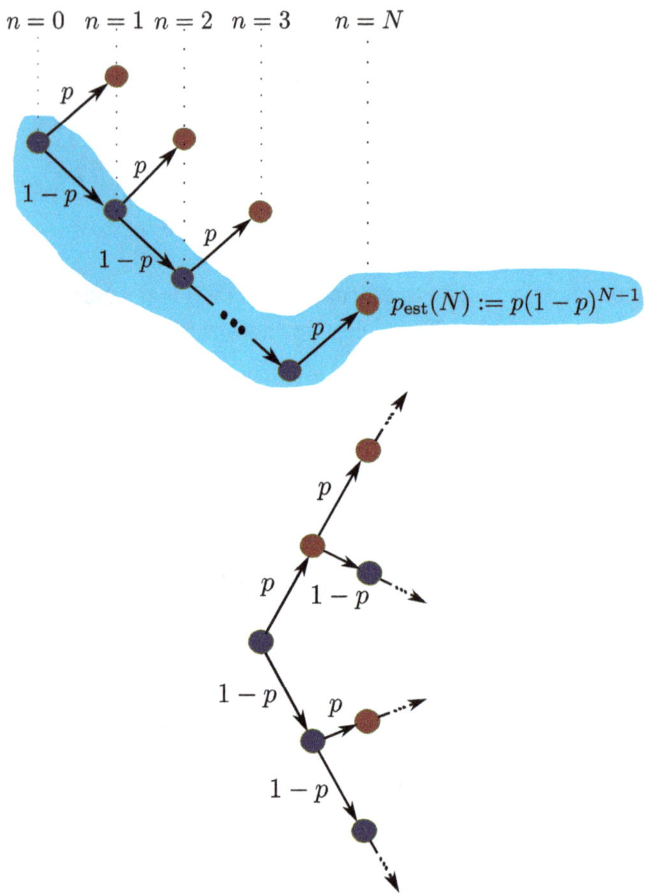


**Figure 6.6:** *Top:* Risk graph neglecting repeating catastrophes as a consequence of an extinction event such as a Nuclear (World) War. *Bottom:* Complete risk graph with repeating catastrophes, i. e., no extinction events.

To answer this question, we first need to derive a reasonable/*explainable*[2] expression estimating the risk of a catastrophe happening in year $n$. Our assumption of this risk process allows us immediately to conclude that in year $n = 1$, the catastrophe either happens with probability $p$ or it does not with probability $1 - p$. The exactly same conclusion holds for year $n = 2$, such that we can immediately obtain the risk graph shown in Figure 6.6. Hence, we can compute the following probability estimate for an extinction catastrophe to happen in year $n$,

$$\begin{aligned} p_{\text{est}}(n) &= p(1 + (1-p) + \cdots + (1-p)^{n-1}) \\ &= p\frac{1-(1-p)^n}{1-(1-p)} \\ &= 1-(1-p)^n . \end{aligned} \tag{6.94}$$

2 Note that the use of *explainable* refers here to Deutsch's philosophically motivated nature of science, see [42].

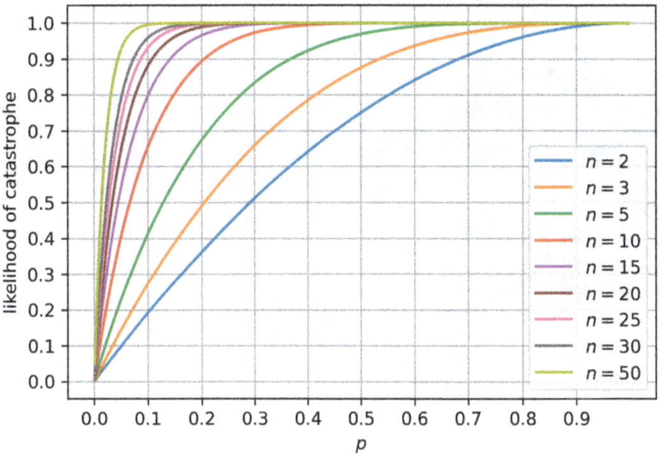


**Figure 6.7:** *Risk graph:* Behavior of the estimated risk of an extinction catastrophe to happen with respect to the probability $p$ of a catastrophe happening during one year and in dependence of a duration in years $n$.

The behavior of this risk estimate (6.94) is visualized in Figure 6.7. The critical year $n_c$, from which onward we have to expect a catastrophic event (that is, $n_c \in \mathbb{N}$ derived from $p_{\text{est}} = 0.5$), then reads,

$$n_c(p) := \left\lfloor \frac{\log(0.5)}{\log(1-p)} \right\rfloor + 1 . \tag{6.95}$$

Hence, for the fractal growth of risk up to year $n = N := 150$ depicted by the graph on the top in Figure 6.8, we propose the following probability weighted *fractal risk dimension*,

$$\begin{aligned} \lim_{n\to\infty} d_{G_n}(p) &:= \lim_{n\to\infty} \frac{\log(\eta_n^w)}{-\log(\ell_n^w)} \\ &= \lim_{n\to\infty} \frac{\frac{1}{n-1}\log(p+n-1)}{\frac{-1}{n-1}\log p - \log(1-p)} = 0 , \end{aligned} \tag{6.96}$$

which is a result of l'Hôpital's rule of the weighted box-number $\eta_n^w$ and of the weighted, longest shortest path belonging to a disastrous event $\ell_n^w$, that is,

$$\begin{aligned} \eta_n^w &:= np + (n-1)(1-p) , \\ \ell_n^w &:= p(1-p)^{n-1} . \end{aligned} \tag{6.97}$$

For finite $n \in \mathbb{N}$, $d_{G_n}(p)$ defines the total number of links with their associated weights divided by the intrinsic length of the longest, shortest, weighted path realizing the catastrophic event to be estimated. As plotted in Figure 6.8, $d_{G_n}(p)$ accounts for the essential properties of a dimension, i. e., it is a nonnegative number characterizing how much of a specific quantity is present in the specific system. As a consequence, the maximal $d_{G_n}(p)$ represents the largest possible weighted risk network for the smallest weighted path to

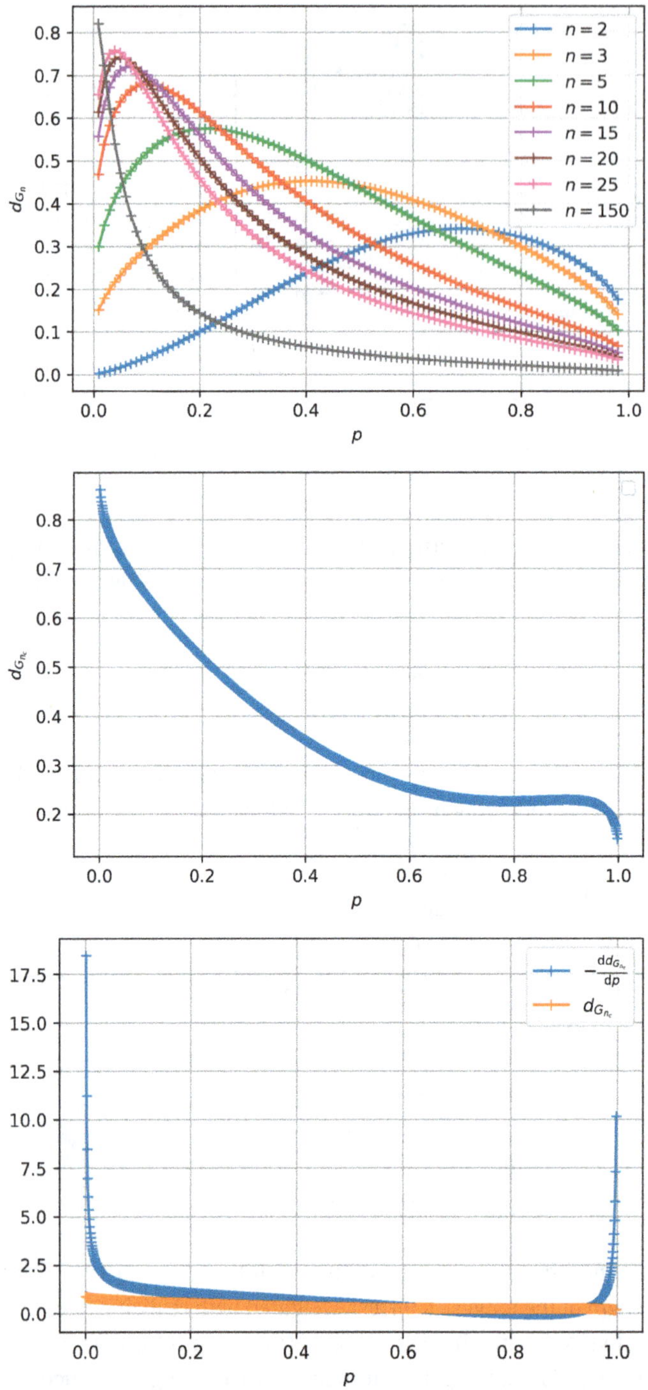


**Figure 6.8:** *Fractal growth of risk and fractal dimension. Top:* Fractal dimension $d_{G_n}$ of the risk graph $G_n(p)$, see (6.96). *Middle:* Critical fractal dimension, that is, $d_{G_n}(p)$ for the critical number of years $n_c(p)$ implying the estimated risk to satisfy $p_{\mathrm{est}}(n_c) > 0.5$. This critical dimension tells us how quickly the risk graph shrinks (i. e., the time to the catastrophe shrinks) under growing likelihood of the catastrophe to happen. *Bottom:* Critical fractal dimension and its negative derivative, which implies a strong decay for $p$ toward zero and for $p$ toward one.

the catastrophe. Hence, the maxima in Figure 6.8 tell us, for which probability $p$ the risk graph is the largest for a particular number of years $n$.[3]

Finally, we can introduce further meaningful fractal dimensions such as $d_{ext}$, which we call *disaster dimension*. It represents the fraction of the disastrous events in a complete risk network that also allows for disasters to repeat. The computation of $d_{ext}$ relies on the following two essential quantities: the total number $n_{tot} = 2^n$ and the number of nodes $n_{ext} = 2(n-1)+1$ defining extinction events, both up to year $n = n_c(p)$. This allows us to immediately define the so-called disaster dimension $d_{ext}$ by,

$$d_{ext} := \frac{\log(n_{ext})}{\log(n_{tot})} = \frac{\log(2\lfloor\frac{\log(0.5)}{\log(1-p)}\rfloor + 1)}{(\lfloor\frac{\log(0.5)}{\log(1-p)}\rfloor + 1)\log(2)}, \tag{6.99}$$

where we used (6.95). In Figure 6.9, we can see for which yearly likelihood $p$ of a disaster to happen, the disastrous event will occupy the largest fraction of its binary risk graph. ◇

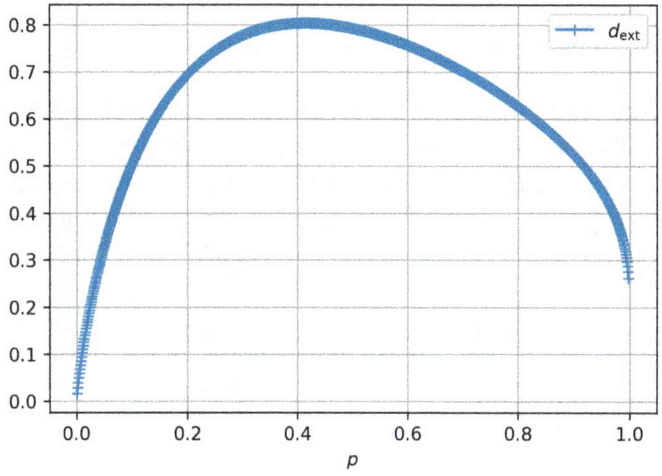


*Figure 6.9: Fractal dimension of disasters in a repeating disaster-network, i. e., the disaster dimension $d_{ext}$:* For specific likelihoods $0 < p < 1$ of a disaster happening during a year, there exists, for a likelihood close to $p$ = 0.4, a maximal fraction $d_{ext}$ of the binary risk graph that is occupied by the disastrous event.

Motivated by discussion in Example 6.8.8, we look at Rosenberg's systematic approach in [131], where he proposes reasonable and convincing concepts and meaningful

3 By looking at ways to account for the probability $p$ in the dimension, we neglect simpler discussions about straightforward box-counting dimensions such as,

$$\text{box}_{G_n}(p) := \frac{\sharp\text{nodes}}{\sharp\text{years}} := \frac{\log(2n+1)}{\log(n)} \overset{(n\to\infty)}{\to} 1, \tag{6.98}$$

where the limit follows by l'Hôpital's rule.

types of fractal dimensions of graphs/networks. In fact, based on our thermodynamic and graph-theoretic approach to state an as general as possible definition of CHeSs (see Definition 2.4.1), general and systematic definitions of fractal dimensions for networks/graphs will immediately represent useful and practically relevant dimensions applicable to CHeSs.

### 6.8.2 Investigating fractal CHeSs

We investigate CHeSs representing fractal structures with a wide range of beneficial properties in applications such as electrochemistry (e. g., electrodes of batteries or fuel cells), biology and plants (e. g., the number of petals or arrangement of leaves), and nautilus shells (e. g., its logarithmic spirals), which are generally approximated by the Fibonacci numbers, see Figure 6.3 and [157]. Here, we want to present an approach for analysing the influence of porosity, surface area, and other geometric properties on the current–voltage characteristics of cathode materials.

### 6.8.3 Cross-shaped fractals: Surface areas, porosities, and capacities in CHeSs

**Porosity.** To develop a general framework and to obtain a first understanding of fractal designs, which generally represent multiple scaled copies of a basic/unit shape arranged in a particular form, we restrict ourselves to very simple unit shapes. That is, for the subsequent definition of fractals, we use a scaling parameter $\lambda > 0$, which will relate the macro- and microscales $L_n$ and $\ell_n$, respectively, for $n \in \mathbb{N}$ as follows,

$$\ell_n = \lambda L_n . \tag{6.100}$$

Let $C_n$ be a cross-shaped fractal of order $n$, i. e., the fractal $C_0$ is obtained after removing a cross of width $\ell_0$ from a square of length $L_0$ at order $n = 0$ and shows four remaining squares of length,

$$L_1 := (L_0 - \ell_0)/2 < L_0 .$$

This immediately leads to the next order $n = 1$ by repeating the above procedure with each of the four smaller squares and resulting in the fractal $C_1$ showing now 16 squares of length,

$$L_2 := (L_1 - \ell_1)/2 < L_1 < L_0 .$$

This procedure leads by induction to the following general formula for the area of the resulting $n$th-order cross-shaped fractal pore space $A_n$,

$$A_n(\lambda) := \lambda(2-\lambda)L_0^2 \sum_{k=0}^{n} (1-\lambda)^{2k}\,. \tag{6.101}$$

Interestingly, the above formula for the area of the 2D-Cantor fractal shows the specific structure of a geometric series. Introducing the variables $q(\lambda) := (1-\lambda)^2$ and $a_0(\lambda) := \lambda(2-\lambda)L_0^2$, we can rewrite (6.101) as follows:

$$A_n(\lambda) := a_0(\lambda)\frac{1-(q(\lambda))^{n+1}}{1-q(\lambda)} \quad \text{for } 0 < \lambda < 1\,. \tag{6.102}$$

We can implement this area formula in Python as follows,

```
def areaA(lmbd,L0,n):

    # Using geometric series: compact formula
    coeff = lmbd*(2-lmbd)*L0**2
    res = coeff*(1-(1-lmbd)**(2*(n+1)))/(1-(1-lmbd)**2)

    return res
```

To obtain an understanding of this area formula for various parameter settings, we can plot the above function for different parameters with the help of the following Python method:

```
fig = plt.figure(dpi=250)
pli =[]
pla = []
for nv in range(10):

    avVSlmbd = areaC(lmbd,L0,nv)
    p = plt.plot(lmbd,avVSlmbd,"*",linestyle="solid")
    pli.append(p[0])
    pla.append("n="+str(nv))

plt.grid()
plt.xlabel(r"$\lambda$")
plt.ylabel(r"$A_n(\lambda)$")
plt.legend(pli,pla,loc="best")
plt.show()
```

In Figure 6.11, we can see how the fractal area of the pore space $A_n(\lambda)$ decreases with increasing the fractal level $n$.

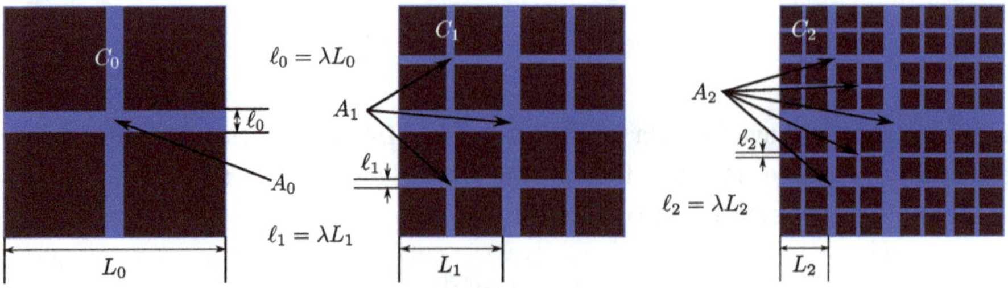


Figure 6.10: *Parameterized 2D Cantor fractal:* Cantor dust $C_n(\lambda)$.

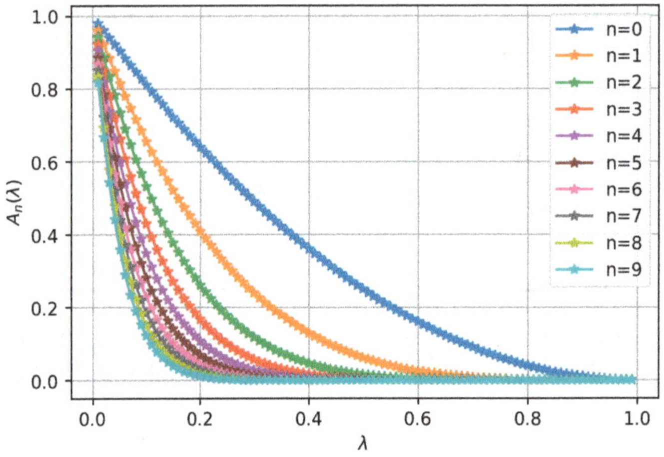


Figure 6.11: *Cross-shaped fractal:* Area $A_n(\lambda)$ versus $\lambda$ and $n$.

**Capacity $C_n(\lambda)$ of the Cantor dust.** The so-called Cantor dust $C_n(\lambda)$ (solid phase) in dependence of the scaling parameter $\lambda$ (see Fig. 6.10) can easily be derived to satisfy the equation,

$$C_n(\lambda) := L_0^2 - A_n(\lambda)\,. \tag{6.103}$$

**The surface area/length of the Cantor dust $C_n(\lambda)$.** For the same fractal $C_n$, let us also compute the surface area $\Gamma_n$ by induction as follows,

$$\begin{aligned}
\Gamma_0 &= 2\cdot 4\cdot L_1\,,\\
\Gamma_1 &= 2\cdot 4^2\cdot L_2\,,\\
&\;\vdots\\
\Gamma_n &= 2\cdot 4^{n+1}\cdot L_{n+1}\,,
\end{aligned} \tag{6.104}$$

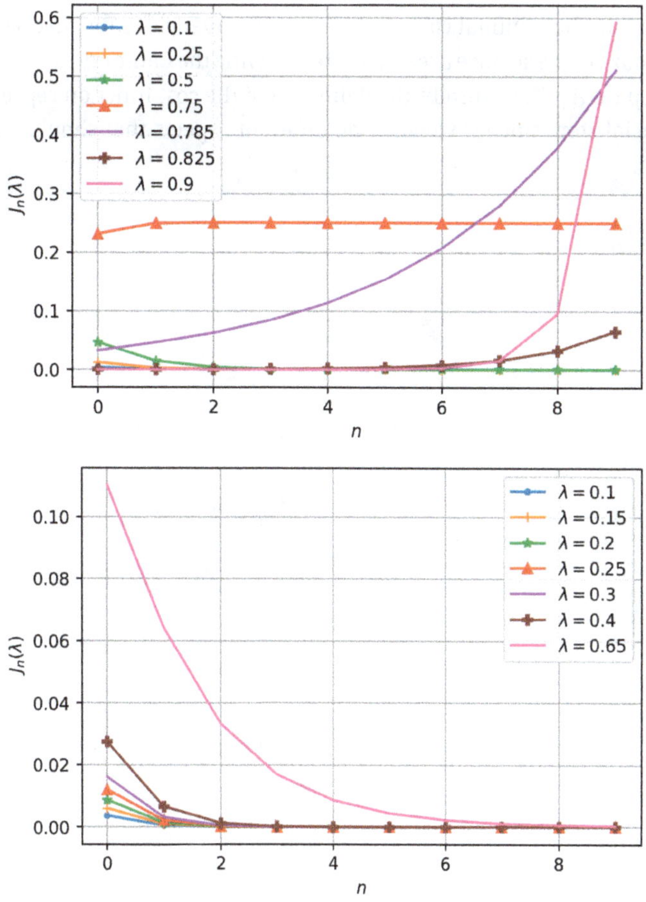


**Figure 6.12:** *Optimal fractal design. Top:* The minimal cost function $J_n(\lambda)$, see (6.106). Note that the plots with $\lambda = 0.785, 0.825, 0.9$ are scaled by the factors $\frac{1}{10}$, $\frac{1}{4.8e3}$, $\frac{1}{3.84e7}$, respectively. *Bottom:* Small $\lambda$-scalings where no scaling factors on the graph have been applied. Here, we can see that for small $\lambda$ (i. e., $\lambda \leq 0.4$), it is recommended to choose the fractal depth/level around $n = 4$ for minimal cost/optimal design. Moreover, for $\lambda$ growing around its mean $\lambda = 0.5$, it is favorable to slowly increase the fractal depth, see the curve for $\lambda = 0.65$.

which leads with the formula $L_n = ((1-\lambda)/2)^{n+1} L_0$ to the surface area,

$$\Gamma_n(\lambda) = 2 \cdot 4^{n+1} \cdot ((1-\lambda)/2)^{n+1} \cdot L_0 . \tag{6.105}$$

**Optimal fractal design: choosing level $n$ and scaling $\lambda$.** For the 2D Cantor fractal depicted in Figure 6.10, we want to investigate the existence of optimal design parameters such as the fractal depth/level $n$ and the scaling $\lambda$. To this end, let us introduce the cost function,

$$J_n(\lambda) := \frac{A_n(\lambda)}{C_n(\lambda) + \Gamma_n^2(\lambda)} , \tag{6.106}$$

where $A_n(\lambda)$ is the pore space defined in (6.102), $C_n(\lambda) := L_0^2 - A_n(\lambda)$ is the area of the Cantor dust/fractal, and $\Gamma_n(\lambda)$ denotes the surface area of the Cantor dust $C_n(\lambda)$. Optimal design parameters such as scaling $\lambda > 0$ and level of fractal depth $n \in \mathbb{N}$ are depicted in Figure 6.12, where we find the minimal costs $J_n(\lambda)$. As a consequence, (6.106) defines optimality here by maximizing the surface area $\Gamma_n(\lambda)$ and growing the Cantor dust $C_n(\lambda)$ while minimizing pore space $A_n(\lambda)$. Naturally, the definition of the cost function represents a delicate and crucial step, since it weights the different aspects that should be favored and disregarded.

# 7 Quantum algorithms describing CHeSs

## 7.1 Introduction

Toward the end of the last decade (2010–2020), there were often multiple news about increases in qubit-size (which we subsequently introduce) per year about companies involved in the development and research of quantum computers such as D-Wave, IBM, Microsoft, Google, etc. In this context, one of the major challenges is the potential loss of safe transactions, of which payments will have an economic impact and hence will be socially most disruptive. Also, the global stability and safety [130] will face uncertainties by the unknown availability of an existing and practicable high-qubit quantum computer [17, 161]. These concerns are well founded on the fact that there already exist algorithmic concepts such as Shor's algorithm to solve the factoring problem, which is still assumed to be intractable for classical computers[1] to decipher encrypted transactions significantly faster and within a reasonable time. For instance, we refer the interested reader to an investigation performed by Kudelski Research [59] in 2021.

At the same time, published research shows that there are still major challenges to overcome, see [72]. For instance, many of these ground breaking algorithms rely on a so-called oracle, whose practical realization or drawbacks are generally not discussed in the literature presenting these algorithms. In fact, such oracle- or query-based algorithms only provide lower bounds on their expected computational costs/complexity.

Before we motivate the promising quantum computational concepts in the context of this book, that is, the efficient and reliable computational realization of descriptions and models for CHeSs, we briefly recall very basic concepts from quantum mechanics and its theory.

## 7.2 Mathematical framework for quantum systems

A very fundamental concept for the theoretical and practical realization of quantum computers is a rigorous understanding about what is a *quantum state* and in particular how can we set, control, and evolve it in a real practical context.

**Quantum State and Quantum Register**

**Definition 7.2.1** (Quantum state and quantum register). A *quantum state* $q$ representing a single qubit exists in a complex Hilbert space $H = [H]^{\otimes 1}$, that is,

$$|q\rangle = c_0\,|0\rangle + c_1\,|1\rangle \in [H]^{\otimes 1}\,, \tag{7.1}$$

1 "intractable" since there is no practical algorithm publicly known up to now despite the fact that there is no proof for this assumption.

https://doi.org/10.1515/9783110579543-007

where $\langle q|q\rangle = c_0^2 + c_1^2 = 1$, and $|0\rangle$ and $|1\rangle$ are the basis states referring to a system that can admit a superposition of two elementary and different discrete states, which we label "0" and "1". Similarly, an *n-qubit quantum state q* reads,

$$|q\rangle = \sum_{i=0}^{2^n-1} c_i\,|i\rangle_n \in [H]^{\otimes n}, \tag{7.2}$$

where,

$$\begin{aligned} |i\rangle_n &= |b_{n-1}b_{n-2}\dots b_0\rangle = |b_{n-1}\rangle\,|b_{n-2}\rangle\dots|b_0\rangle \\ &= |b_{n-1}\rangle\otimes|b_{n-2}\rangle\otimes\dots\otimes|b_0\rangle \in [H]^{\otimes n}, \end{aligned} \tag{7.3}$$

denotes the standard orthonormal basis of the $n$-qubit Hilbert space $[H]^{\otimes n}$, for $0 \le i \le 2^n - 1$, with the binary representation $i = \sum_{k=0}^{n-1} b_k 2^k \in \mathbb{N}$, $b_k \in \{0,1\}$, and where the coefficients, $c_i \in \mathbb{C}$, again satisfy the normalization condition $\sum_{i=0}^{2^n-1} c_i^2 = 1$. $[H]^{\otimes n}$ is the product Hilbert space of $n$ single qubit Hilbert spaces $H = [H]^{\otimes 1}$.

The $n$ qubits represented in the standard orthonormal basis $|i\rangle_n$ for $1 \le i \le 2^n - 1$ with $i \in \mathbb{N}$, on which a quantum computer performs its operations, are called *a quantum register*.

**Remark 7.2.2** (Number of basic states). In Definition 7.2.1, we follow the classical and generally applied convention of restricting ourselves to two basic states "0" and "1". However, we note that we could also build a quantum computer relying on more than just two basic states, which is referred to as *qudit-based quantum computing*, but without necessarily gaining a clear advantage. Motivations for quidit-based computing are given in [173]. ◇

## 7.3 Solving multiscale problems on a quantum computer

We consider a multiscale problem consisting of a charged composite material $\mathcal{M}^\epsilon$ showing two different electric permittivities $\varepsilon_0$ and $\varepsilon_1$ belonging to material phases $Y_0$ and $Y_1$, respectively, see Section 6.3. The distribution of charges in the composite $\mathcal{M}^\epsilon$ is given by a charge density $\rho(\mathbf{x})$. The parameter $\epsilon$ refers to the heterogeneity of the composite $\mathcal{M}^\epsilon$ and generally is defined by $\epsilon := \frac{l}{L} > 0$, where $l$ is the microscopic length of a (quadratic) reference cell, and $L$ denotes the macroscopic length of the composite $\mathcal{M}^\epsilon$. In fact, $\mathcal{M}^\epsilon$ is a periodic covering by placing characteristic reference cells next to each other. For random materials, such a reference cell/representative volume element can be obtained via a suitable statistical averaging, for instance. Now, we are interested in the electric potential $\phi^\epsilon$ that solves the following *microscopic equation*,

$$-\mathrm{div}\left(\varepsilon\left(\frac{\mathbf{x}}{\epsilon}\right)\nabla\phi^\epsilon(\mathbf{x})\right) = \rho(\mathbf{x}), \quad \text{in } \mathcal{M}^\epsilon, \tag{7.4}$$

where the composite introduces a heterogeneity via the electric permittivity,

$$\varepsilon^{\epsilon}(\mathbf{x}) := \varepsilon\left(\mathbf{x}, \frac{\mathbf{x}}{\epsilon}\right) := \varepsilon(\mathbf{y}) := \varepsilon_0\chi_{Y_0}(\mathbf{y}) + \varepsilon_1\chi_{Y_1}(\mathbf{y}),$$

where $Y_0$ and $Y_1$ are materials with electric permittivities $\varepsilon_0$ and $\varepsilon_1$, respectively.

Naturally, due to the complexity induced by multiple domains resolving different scales, such problems are generally solved numerically. However, we point out that for strongly heterogeneous problems, that is, for $\epsilon \ll 1$, this leads to very high-dimensional problems due to the general requirement of $h \ll \epsilon$, where $h > 0$ is the mesh/grid size used in the corresponding numerical computation. This latter requirement means that a reliable resolution of characteristic heterogeneities in composites imposes a mesh constraint where the mesh size $h > 0$ needs to be chosen much smaller than the heterogeneity $\epsilon > 0$. As a consequence, classical numerical discretization strategies for multiscale problems intrinsically suffer from the curse of dimensionality.

Hence, since quantum computing promises to offer a new strategy for attacking classical high-dimensional problems, once quantum computers reach a certain threshold in the number of available qubits, we present a promising approach based on recent developments in quantum computing [70]. In particular, we motivate that the topics on homogenization, as discussed in Section 6.3, especially on two-scale convergence (see Section 10), will still remain equally relevant, if not more so, even after achieving quantum supremacy.[2] The reason is that the so-called two-scale limit of a multiscale problem such as (7.4) is a problem that allows for more stable computations, but for the price of still increasing the space dimension to include the macro- and the microscale at the same time, see Section 7.3.2 for formal details.

### 7.3.1 Quantum algorithms required for multiscale problems

There are various fundamental classical tools and algorithms that have a relatively simple counterpart in the quantum computing context. In the subsequent sections, we will recall and motivate the Quantum Fourier Transform (QFT), the Quantum Phase Estimation (QPE), and the Quantum Linear Systems Solver (QLSS).

#### 7.3.1.1 From classical to quantum Fourier transform (QFT)

An intuitive approach to motivate the algorithm for the QFT is to recall the classical Fourier transform, $\tilde{f}(\xi) := \mathcal{F}(f(x))$, of a periodic and square-integrable function $f \in L^2_\sharp(0, L)$.

2 The term *quantum supremacy* seems to go back to John Preskill (Caltech) and describes the engineering and computational complexity aspects of demonstrating that for a specific problem, where classical computers suffer from computational times going beyond practical applicability, quantum computers can solve it in practically useful time.

**Fourier Transform (FT).** A periodic function of period $L > 0$ can be represented by the Fourier series

$$f(x) = \frac{a_0}{2} + \sum_{n=1}^{\infty}\left( a_n \cos\left(\frac{2\pi n x}{L}\right) + \sum_{n=1}^{\infty} b_n \sin\left(\frac{2\pi n x}{L}\right)\right), \tag{7.5}$$

where

$$\begin{cases} a_n := \frac{2}{L}\int_0^L f(x)\cos\left(\frac{2\pi n x}{L}\right)dx\,, \\ b_n := \frac{2}{L}\int_0^L f(x)\sin\left(\frac{2\pi n x}{L}\right)dx\,. \end{cases} \tag{7.6}$$

Using the Euler formula $\exp(\mathrm{i}x) = \cos(x) + \mathrm{i}\sin(x)$, $a_n = a_{-n}$, and $b_n = -b_{-n}$, allows us to rewrite (7.5)–(7.6) as follows,

$$\begin{cases} f(x) = \sum_{n=-\infty}^{\infty} h_n \exp\left(\frac{2\pi \mathrm{i} n x}{L}\right), \\ h_n = \frac{1}{L}\int_0^L f(x)\exp\left(-\frac{2\pi \mathrm{i} n x}{L}\right)dx\,. \end{cases} \tag{7.7}$$

Finally, the continuous Fourier transform (FT) can be obtained with the help of the limit as $L \to \infty$,

$$\begin{cases} \tilde{f}(\xi) := \mathcal{F}(f) = \frac{1}{2\pi}\int_{-\infty}^{\infty} f(x)\exp(-\mathrm{i}\xi x)\,dx\,, \\ \mathcal{F}^{-1}(\tilde{f}(\xi)) = \int_{-\infty}^{\infty} \tilde{f}(\xi)\exp(-\mathrm{i}\xi x)\,d\xi\,. \end{cases} \tag{7.8}$$

**Discrete Fourier Transform (DFT).** The DFT is a result of a suitable numerical approximation of the Fourier transform (FT). It turns out that the FT can be discretized with the help of the trapezoidal quadrature rule, i. e.,

$$\int_{x_0}^{x_N} f(x)\,dx = \frac{(f(x_N) + f(x_0))\mathfrak{h}}{2} + \mathfrak{h}\sum_{i=1}^{N-1} f(x_i) + e_T\,, \tag{7.9}$$

where $x_i = x_0 + i\mathfrak{h}$ for $i \in \{0, 1, 2, \ldots, N\}$, and $e_T$ is the error of the trapezoidal rule. This error can be estimated by,

$$|e_T| \leq \frac{\mathfrak{h}^2}{8}\int_{x_0}^{x_N} |f''(x)|\,dx\,, \tag{7.10}$$

for functions $f \in C^2([x_0, x_N])$. The parameter $\mathfrak{h} > 0$ is a constant mesh size. In fact, for periodic functions $f(x)$, i. e., $f \in C_\sharp(x_0, x_N)$, the trapezoidal rule simplifies to,

$$\int_{x_0}^{x_N} f(x)\,dx = \mathfrak{h}\sum_{i=1}^{N-1} f(x_i) + e_T\,. \tag{7.11}$$

As a consequence, the *DFT* simply represents the trapezoidal approximation of $(7.7)_2$, that is,

$$\begin{aligned} h_n &= \frac{1}{L}\int_{x_0}^{x_N} f(x)\exp\left(-\frac{i2\pi nx}{L}\right)dx \\ &\approx \frac{1}{N}\sum_{k=1}^{N-1} f(x_k)\exp\left(-\frac{i2\pi nk}{N}\right) \\ &=: \tilde{h}_n\,, \end{aligned} \tag{7.12}$$

since $x_k = x_0 + k\mathfrak{h}$ and $\mathfrak{h} := \frac{L}{N}$. For numerical purposes, it is convenient to set $f_i := f(x_i)$ and to introduce the vector notations,

$$f^{(N)} := \begin{bmatrix} f_0 \\ f_1 \\ \vdots \\ f_{N-1} \end{bmatrix} \in \mathbb{C}^N, \quad \tilde{h}^{(N)} := \begin{bmatrix} \tilde{h}_0 \\ \tilde{h}_1 \\ \vdots \\ \tilde{h}_{N-1} \end{bmatrix} \in \mathbb{C}^N, \tag{7.13}$$

and to write the numerical approximation as follows,

$$\tilde{h}^{(N)} = \frac{1}{N}\hat{\mathrm{W}}f^{(N)}, \tag{7.14}$$

where $\hat{\mathrm{W}}$ is the matrix,

$$\begin{aligned} \hat{\mathrm{W}} &:= \{w_{nk}\}_{n,k=0,1,\dots,N-1} := \{w^{nk}\}_{n,k=0,1,\dots,N-1} \\ &= \begin{bmatrix} 1 & 1 & 1 & \dots & 1 \\ 1 & w & w^2 & \dots & w^{(N-1)} \\ 1 & w^2 & w^4 & \dots & w^{2(N-1)} \\ \vdots & \vdots & \vdots & \ddots & \vdots \\ 1 & w^{(N-1)} & w^{(N-1)2} & \dots & w^{(N-1)(N-1)} \end{bmatrix}, \end{aligned} \tag{7.15}$$

for the parameter $w := \exp(-\frac{i2\pi L}{N})$.

Note that based on these considerations, we can define the *Discrete Fourier Transform (DFT)* by the *unitary* map,

$$\begin{aligned} \mathcal{F}_d : \mathbb{C}^N &\to \mathbb{C}^N, \\ f^{(N)} &\mapsto \frac{1}{\sqrt{N}}\hat{\mathrm{W}}f^{(N)}, \end{aligned} \tag{7.16}$$

which can be easily verified by computing,

$$\mathcal{F}_d \overline{\mathcal{F}}_d^T = \frac{1}{\sqrt{N}} \hat{\mathrm{W}} \left( \frac{1}{\sqrt{N}} \hat{\mathrm{W}} \right)' = \hat{\mathrm{I}} . \tag{7.17}$$

**Remark 7.3.1** (Complexity of the DFT). We recall that the DFT is rather expensive, i. e., it requires $N^2$ complex multiplications. To this end, the so-called Fast Fourier Transforms (FFTs) were developed with a complexity governed by $N \log N$ for $N$ large [38]. ◇

**Quantum Fourier Transform (QFT).** We recall that quantum operators need to be *reversible*, and therefore *unitary* structures are well suited for this purpose. Recalling that the classical DFT is unitary, we can state its quantum analog by looking at (7.12).

**Quantum Fourier Transform (QFT)**

**Definition 7.3.2** (Quantum Fourier transform (QFT)). Let $|j\rangle_n \in [H]^{\otimes n}, j \in \mathbb{N}$ with $0 \le j \le N-1$ and $N := 2^n$, be the standard orthonormal basis $|j\rangle_n = |j_{n-1}j_{n-2}\cdots j_0\rangle$ of the $n$-qubit Hilbert space $[H]^{\otimes n}$ with binary representation $j = \sum_{k=0}^{N-1} j_k 2^k$. We then define the *Quantum Fourier Transform (QFT)* by the map,

$$\begin{aligned} \mathcal{F}_q : [H]^{\otimes n} &\rightarrow [H]^{\otimes n}, \\ |j\rangle_n &\mapsto \frac{1}{\sqrt{N}} \sum_{k=0}^{N-1} \exp\left( -\frac{\mathrm{i}2\pi jk}{N} \right) |k\rangle_n . \end{aligned} \tag{7.18}$$

◇

Let us recall how the QFT-operator $\mathcal{F}_q$ operates in a $n$-qubit register.

**Remark 7.3.3** (Sign of QFT). The sign of the QFT in the exponential phase appears to vary in the literature, depending on whether the definition should correspond to the classical DFT or not. ◇

Let us recall how the QFT-operator $\mathcal{F}_q$ operates on an $n$-qubit register, see [179], for instance.

**Properties of the QFT operator**

**Lemma 7.3.4** (Properties of the QFT operator $\mathcal{F}_q$). *We assume to be working with a register of n qubits and that $|j\rangle_n = |j_{n-1}j_{n-2}\cdots j_0\rangle$ represents for $0 \le j \le N-1$ and $N := 2^n$, the canonical orthonormal basis in this register, where $j_k \in \{0,1\}$ for $0 \le k \le n-1$. The QFT-operator $\mathcal{F}_q$ has the following properties:*

(i) *The operator $\mathcal{F}_q$ admits on an n-qubit register the following matrix form,*

$$\hat{\mathrm{F}}_q = \{f_{ij}^q\}_{i,j=0}^{N-1}, \tag{7.19}$$

*where $f_{ij}^q := \frac{1}{\sqrt{N}} \exp(\mathrm{i}2\pi ij/N)$.*

(ii) *The operator $\mathcal{F}_q$ is unitary, i. e., in an n-qubit register, we have $\hat{\mathrm{F}}_q \overline{\hat{\mathrm{F}}}_q^T = \hat{\mathrm{I}} \in \mathbb{R}^{N \times N}$.*

(iii) *The QFT-operator $\mathcal{F}_q$ satisfies the following identity,*

$$\begin{aligned} \mathcal{F}_q |j\rangle_n = \frac{1}{\sqrt{N}} \big(|0\rangle + \exp(\mathrm{i}2\pi 0._{[2]}j_0)\,|1\rangle\big) \otimes \big(|0\rangle + \exp(\mathrm{i}2\pi 0._{[2]}j_1 j_0)\,|1\rangle\big) \otimes \ldots \\ \otimes \big(|0\rangle + \exp(\mathrm{i}2\pi 0._{[2]}j_{n-1}\cdots j_1 j_0)\,|1\rangle\big), \end{aligned} \tag{7.20}$$

*where the binary representation* $j = \sum_{l=0}^{n-1} j_l 2^l$ *is written in nonperiodic fractional form, i. e.,*

$$\begin{aligned} \left]\frac{2^{n-1}}{N} j\right[ &=: 0._{[2]} j_0 \,, \\ \left]\frac{2^{n-2}}{N} j\right[ &=: 0._{[2]} j_0 j_1 \,, \\ &\cdots \\ \left]\frac{1}{N} j\right[ &=: 0._{[2]} j_{n-1} \cdots j_{n-2} \cdots j_0 \,, \end{aligned} \tag{7.21}$$

*and for a fractional binary number* $b = b^{m-1} b^{m-2} \dots b^0 ._{[2]} b_{n-1} b_{n-2} \dots b_0$*, the operator* $]b[$ *is defined by solely picking the fractional part, that is,*

$$]b[ := 0._{[2]} b_{n-1} b_{n-2} \dots b_0 \,. \tag{7.22}$$

*Proof.* (i) After looking at the definition of the QFT operator (7.18) and using the fact that $|i\rangle_n$ and $|j\rangle_n$ represent standard orthonormal bases of an $n$-qubit quantum register, we can directly compute,

$$\begin{aligned} (\hat{\mathrm{F}}_q)_{ji} = \langle i| \mathcal{F}_q |j\rangle_n &= \frac{1}{\sqrt{N}} \sum_{k=0}^{N-1} \langle i| \exp(\mathrm{i}2\pi jk/N) |k\rangle_n \\ &= \frac{1}{\sqrt{N}} \sum_{k=0}^{N-1} \exp(\mathrm{i}2\pi jk/N) \delta_{ik} \\ &= \frac{1}{\sqrt{N}} \exp(\mathrm{i}2\pi ji/N) \in \mathbb{C} \,, \end{aligned} \tag{7.23}$$

for all $0 \le i, j \le N-1$ where $i, j \in \mathbb{N}$, so that $\hat{\mathrm{F}}_q \in \mathbb{C}^{N\times N}$.

(ii) Thanks to (i), we immediately obtain,

$$(\hat{\mathrm{F}}_q \overline{\hat{\mathrm{F}}}_q^T)_{ji} = \frac{1}{N} \sum_{k=0}^{N-1} \exp(\mathrm{i}2\pi jk/N) \exp(-\mathrm{i}2\pi ki/N) = \frac{1}{N} \sum_{k=0}^{N-1} \exp(\mathrm{i}2\pi k(j-i)/N) \,. \tag{7.24}$$

We consider the following two cases:

*Case 1:* ($j = i$) We immediately obtain the claim.

*Case 2:* ($j \ne i$) Introducing the parameter $l = j - i < N$ allows us to rewrite (7.24) as follows,

$$(\hat{\mathrm{F}}_q \overline{\hat{\mathrm{F}}}_q^T)_{ji} = \frac{1}{N} \sum_{k=0}^{N-1} \exp(\mathrm{i}2\pi kl/N) \,, \tag{7.25}$$

which after setting $q = \exp(\mathrm{i}2\pi k/N)$ and using the formula for geometric series,[3] immediately gives the following expression,

3 Recall the formula for finite geometric series $\sum_{l=0}^{L-1} q^l = \frac{q^L - 1}{q-1}$.

$$(\hat{\mathrm{F}}_q \overline{\hat{\mathrm{F}}}_q^{T})_{ji} = \frac{1}{N}\sum_{l=0}^{N-1} z^l = \frac{1}{N}\frac{\exp(\mathrm{i}2\pi k) - 1}{\exp(\mathrm{i}2\pi k/N) - 1}\,. \tag{7.26}$$

Clearly, in (7.26), the numerator vanishes, and hence the claim follows.

(iii) The verification of (7.20) relies on rewriting Definition 7.3.2 of the QFT operation $\mathcal{F}_q$,

$$\begin{aligned}
\mathcal{F}_q\,|j\rangle_n &= \frac{1}{\sqrt{N}}\sum_{k=0}^{N-1}\exp(\mathrm{i}2\pi]kj/N[)\,|k\rangle_n = \frac{1}{\sqrt{N}}\sum_{k=0}^{N-1}\exp\left(\mathrm{i}2\pi\left]j\left(\sum_{l=0}^{n-1}k_l 2^{l-n}\right)\right[\right)|k\rangle_n \\
&= \frac{1}{\sqrt{N}}\sum_{k_0=0}^{1}\sum_{k_1=0}^{1}\cdots\sum_{k_{n-1}=0}^{1}\prod_{l=0}^{n-1}\mathrm{e}^{\mathrm{i}2\pi]jk_l 2^l/N[}\,|k_{n-1}k_{n-2}\dots k_0\rangle \\
&= \frac{1}{\sqrt{N}}\sum_{k_{n-1}=0}^{1}\mathrm{e}^{\mathrm{i}2\pi]jk_{n-1}2^{n-1}/N[}\,|k_{n-1}\rangle \otimes \sum_{k_{n-2}=0}^{1}\mathrm{e}^{\mathrm{i}2\pi]jk_{n-2}2^{n-2}/N[}\,|k_{n-2}\rangle \otimes \dots \\
&\quad \otimes \sum_{k_0=0}^{1}\mathrm{e}^{\mathrm{i}2\pi]jk_0 2^0/N[}\,|k_0\rangle \\
&= \frac{1}{\sqrt{N}}\left(|0\rangle + \mathrm{e}^{\mathrm{i}2\pi]j2^{n-1}/N[}\,|1\rangle\right) \otimes \left(|0\rangle + \mathrm{e}^{\mathrm{i}2\pi]j2^{n-2}/N[}\,|1\rangle\right) \otimes \dots \\
&\quad \otimes \left(|0\rangle + \mathrm{e}^{\mathrm{i}2\pi]j2^0/N[}\,|1\rangle\right),
\end{aligned} \tag{7.27}$$

where we used the fact that $0 \le k \le N-1$ has the binary representation $k = \sum_{l=0}^{N-1} k_l 2^l$. Hence, the claim follows with the nonperiodic representation of the phase angle (7.22). □

For applications, it is important how a unitary operator $\mathcal{U}$ such as the QFT $\mathcal{F}_q$ can be represented with the help of universal[4] quantum gates. Therefore, Figure 7.1 shows a possible quantum circuit implementing the QFT operator $\mathcal{F}_q$.

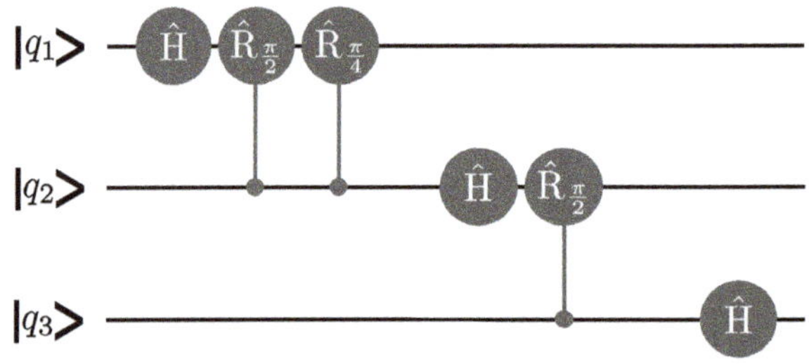


**Figure 7.1:** *Three-qubit QFT circuit not restricted to sole, nearest-neighbor interactions:* First, a *Hadamard gate* $\hat{\mathrm{H}}$ operates on each qubit. Then, the so-called *phase shift operators* realized by *phase shift gates* $\hat{\mathrm{R}}_\phi$ for rotations/phase shifts of angle $\phi$ follow.

4 Universality refers in classical and in quantum computing to the fact that a specific and limited class of gates are able to perform all possible operations.

#### 7.3.1.2 Quantum phase estimation (QPE)

Before we explain the technical details of QPE, we note that QPE represents an eigenvalue decomposition of a Hermitian[5] matrix $\hat{\mathrm{A}} \in \mathbb{C}^{N\times N}$, $N \in \mathbb{N}_{>0}$. Thus, the algorithm has to identify for a given input vector $\mathbf{v} \in \mathbb{C}^N$ (not necessarily the eigenstate), the eigenvalues $\lambda_i \in \mathbb{R}$, $1 \leq i \leq N$,[6] that satisfy the equation,

$$\hat{\mathrm{A}}\mathbf{v}_i = \lambda_i \mathbf{v}_i \quad \text{for } 1 \leq i \leq N\,, \tag{7.28}$$

where $\mathbf{v}_i$ for $1 \leq i \leq N$ are the associated eigenvectors. Note that subsequently, we generally use the notation $\mathcal{B}$ to represent abstract operators. The associated ($l$-dimensional, $l \in \mathbb{N}$) approximations or special cases are represented in matrix form, i. e., $\hat{\mathrm{B}} \in \mathbb{C}^{l\times l}$. Since gate-based quantum computers rely on unitary operators $\mathcal{U}$, that is, $\mathcal{U}\mathcal{U}^* = \mathcal{I}$, we will map a Hermitian operator $\mathcal{A}$ realized by the unitary operator,

$$\mathcal{U} := \exp(\mathrm{i}\mathcal{A})\,, \tag{7.29}$$

unless we are already given a unitary operator $\mathcal{A}$ from the beginning. The eigenvalues $\lambda$ of unitary tensors $\hat{\mathrm{U}}$ satisfy $|\lambda|^2 = \lambda\overline{\lambda} = 1$, which is also motivated by,

$$\mathbf{v} = \overline{\hat{\mathrm{U}}}^T \hat{\mathrm{U}}\mathbf{v} = \lambda \overline{\hat{\mathrm{U}}}^T \mathbf{v} = \lambda\overline{\lambda}\mathbf{v}\,. \tag{7.30}$$

The subsequent result presents a possible quantum approach to obtain spectral characteristics of unitary operators.

**Quantum Phase Estimation (QPE)**

**Lemma 7.3.5** (Quantum Phase Estimation (QPE)). *Suppose that we are working with a register of $n \in \mathbb{N}_{>0}$ qubits. Let a Hermitian operator $\mathcal{A}$ be realized by a tensor/matrix $\hat{\mathrm{A}} \in \mathbb{C}^{N\times N}$ for $N := 2^n$, and let $|\mathbf{v}\rangle_n = \sum_{i=0}^{N-1} v_i\,|i\rangle_n \in [H]^{\otimes n}$ be an arbitrary quantum state, i. e., $1 = \sum_{i=0}^{N-1} v_i^2$.*

*There exists a QPE-operator $\mathcal{F}_{\mathrm{QPE}}$ such that,*

$$\mathcal{F}_{\mathrm{QPE}}(\hat{\mathrm{A}}, |\mathbf{v}\rangle_n) = \varphi_n\,, \tag{7.31}$$

*where $\varphi_n = 0._{[2]}j_{n-1}j_{n-2}\cdots\cdots j_0 = \sum_{i=0}^{n-1} j_{n-i-1}2^{-i-1}$ is an $n$-digit binary approximation of the phase $\varphi$ such that,*

$$\lambda = e^{\mathrm{i}2\pi\varphi}\,, \tag{7.32}$$

*which represents the most likely eigenvalue of $\mathcal{U} = \exp(\mathrm{i}\mathcal{A})$ associated with the given input state vector $|\mathbf{v}\rangle_n$.*

**Remark 7.3.6** (Most likely eigenvalues). In fact, a quantum computing algorithm realizing the QPE operator $\mathcal{F}_{\mathrm{QPE}}$ exploits the fact that an arbitrary quantum state $|\mathbf{v}\rangle_n \in [H]^{\otimes n}$ can be represented as a linear combination of eigenfunctions $|\mathbf{v}_i\rangle_n \in [H]^{\otimes n}$. ◇

5 A Hermitian matrix $\hat{\mathrm{A}}$ is defined by the property $\overline{\hat{\mathrm{A}}}^T = \hat{\mathrm{A}}$.

6 Note that the eigenvalues of a Hermitian matrix are real.

*Proof.* Property (7.30) allows us to represent the eigenvalues of unitary operators $\hat{\mathrm{U}}$ with the help of the phase variable $\varphi$ by (7.32). A key idea in the construction of the QPE operator $\mathcal{F}_{\mathrm{QPE}}$ is the smart representation of the eigenvalues $\lambda$ by rewriting the exponent $\varphi$ in binary form as in (7.21), see also [72] for an applied overview. Naturally, the quality of the approximation is governed by the number of bits, $n \in \mathbb{N}$, appearing after the $\cdot_{[2]}$-point in (7.21). We construct $\mathcal{F}_{\mathrm{QPE}}$ in four essential steps.

**Step (i):** *Assumption and initial states.* For simplicity, suppose that we know the eigenstate $|\boldsymbol{v}\rangle_n \in [H]^{\otimes n}$ of the observable $\mathcal{U}$, i. e., the eigenfunction of the unitary operator $\mathcal{U}$. In fact, this assumption does not represent a restriction, since an arbitrary state $|\mathbf{v}\rangle_n \in [H]^{\otimes n}$ can be decomposed with the help of the eigenbasis $|\boldsymbol{v}_j\rangle_n$, $1 \leq j \leq l$, associated with the operator $\mathcal{U}$, that is,

$$|\mathbf{v}\rangle_n = \sum_{j=0}^{N-1} a_j \left|\boldsymbol{v}_j\right\rangle_n , \quad a_j \in \mathbb{C} . \tag{7.33}$$

As a frequent practice in quantum algorithms, we choose $|\mathbf{v}\rangle_n = |0\rangle_n$. Hence, this allows us to identify the most likely result as the one for which the coefficients $|a_j| = \left|\langle v|\boldsymbol{v}_j\rangle_n\right|$ in the expression,

$$|\langle 0|\boldsymbol{v}\rangle_n| = \sum_{j=0}^{N-1} a_j \left\langle 0|\boldsymbol{v}_j\right\rangle_n , \tag{7.34}$$

are the largest. Now, we will use the above mentioned eigenstate $|\boldsymbol{v}\rangle_n$ associated to the observable $\mathcal{U}$ as the initial state of a second register used in the QPE algorithm. The first register represents the so-called control/phase register of $n$ qubits (note that $n$ is the number of bits in the binary representation of the phase $\varphi$), where each qubit is prepared in the equal superposition state $\frac{1}{\sqrt{2}}(|0\rangle + |1\rangle)$.

**Step (ii):** *QPE operation in the two-qubit case.* Recall that we will use two registers, a first $n$-qubit register from which we want to extract the phase and a second register being without loss of generality an eigenvector to $\mathcal{U} = \exp(i\mathcal{A})$. This implies the following initial state configuration representing the first register by a simple qubit for simplicity, i. e.,

$$\frac{1}{\sqrt{2}}(|0\rangle + |1\rangle) \otimes |\boldsymbol{v}\rangle_n = \frac{1}{\sqrt{2}}(|0\rangle\,|\boldsymbol{v}\rangle_n + |1\rangle\,|\boldsymbol{v}\rangle_n) . \tag{7.35}$$

With the unitary operator $\mathcal{U}$, for which we want to compute its eigenvalues, we apply a controlled-$\mathcal{U}$ operation to the initial state (7.35) of the second register. Rewriting the resulting term as follows,

$$\begin{aligned} \frac{1}{\sqrt{2}}(|0\rangle\,|\boldsymbol{v}\rangle_n + |1\rangle\,\mathcal{U}\,|\boldsymbol{v}\rangle_n) &= \frac{1}{\sqrt{2}}(|0\rangle\,|\boldsymbol{v}\rangle_n + \exp(\mathrm{i}2\pi\varphi)\,|1\rangle\,|\boldsymbol{v}\rangle_n) \\ &= \frac{1}{\sqrt{2}}(|0\rangle + \exp(\mathrm{i}2\pi\varphi)\,|1\rangle) \otimes |\boldsymbol{v}\rangle_n , \end{aligned} \tag{7.36}$$

allows us to immediately see that the eigenvalue of the unitary observable $\mathcal{U}$ can be collected in the first register in the form of the phase variable j.

Exactly at this stage the binary representation (7.21) and the approximation $\varphi_n$ (of $\varphi$) becomes useful in the sense that it allows us to extract bit by bit with controlled-$\mathcal{U}^{2^l}$ operations for $l = 0, 1, \ldots, n-1$. For instance, applying a controlled-$\mathcal{U}^2$ operation on (7.35) immediately gives,

$$\begin{aligned}\frac{1}{\sqrt{2}}(|0\rangle\,|\mathbf{v}\rangle_n + |1\rangle\,\mathcal{U}^2\,|\mathbf{v}\rangle_n) &= \frac{1}{\sqrt{2}}(|0\rangle\,|\mathbf{v}\rangle_n + \exp(\mathrm{i}2\pi 2\mathrm{j})\,|1\rangle\,|\mathbf{v}\rangle_n) \\ &= \frac{1}{\sqrt{2}}(|0\rangle + \exp(\mathrm{i}2\pi 0._{[2]}j_{n-2}\ldots j_0)\,|1\rangle) \otimes |\mathbf{v}\rangle_n \,. \end{aligned} \tag{7.37}$$

Therefore, a general, controlled-$\mathcal{U}^{2^k}$ operation on (7.35) for $k = 0, 1, \ldots, n-1$ leads to,

$$\frac{1}{\sqrt{2}}(|0\rangle\,|\mathbf{v}\rangle_n + |1\rangle\,\mathcal{U}^{2^k}\,|\mathbf{v}\rangle_n) = \frac{1}{\sqrt{2}}(|0\rangle + \exp(\mathrm{i}2\pi 0._{[2]}j_{n-1-k}\ldots j_0)\,|1\rangle) \otimes |\mathbf{v}\rangle_n \,. \tag{7.38}$$

**Step (iii):** *Extending the 1-qubit control register to m qubits.* Let us consider $m$ qubits in our first, control/phase register. Calculations (7.37) and (7.38) show, that the $k$-th qubit is related to the control $\mathcal{U}^{2^k}$. Hence, we can write the state in the first register after these controlled operations as follows,

$$\begin{aligned}\frac{1}{\sqrt{2}}(|0\rangle + \exp(\mathrm{i}2\pi 0._{[2]}j_{n-1}j_{n-2}\ldots j_0)\,|1\rangle) \otimes \frac{1}{\sqrt{2}}(|0\rangle + \exp(\mathrm{i}2\pi 0._{[2]}j_{n-2}\ldots j_1 j_0)\,|1\rangle) \otimes \ldots \\ \cdots \otimes \frac{1}{\sqrt{2}}(|0\rangle + \exp(\mathrm{i}2\pi 0._{[2]}j_0)\,|1\rangle)\,. \end{aligned} \tag{7.39}$$

**Step (iv):** *Phase extraction from the first control/phase register.* Fortunately, the structure of (7.39) in (iii) allows us to apply the QFT to extract the necessary phase parameters from the binary approximation (7.21), which defines the eigenvalue according to (7.32). In fact, applying the inverse QFT $\mathcal{F}_q^* := \overline{\mathcal{F}_q}^T$ to the controlled first register (7.39) immediately gives us the phase by the tensorproduct,

$$|j_{n-1}\rangle \otimes |j_{n-2}\rangle \otimes \cdots \otimes |j_0\rangle \in [H]^{\otimes n}\,, \tag{7.40}$$

which gives $\varphi_n = 0._{[2]}j_{n-1}\ldots j_0$ after a measurement in the standard basis, i. e., extracting the most likely bits associated with $\varphi_n$.

Finally, the operations of steps (i)–(iv) can be visualized with the help of quantum circuits representing the QPE algorithm, see Figure 7.2. □

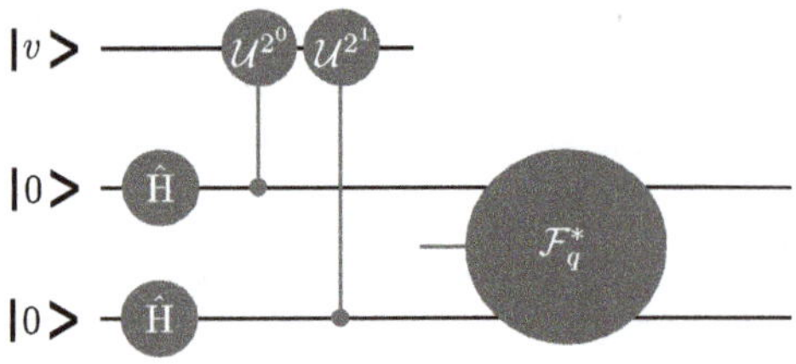


Figure 7.2: *Three-qubit QPE circuit:* Computation of the quantum phase $\varphi$ (see equation (7.32)) for two binary digits represented by $\varphi_1$ and $\varphi_0$. First, a *Hadamard gate* $\hat{H}$ is applied to the two qubits representing $\varphi_1$ and $\varphi_0$. Subsequently, powers of the given unitary operator $\mathcal{U} = \exp(\hat{\mathcal{A}})$ are applied to an arbitrary quantum state $|\mathbf{v}\rangle_n$ and these two qubits.

#### 7.3.1.3 Quantum linear systems solver (QLSS): the Harrow–Hassidim–Lloyd (HHL) algorithm

Almost every science, engineering, and computational model relies at some point on the solution of a linear system of equations. The essential mathematical formulation of such a coupled linear system can be stated with the help of an intrinsic matrix $\hat{A} \in \mathbb{C}^{M\times N}$ and a vector $\mathbf{b} \in \mathbb{C}^N$ with the goal of finding a solution $\mathbf{x} \in \mathbb{C}^{M\times N}$ for $M, N \in \mathbb{N}_{>0}$, such that the following equation is satisfied,

$$\hat{A}\mathbf{x} = \mathbf{b}\,. \tag{7.41}$$

For simplicity, we subsequently discuss the case of Hermitian square matrices $\hat{A} \in \mathbb{C}^{N\times N}$, i. e., matrices satisfying $\hat{A}^* = \hat{A}$. Note that if $\hat{A}$ is arbitrary, then we can define a Hermitian representation by,

$$\tilde{\hat{A}} := \begin{bmatrix} \hat{0} & \overline{\hat{A}}^T \\ \hat{A} & \hat{0} \end{bmatrix}, \tag{7.42}$$

and immediately also a unitary matrix by $\hat{U} := \exp(i\hat{A})$ as required for quantum computations (referred to as Hamiltonian encoding). A quantum way of solving a linear system can be obtained by a quantum operator defined as follows.

**Harrow–Hassidim–Lloyd (HHL) operator**

**Lemma 7.3.7** (Harrow–Hassidim–Lloyd (HHL) operator, $\mathcal{F}_{\text{HHL}}$). *Suppose that we have a register of n qubits available as well as an ancilla and m worker qubits where $m \in \mathbb{N}_{>0}$ denotes the binary approximation level of $\hat{A}$'s eigenvalues. Let $\hat{A} \in \mathbb{C}^{M\times N}$ and $\mathbf{b} \in \mathbb{C}^N$ be given for $M, N \in \mathbb{N}_{>0}$ with $N := 2^n$. There exists a quantum operator $\mathcal{F}_{\text{HHL}}(\hat{A}, |\mathbf{b}\rangle_n) = |\mathbf{x}\rangle_n$, where the following representations hold,*

$$\begin{aligned} |\mathbf{b}\rangle_n &:= \sum_{i=0}^{N-1} b_i\,|i\rangle_n \Big/ \sqrt{\sum_{l=0}^{N-1} b_l^2}\,, \\ |\mathbf{x}\rangle_n &:= \sum_{i=0}^{N-1} x_i\,|i\rangle_n \Big/ \sqrt{\sum_{l=0}^{N-1} x_l^2}\,, \end{aligned} \tag{7.43}$$

*such that for a given accuracy $\epsilon > 0$, the image $\mathbf{x} := [x_0, x_1, \ldots, x_{N-1}]' \in \mathbb{C}^N$ associated to $|\mathbf{x}\rangle_n$ satisfies,*

$$\|\mathbf{x} - \tilde{\mathbf{x}}\| \le \epsilon\,, \tag{7.44}$$

*where $\tilde{\mathbf{x}}$ is the classical solution of $\hat{A}\tilde{\mathbf{x}} = \mathbf{b}$.*

*Proof.* We prove the existence of $\mathcal{F}_{\text{HHL}}$ by providing its explicit construction following [72] in the context of a quantum computer showing three basic quantum registers, that is, the so-called ancilla (auxiliary) qubits $|q_A\rangle$, worker qubits $|q_W\rangle$, and input/output qubits $|q_{IO}\rangle$, such that we get,

$$q = |q_A\rangle \otimes |q_W\rangle \otimes |q_{IO}\rangle \;. \tag{7.45}$$

We start with an initial quantum state $q_0 = |0\rangle \otimes |0\rangle \otimes |\mathbf{b}\rangle_n$, where the ancilla and worker qubits are in the zero state.

**Step 1: (QPE)** We can apply the quantum operator $\mathcal{F}_{\text{QPE}}$ to this initial state and collect the eigenvalues in the worker qubits $q_W$, that is, we end up with a subsequent quantum state $|q_1\rangle$ of the form,

$$|q_1\rangle = |0\rangle \otimes \left|\tilde{\lambda}_i\right\rangle \otimes |\mathbf{v}_i\rangle_n \;, \tag{7.46}$$

where we assumed without loss of generality and for simplicity of the representation, that the vector **b** is an eigenvector of the Hermitian matrix Â, see (7.42) in the case of non-Hermitian Â. Note that $\tilde{\lambda}_i$ denotes the binary approximation of $\lambda_i$, which subsequently translates into the precision (7.44).

**Step 2: ($Y$-rotation)** By defining the angle,

$$\alpha_i = \arccos\left(\frac{\eta}{\tilde{\lambda}_i}\right), \tag{7.47}$$

for a hyperparameter[7] $\eta$, which accounts for the Hamiltonian simulation strategy, i. e., $e^{i\hat{A}t}$. Now applying the $e^{-iY\alpha_i}$-rotation leads to the following quantum state,

$$|q_2\rangle = \left(\sqrt{1-\eta^2/\tilde{\lambda}_i^2}\,|0\rangle + \eta/\tilde{\lambda}_i\,|1\rangle\right) \otimes \left|\tilde{\lambda}_i\right\rangle \otimes |\mathbf{v}_i\rangle \;. \tag{7.48}$$

Note that the ancilla register contains a useful representation of the eigenvalues, since the Hermitian nature of Â implies the existence of an orthogonal basis of eigenvectors $\mathbf{v}_i$, such that any given right-hand side **b** can be rewritten as follows,

$$|\mathbf{b}\rangle = \sum_{l=0}^{N-1} \beta_l\,|\mathbf{v}_l\rangle \;. \tag{7.49}$$

Herewith, (7.48) can be rewritten in a more general form,

7 An intrinsic parameter of the algorithm and not of the underlying physical problem. Hence it is not governed by actual (statistical) data.

$$|q_3\rangle = \sum_{l=0}^{N-1} \beta_l \left(\sqrt{1 - \eta^2/\tilde{\lambda}_l^2}\,|0\rangle + \eta/\tilde{\lambda}_l\,|1\rangle\right) \otimes \left|\tilde{\lambda}_l\right\rangle \otimes |\mathbf{v}_l\rangle_n\,, \tag{7.50}$$

which also motivates our wanted target representation $\sum_{l=0}^{N-1} \beta_l \frac{\eta}{\tilde{\lambda}_l} |\mathbf{v}_l\rangle_n = |\mathbf{x}\rangle_n$. Since we do not need values in the worker-qubit register anymore, we can uncompute them by resetting them to the standard zero quantum state $|0\rangle$.

**Step 3: (Measurement and postselection)** Since the quantum state (7.50) contains all the information necessary for a quantum solution to linear system problems, we simply need to measure the ancilla qubit register and postselect on the state $|1\rangle$-result such that we obtain the following quantum state,

$$|q_4\rangle = \sum_{l=0}^{N-1} \beta_l \frac{\eta}{\tilde{\lambda}_l} |\mathbf{v}_l\rangle_n \approx |\mathbf{x}\rangle_n\,, \tag{7.51}$$

where the approximation, as indicated in (7.44), accounts for the fact that the eigenvalues $\tilde{\lambda}_l$ are binary approximations of the true eigenvalues up to a certain precision. This proves the lemma. □

### 7.3.2 Multiscale problems' benefit from quantum computers

For the subsequent argument, we restrict ourselves to a stationary and elliptic multiscale problem, which still enjoys a wide range of applications such as describing flow or conductivity in highly heterogeneous materials/composites such as soil, for instance. In fact, we motivate that despite the practically relevant simplifications achieved by homogenization/systematic upscaling such as by the two-scale convergence method [3], the resulting limiting problem still represents a computationally demanding problem with the advantage that we do not have to fully resolve the microscale anymore with a computational mesh much smaller than the intrinsic heterogeneity of the underlying heterogeneous material.

We denote our heterogeneous material by $\mathcal{D} \subset \mathbb{R}^d$, which is a bounded open set, and $d > 0$ is the dimension of space. For a given charge density $\rho(\mathbf{x})$, we are looking for solutions $\phi^\epsilon(\mathbf{x})$ of the following classical problem,

$$\begin{cases} \operatorname{div}(\hat{\mathrm{A}}(\mathbf{x}, \mathbf{x}/\epsilon)\nabla\phi^\epsilon(\mathbf{x})) = \rho(\mathbf{x})\,, & \text{in } \mathcal{D}\,, \\ \phi^\epsilon = 0\,, & \text{on } \partial\mathcal{D}\,, \end{cases} \tag{7.52}$$

where the tensor $\hat{\mathrm{A}}$ represents the specific material conductivities. More precisely, we make the following assumptions on the given data $\rho$ and $\hat{\mathrm{A}}$ in the multiscale problem (7.52).

**Assumptions.**

**(A1)** *The right-hand side (e. g., charge density)* $\rho \in L^2(\mathcal{D})$.

**(A2)** *The conductivity tensor* $\hat{\mathrm{A}}(\mathbf{x},\mathbf{y})$ *is* $\mathbf{y}$*-periodic and strongly elliptic, i. e., there exist constants* $b \geq a > 0$ *such that,*

$$a|\zeta|^2 \leq \sum_{k,l=1}^{d} a_{kl}\zeta_k\zeta_l \leq b|\zeta|^2 . \tag{7.53}$$

**(A3)** *The conductivity* $\hat{\mathrm{A}}(\mathbf{x},\mathbf{y})$ *is an admissible test function, e. g.,*

$$\hat{\mathrm{A}} \in \left[C(\mathcal{D}; L^{\infty}_{\sharp}(Y))\right]^{d\times d} ,$$

*see* [3] *for a weaker admissibility requirement albeit more difficult to verify in practice.* ◇

We can easily verify weak solutions $\phi^\epsilon \in H^1_0(\mathcal{D})$ for (7.52). In fact, for problem (7.52) together with the assumptions **(A1)–(A3)**, the two-scale convergence allows us to rigorously establish the following two-scale limit (i. e., passing to the limit as $\epsilon \to 0$) (see Theorem 2.3 in [3]),

$$\begin{cases} -\mathrm{div}_{\mathbf{y}}\big(\hat{\mathrm{A}}(\mathbf{x},\mathbf{y})(\nabla_{\mathbf{x}}\phi(\mathbf{x}) + \nabla_{\mathbf{y}}\phi^1(\mathbf{x},\mathbf{y}))\big) = 0 , & \text{in } \mathcal{D}\times Y , \\ -\mathrm{div}_{\mathbf{x}}\big(\int_Y \hat{\mathrm{A}}(\mathbf{x},\mathbf{y})(\nabla_{\mathbf{x}}\phi(\mathbf{x}) + \nabla_{\mathbf{y}}\phi^1(\mathbf{x},\mathbf{y}))\,d\mathbf{y}\big) = \rho(\mathbf{x}) , & \text{in } \mathcal{D} , \\ \phi(\mathbf{x}) = 0 , & \text{on } \partial\Omega , \\ \phi^1(\mathbf{x},\mathbf{y}) , & \text{is } Y\text{-periodic.} \end{cases} \tag{7.54}$$

Note that multiscale modeling relying on a classical computer preferably aims to decouple the macroscale from the microscale by averaging out the microscopic information and providing it to the macroscopic formulation in the form of effective model parameters such as effective diffusion or conductivity coefficients; see the sections on upscaling (Chapter 6) and homogenization (Chapter 10).

Our key motivation for the general usefulness of quantum computing for multiscale modeling relies on the fact that upscaled effective models obtained by homogenization and related upscaling techniques reduce to solving linear systems of equations after choosing a specific numerical discretization strategy. To simplify the argument, we consider a separable material-specific tensor $\hat{\mathrm{A}}(\mathbf{x},\mathbf{y})$ (as motivated by Harbrecht and Schwab [69]), i. e., $\hat{\mathrm{A}}(\mathbf{x},\mathbf{y}) = \hat{\mathrm{A}}_1(\mathbf{x})\hat{\mathrm{A}}_2(\mathbf{y})$. Since this separability property is generally not available in practical examples, we note that it can be approximated via an expansion of the form,

$$\hat{\mathrm{A}}(\mathbf{x},\mathbf{y}) = \sum_{l=1}^{L} \hat{\mathrm{A}}^l_1(\mathbf{x})\hat{\mathrm{A}}^l_2(\mathbf{y}) . \tag{7.55}$$

Hence, we continue our argument under the assumption of a separable tensor, i. e., $\hat{A}(\mathbf{x},\mathbf{y}) = \hat{A}_1(\mathbf{x})\hat{A}_2(\mathbf{y})$.

We choose the finite element spaces $V_0^h \subset H_0^1(\mathcal{D})$, $V^{h_0} \subset H^1(\mathcal{D})$, and $V^{h_1} \subset H_\sharp^1(\mathcal{Y})$ for mesh sizes $h, h_0, h_1 > 0$, and for linear basis functions $P_1(K)$ defined on triangles $K \in \mathcal{T}_\delta^D$, where $\delta \in \{h, h_0, h_1\}$ is the mesh size of the specific triangulation $\mathcal{T}_\delta^D$ applied to the domain $D \in \{\mathcal{D}, Y\}$ to discretize the two-scale limit (7.54), that is,

$$\begin{cases} V_0^h := \operatorname{span}\{\varphi \in C(\overline{\mathcal{D}}) \mid \varphi|_K \in P_1(K), K \in \mathcal{T}_h^{\mathcal{D}}\}, \\ V^{h_0} := \operatorname{span}\{\varphi \in C(\overline{\mathcal{D}}) \mid \varphi|_K \in P_1(K), K \in \mathcal{T}_{h_0}^{\mathcal{D}}\}, \\ V^{h_1} := \operatorname{span}\{\varphi \in C(\overline{Y}) \mid \varphi|_K \in P_1(K), K \in \mathcal{T}_{h_1}^{Y}\}. \end{cases} \tag{7.56}$$

Hence, $V_0^h$, $V^{h_0}$, and $V^{h_1}$ discretize the Sobolev spaces $H_0^1(\mathcal{D})$, $H^1(\mathcal{D})$, and $H_\sharp^1(\mathcal{Y})$, respectively. In particular, $V^{h_0}$ resolves any constraints different from homogeneous Dirichlet conditions imposed on the boundary.

Therefore, for the discrete charge density $\rho_h \in V_0^h$ given by,

$$\rho_h(\mathbf{x}) := \sum_{K \in \mathcal{T}_h^{\mathcal{D}}} \rho_K \varphi_K(\mathbf{x}) \in V_0^h, \tag{7.57}$$

we can look for numerical solutions,

$$\phi_h(\mathbf{x}) = \sum_{K \in \mathcal{T}_h^{\mathcal{D}}} \phi_K \varphi_K(\mathbf{x}) \in V_0^h,$$

and,

$$\phi_h^1(\mathbf{x}) = \sum_{K \in \mathcal{T}_h^{\mathcal{D}}} \sum_{K' \in \mathcal{T}_{h_1}^{Y}} \phi_{K,K'} \varphi_K(\mathbf{x}) \otimes \psi_{K'}(\mathbf{y}) \in V_{h_0} \otimes V_{h_1},$$

solving the following discrete linear system approximating (7.54),

$$\hat{A} \begin{bmatrix} \phi_h(\mathbf{x}) \\ \phi_h^1(\mathbf{x},\mathbf{y}) \end{bmatrix} = \begin{bmatrix} \rho_h(\mathbf{x}) \\ 0 \end{bmatrix}, \tag{7.58}$$

where the system matrix $\hat{A}$ is defined for the number of basis elements $N_x$ on the macroscale $\mathcal{D}$ and $N_y$ on the microscale $\mathcal{Y}$ by,

$$\hat{A} := \begin{bmatrix} a\hat{A}_1^{\nabla,\nabla} & \sum_{i=1}^d \hat{b}^i \otimes \hat{A}_1^{\nabla,i} \\ (\sum_{i=1}^d \hat{b}^i \otimes \hat{A}_1^{\nabla,i})^T & \hat{A}_2^{\nabla,\nabla} \otimes \hat{A}_1 \end{bmatrix} \in \mathbb{R}^{(dN_x+dN_y dN_x)\times(dN_x+dN_y dN_x)}, \tag{7.59}$$

and its components satisfy,

$$\hat{b}^i := \{b^i_{k_1,k_2,\dots,k_d}\}_{1\le k_i \le N_y},$$
$$\hat{A}_1 := \{A^1_{k_1,k_2,\dots,k_d;l_1,l_2,\dots,l_d}\}_{1\le k_i,l_i \le N_x},$$

$$\hat{A}_1^{\nabla,\nabla} := \{A^{1,\nabla,\nabla}_{k_1,k_2,\dots,k_d;l_1,l_2,\dots,l_d}\}_{1\le k_i,l_i\le N_x},$$
$$\hat{A}_1^{\nabla,i} := \{A^{1,\nabla,i}_{k_1,k_2,\dots,k_d;l_1,l_2,\dots,l_d}\}_{1\le k_i,l_i\le N_x},$$

and,

$$\hat{A}_2^{\nabla,\nabla} := \{A^{2,\nabla,\nabla}_{k_1,k_2,\dots,k_d;l_1,l_2,\dots,l_d}\}_{1\le k_i,l_i\le N_x},$$

where,

$$\begin{aligned}
a &:= \int_Y A_2(\mathbf{y})\,d\mathbf{y}\,,\\
b^i_{k_1,k_2,\dots,k_d} &:= \int_Y A_2(\mathbf{y})\frac{\partial}{\partial y_i}(\psi_{k_1}(y_1)\otimes\psi_{k_2}(y_2)\otimes\cdots\otimes\psi_{k_d}(y_d))\,d\mathbf{y}\,,\\
A^1_{k_1,k_2,\dots,k_d;l_1,l_2,\dots,l_d} &:= \int_{\mathcal{D}} A_1(\mathbf{x})(\varphi_{k_1}(x_1)\otimes\varphi_{k_2}(x_2)\otimes\cdots\otimes\varphi_{k_d}(x_d))\\
&\quad(\varphi_{l_1}(x_1)\otimes\varphi_{l_2}(x_2)\otimes\cdots\otimes\varphi_{l_d}(x_d))\,d\mathbf{x}\,,\\
A^{1,\nabla,\nabla}_{k_1,k_2,\dots,k_d;l_1,l_2,\dots,l_d} &:= \sum_{i=1}^d\int_{\mathcal{D}} A_1(\mathbf{x})\frac{\partial}{\partial x_i}(\varphi_{k_1}(x_1)\otimes\varphi_{k_2}(x_2)\otimes\cdots\otimes\varphi_{k_d}(x_d))\\
&\quad\frac{\partial}{\partial x_i}(\varphi_{l_1}(x_1)\otimes\varphi_{l_2}(x_2)\otimes\cdots\otimes\varphi_{l_d}(x_d))\,d\mathbf{x}\,,\\
A^{1,\nabla,i}_{k_1,k_2,\dots,k_d;l_1,l_2,\dots,l_d} &:= \int_{\mathcal{D}} A_1(\mathbf{x})\frac{\partial}{\partial x_i}(\varphi_{k_1}(x_1)\otimes\varphi_{k_2}(x_2)\otimes\cdots\otimes\varphi_{k_d}(x_d))\\
&\quad(\varphi_{l_1}(x_1)\otimes\varphi_{l_2}(x_2)\otimes\cdots\otimes\varphi_{l_d}(x_d))\,d\mathbf{x}\,,\\
A^{2,\nabla,\nabla}_{k_1,k_2,\dots,k_d;l_1,l_2,\dots,l_d} &:= \sum_{i=1}^d\int_Y A_2(\mathbf{y})\frac{\partial}{\partial y_i}(\psi_{k_1}(y_1)\otimes\psi_{k_2}(y_2)\otimes\cdots\otimes\psi_{k_d}(y_d))\\
&\quad\frac{\partial}{\partial y_i}(\psi_{l_1}(y_1)\otimes\psi_{l_2}(y_2)\otimes\cdots\otimes\psi_{l_d}(y_d))\,d\mathbf{y}\,.
\end{aligned}\tag{7.60}$$

Finally, we recall that a standard one-dimensional linear basis element is defined by,

$$\zeta_i(x) := \begin{cases}\frac{x-x_{i-1}}{\delta} & \text{if } x_{i-1}\le x\le x_i\,,\\ \frac{x_{i+1}-x}{\delta} & \text{if } x_i\le x\le x_{i+1}\,,\\ 0 & \text{else}\,.\end{cases}\tag{7.61}$$

The form of the numerical multiscale problem (7.58) obtained by a fairly standard discretization of the two-scale limit/upscaled effective macroscopic description (7.54) provides a general linear problem, to which we can apply the quantum operator $\mathcal{F}_{HHL}$ proposed by Harrow, Hassidim, and Lloyd as discussed in Section 7.3.1.3.

Hence, thanks to Lemma 7.3.7, we immediately obtain the following main result:

**Quantum Computing Algorithm (QCA) solving Upscaled Elliptic Equations (UEE)**

**Lemma 7.3.8.** *Suppose that we have access to an n-qubit register as well as an ancilla and m worker qubits where $m \in \mathbb{N}_{>0}$ denotes the binary approximation level of $\hat{A}$'s eigenvalues. Let $\hat{A}$ be defined as in (7.59), and let* **b** *be the right-hand side in (7.58). There exists a quantum operator $\mathcal{F}_{\mathrm{HHL}}(\hat{A}, |\mathbf{b}\rangle_n) = |\mathbf{x}\rangle_n$, where the following representations hold,*

$$|\mathbf{b}\rangle_n := \sum_{i=0}^{M-1} b_i \,|i\rangle_n \Big/ \sqrt{\sum_{l=0}^{M-1} b_l^2}\,,$$
$$|\mathbf{x}\rangle_n := \sum_{i=0}^{M-1} x_i \,|i\rangle_n \Big/ \sqrt{\sum_{l=0}^{M-1} x_l^2}\,, \tag{7.62}$$

*such that for a given accuracy $\epsilon > 0$, the image $\mathbf{x} := [x_0, x_1, \ldots, x_{M-1}]' \in \mathbb{C}^M$, $M := 2^{dN_x + dN_y dN_x}$, associated to $|\mathbf{x}\rangle_n$ satisfies,*

$$\|\mathbf{x} - \tilde{\mathbf{x}}\| \le \epsilon\,, \tag{7.63}$$

*where $\tilde{\mathbf{x}}$ is the classical solution of $\hat{A}\tilde{\mathbf{x}} = \mathbf{b}$.*

**Remark 7.3.9.** Lemma 7.3.8 guarantees that solutions $|\mathbf{x}\rangle$ approximate the corresponding classical numerical solution $\tilde{\mathbf{x}} := [\phi_h(\mathbf{x}), \phi_h^1(\mathbf{x}, \mathbf{y})]'$ with a given accuracy $\epsilon > 0$. Moreover, the numerical solution $|\mathbf{x}\rangle_n$ is a quantum representation of the linear finite element tensor product approximation of the solution $[\phi(\mathbf{x}), \phi^1(\mathbf{x}, \mathbf{y})]'$ of the upscaled/homogenized equation (7.54). ◇

The considerations in this section clearly demonstrate that the fundamental techniques for multiscale modeling and upscaling of microscopic scientific model formulations remain of general importance in a postquantum computing era. In particular, the availability of quantum computers will allow us to efficiently compute rigorously upscaled, nonlinear, and coupled systems of equations, which require us to solve for additional auxiliary equations involving micro- and macroscale quantities next to the effective macroscopic equation showing a (reference cell)-avarage of terms depending on both, the micro- and the macroscale, see (7.54).

# 8 Electrochemical CHeSs

## 8.1 Energy storage and fuel cell systems

Before delving into specific examples of batteries and fuel cells for energy storage, let us briefly discuss, why these electrochemical devices are examples of CHeSs. The very intuitive and straightforward motivation is that electrochemical energy storage systems represent in general a system of complex systems due to their composition of interlinked subsystems such as anode, cathode, and separator under a macroscopic point of view. In fact, the number of linked systems depends on the level of detail one is interested in. A more microscopic view would also account for the components building the electrodes such as active intercalation hosts, carbon black, and electrolyte, or the composition of the electrolyte forming the separator, for instance.

### 8.1.1 CHeSs describing energy storage systems

The term *energy storage system* is very general in the sense that it accounts for any system that is able to store energy in the form of heat (e. g., media/composites efficiently storing heat[1]), a mechanical process (e. g., hydro power station), charged species (e. g., batteries, solar cells, fuel cells), and corresponding physical, chemical, or even biological opportunities. We restrict our considerations to electrochemical storage systems such as batteries and fuel cells.

#### 8.1.1.1 Six commonly used Li-battery designs

The main components of batteries are the following: anodic and cathodic current collectors, anode (i. e., negative electrode), electrolyte and separator,[2] and cathode (i. e., positive electrode). We further present a summary of performance characteristics of six different electrode–electrolyte combinations as published online, see Figure 8.2. Common to all these six battery designs is the fact that $Li^+$-ions represent the charge transport in the electrolyte phase, see Figure 8.1.

**Lithium Cobalt Oxide (LCO) Battery:** LCO batteries consist of graphite ($LiC_6$) and lithium cobalt oxide ($LiCoO_2$) as cathode and anode materials, respectively.

**Lithium Manganese Oxide (LMO) Battery:** LMO batteries have graphite ($LiC_6$) as anode material and commonly lithium manganese oxide (the spinel $LiMn_2O_4$) as the cathode material.

1 Efficient heat storage is still an active research field.

2 Often, the electrolyte is already considered as a separator, since it prevents the movement of electrons.

https://doi.org/10.1515/9783110579543-008

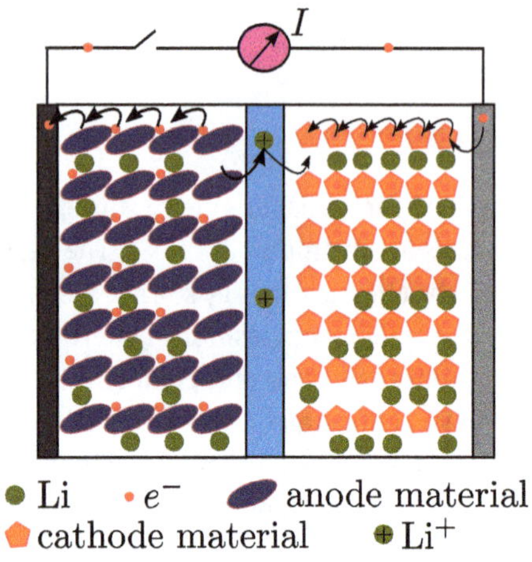


Figure 8.1: *General design of a Li-ion battery:* The anode material (negative electrode) releases an electron $e^-$ into the external circuit while giving a positively charged Li-ion $Li^+$ free to the electrolyte/separator (blue square). The electron leaves the battery via the dark grey current collector (which generally is aluminum (Al)) of the anode. While the electron moves via an external circuit driving a device under a certain current $I$ to the cathodic current collector (which one frequently chooses to be copper (Cu)), the Li-ion goes through the electrolyte to the cathode, into which it intercalates by consuming the electron $e^-$ from the external circuit. Naturally, there are primary useful anodic and cathodic reactions, along with various secondary side reactions at different interfaces such as the "current collector"–"electrode" or "electrode"–"electrolyte" interfaces.

**Lithium Iron Phosphate (LFP) Battery:** LFP batteries use iron phosphate ($FePO_4$) and graphitic carbon ($LiC_6$) as the cathode and anode materials, respectively.

**Nickel Manganese Cobalt (NMC) Battery:** NMC batteries use lithium nickel manganese cobalt oxide ($LiNiMnCoO_2$) as the cathode material and again graphite ($LiC_6$) as the anode material.

**Nickel Cobalt Aluminium (NCA) Battery:** NCA batteries consist of lithium nickel cobalt aluminum oxide ($LiNiCoAlO_2$) and graphite ($LiC_6$) as the cathode and anode materials, respectively.

**Lithium Titanium Oxide (LTO) Battery:** LTO batteries differ from the previous examples by replacing the anode material with titanate (e. g., $Li_2TiO_3$), in which case lithium manganese oxide (NMC) can be chosen as the cathode material.

### 8.1.2 Fuel cell systems representing electrochemical CHeSs

The design of fuel cells is very close to batteries in that they also consist of two electrodes (i. e., an anode and a cathode) and a separator (electrolyte). Unlike batteries, the

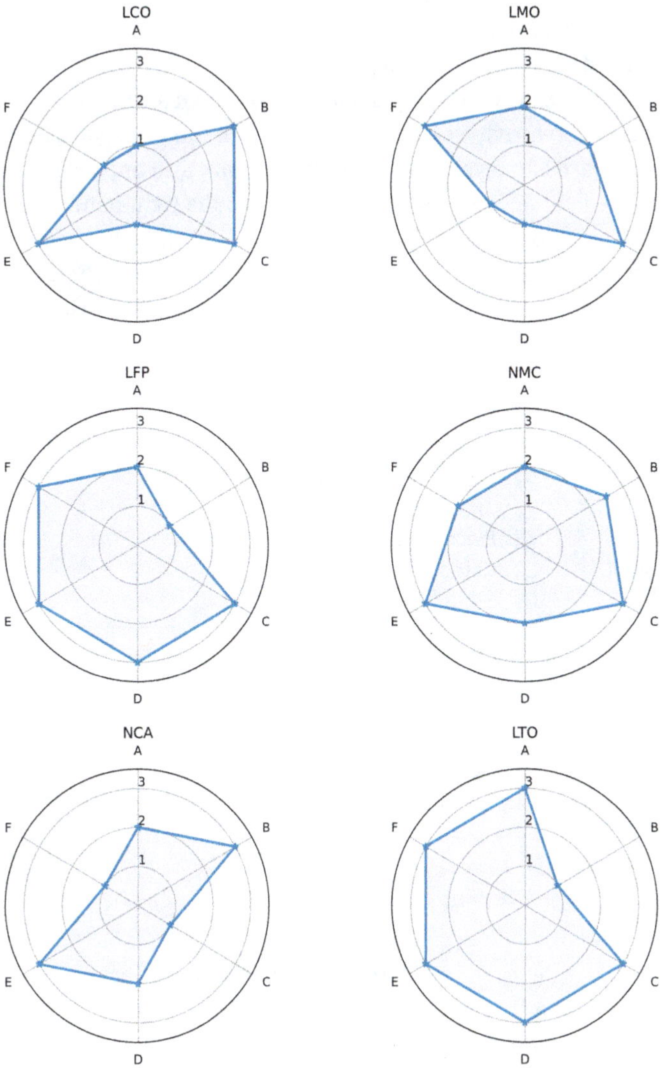


**Figure 8.2:** *Comparison of six different Li-battery types:* A, B, C, D, E, and F denote *specific power*, *specific energy*, *cost*, *life span*, *performance*, and *safety*, respectively. (Source of image: https://dragonflyenergy.com/types-of-lithium-batteries-guide/. Date of access: 31 October 2023..)

catalytic performance of the anode is improved with the help of expensive catalysts such as palladium, platinum, and even ruthenium (Ru). A primary driver for interest in fuel cell technology is the operational time, which is generally longer for fuel cells as demonstrated by NASA for its extended missions to space [112]. In fact, the fuel cells installed by NASA into the space shuttle produce water as a byproduct for consumption/use.

#### 8.1.2.1 Direct methanol fuel cell (DMFC)

Consider a rather standard design of a DMFC depicted in Figure 8.3. Frequently used anode materials are graphene oxide, polyvinylpyrrolidone, and even a combination of both. Additionally, to increase reactivity, catalysts like platinum (Pt) nanoparticles or a combination of Pt and Ru are employed. The same catalysts are also applied on the cathode side. In fact, on the left-hand side of the anode and on the right-hand side of the cathode, there are gas diffusion layers. Between the anode and cathode, there is a *polymer electrolyte membrane* as electrolyte. Hence, the DMFC is close to a *Proton Exchange Membrane (PEM)* fuel cell albeit it can draw hydrogen from (inexpensive and easily transported and stored) pure methanol via a Pt-Ru-catalyst.

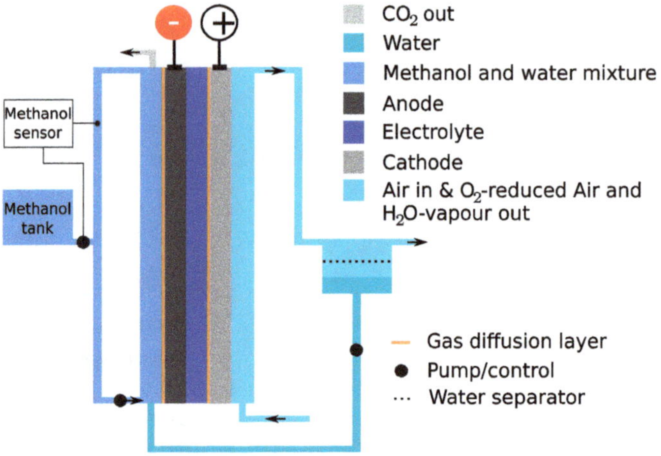


**Figure 8.3:** *Direct Methanol Fuel Cell design:* The image does not explicitly show the presence of gas diffusion layers on the left-hand side of the anode and on the right-hand side of the cathode.

#### 8.1.2.2 Proton exchange membrane fuel cell (PEMFC)

A general design of a PEMFC is depicted in Figure 8.4. We immediately recognize that the *Gas Diffusion Layer (GDL)* is a common feature to both DMFC and PEMFC. Its primary purpose is to reduce degradations in transport of reactants to the CL, e. g., water

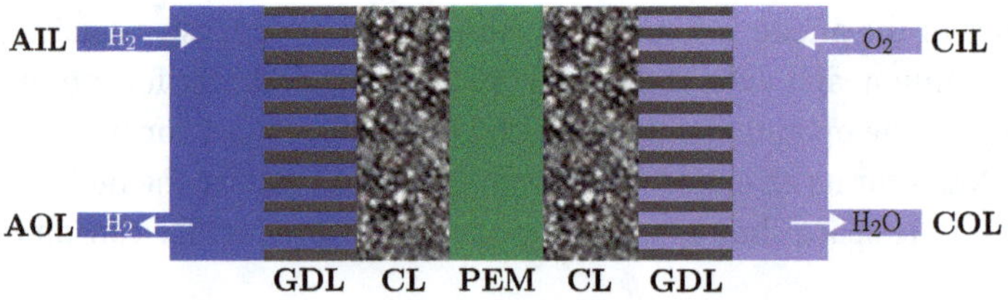


**Figure 8.4:** *Proton Exchange Membrane Fuel Cell design:* Anodic InLet (AIL), Anodic OutLet (AOL), Gas Diffusion Layer (GDL), Catalyst Layer (CL), Proton Exchange Membrane (PEM), Cathodic InLet (CIL), and Cathodic OutLet (COL). The electrons flow from the anodic CL over an external circuit and come back to the cathodic CL after powering an external device.

management. The catalytic performance is primarily defined by the geometric and material/chemical properties of the CL design. In this context, an effective macroscopic catalyst layer description has been reliably and systematically derived in [140, 141] using the upscaling techniques presented in this book by allowing to account for the underlying chemical and geometric characteristics of the employed materials.

#### 8.1.2.3 Solid oxide fuel cell (SOFC)

The general design of a SOFC shown in Figure 8.5 is simpler than that of PEMFCs, since no GDLs are required. The high temperature (~600 °C) generated by the SOFC makes it also interesting for future commercial, governmental, industrial, and residential applications. However, the Wilton E. Scott Institute of Carnegie Mellon University (see [76]) estimates that the U.S. Department of Energy's targets on costs, such as \$ 900/kW$_{net}$ and a stack voltage degradation rate of 0.2 %/1000 h, will be reached between 2035 and 2050.

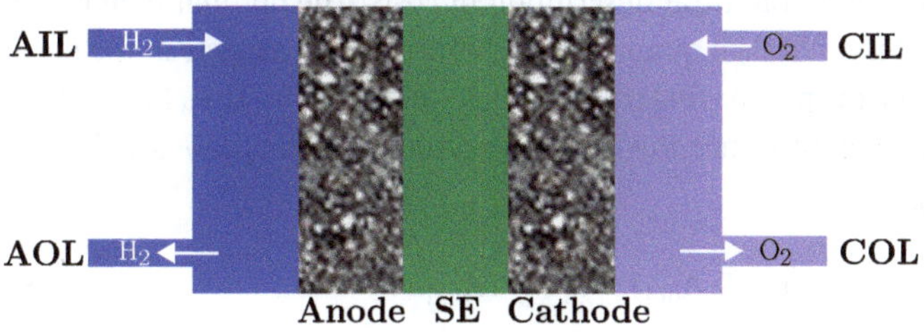


Figure 8.5: *Solid Oxide Fuel Cell design:* Solid Electrolyte (SE).

## 8.2 Optimal material design as interacting CHeSs

The goal of this section is to exploit the tools established in Section 2.8 for optimizing material design such as composition. For simplicity, we focus on optimizing the composition of two types of matter showing competing properties such as different conductivities and geometric characteristics. In fact, we look at the material characteristics of two different materials $M_1$ and $M_2$, see Table 8.1.

Table 8.1: *Optimizing material conductivity.* The conductivities $\sigma_i$, $i \in \{1,2\}$, are homogeneous across the same material and hence represent conductivity densities along the electron flow (1D in general).

| **Property** | $M_1$ | $M_2$ | **Physical Dimension** |
|---|---|---|---|
| Volume | $V_1$ | $V_2$ | $[V_i]_d = [\mathrm{m}^3]$ for $i \in \{1,2\}$ |
| Conductivity | $\sigma_1$ | $\sigma_2$ | $[\sigma_i]_d = [[\mathrm{kg}]^{-1}[\mathrm{m}]^{-3}[\mathrm{s}]^3[\mathrm{A}]^2] = [\mathrm{S}]/[\mathrm{m}]$ for $i \in \{1,2\}$ |
| Costs | $c_1$ | $c_2$ | $[c_i]_d = [\$]$ |
| $\sigma$-Proportionality | $q_1$ | $q_2$ | $q_1 + q_2 = 1$ |
| $V$-Proportionality | $p_1$ | $p_2$ | $p_1 + p_2 = 1$ |

Recall that voltage can be expressed in SI base units [m], [kg], [s], and [A] by,

$$[V] = \frac{[E_{\text{pot}}]_d}{[Q]_d} = \frac{[J]}{[C]} = \frac{[\text{kg}][\text{m}^2][\text{s}^{-2}]}{[\text{A}][\text{s}]} = [\text{A}][\Omega]\,,$$

since a resistor $R$ ($[R]_d = [\Omega]$), a current $I$ ($[I]_d = [\text{A}]$), and a voltage $U$ ($[U]_d = [\text{V}]$) are related to each other via Ohm's law. A parameter that accounts for the operational setup of our conducting materials depends on the current $I$ applied across it, the surface area $A_{\text{nfd}}$ ($[A_{\text{nfd}}]_d = [\text{m}]^2$) orthogonal to the flow direction of the current, and the cost of (electric) energy $c_{\text{en}}$ ($[c_{\text{en}}]_d = \frac{[\$]}{[\text{W}]}$), that is,

$$a := \frac{I^2 c_{\text{en}}}{L^i}\,, \tag{8.1}$$

where $L^i$ ($[L^i]_d = [\text{m}]$) is the length of the conductor along the direction of electron flow. The latter parameter (8.1) allows us to write terms into the dimension of a specific currency and hence into a natural description of costs. The optimal material design/composition requires a physical quantity that defines, which aspects in the design we want to favor, such as low purchasing prices or high performance (i. e., high conductivity). For this purpose, we introduce the following system specific goal/cost/objective function,

$$G(\sigma, c, L, A_{\text{nfd}}) := -\frac{a}{\sigma} - cA_{\text{nfd}}L\,. \tag{8.2}$$

For simplicity and without loss of generality for our discussion, we introduce,

$$\textbf{Assumption (MDI)} \qquad \begin{cases} I = 1\,, \\ c_{\text{en}} = 1\,. \end{cases} \tag{8.3}$$

A plot of the objective function (8.2) shows that its characteristic shape does not visibly change except for very small deviations in the corresponding values, see Figure 8.6. Now, we are able to state our main interest based on the above conductivity characterization of materials:

**Optimal Design of Electron Conducting Materials (ODoECM):** *We want to maximize the material conductivity $\sigma$ by taking into account material costs $c$, operational aspects $a$, and geometric design properties via $A_{\text{nfd}}$ and $L$ as combined in the general objective function $G$ in* (8.2). ◇

To achieve our goal defined in **(ODoECM)**, we apply the framework of *reasonably interacting CHeSs* to the design of conducting materials.

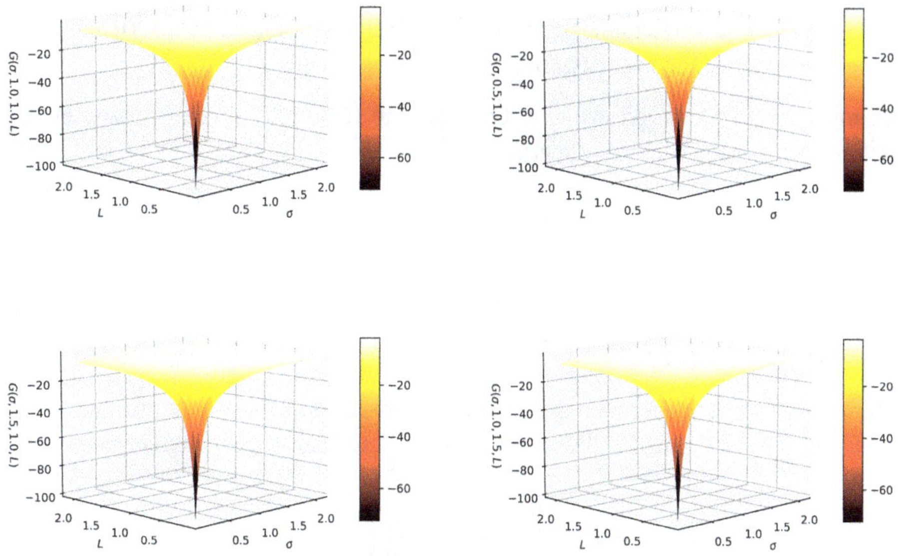


Figure 8.6: *Objective function* $G(\sigma, c, L, A_{\text{nfd}})$*:* There is no visible difference between various different parameter settings.

---

**Optimizing Material Design as an Interacting CHeS (OMDIC)**

To *optimize material design* based on a characteristic *objective/goal function*, rewrite the problem as an interaction problem of two systems, where system $\mathcal{S}_1$ represents the available options for the geometric design (e. g., $A_{\text{nfd}}$ and $L$), and system $\mathcal{S}_2$ controls/chooses the material properties (e. g., $\sigma$ and $c$). The solution to the optimal material design problem **(ODoECM)** then consists of the *reasonable actions/optimal trade off* between both systems $\mathcal{S}_1$ and $\mathcal{S}_2$ under the given circumstances. This leads to two possible solutions:

**[PD]** **Pure Design:** The interacting systems show a *saddle point*, which implies a *pure design* and hence an *equilibrium*.

**[MD]** **Mixed Design:** In the absence of a *saddle point*, the optimal material design is a mixture of all the available material characteristics, i. e., a *mixed design*. This represents a *mixed interaction equilibrium*.

---

**Remark 8.2.1** (Optimization as a game). Note that the key observation in our formulation of the above method **(OMDIC)** relies on the general fact that optimization problems depend on competing quantities such as design parameters versus costs and performance of materials.[3] Therefore, it can be directly translated into a *game*, i. e., a *reasonably interacting CHeS*, in which we aim to look for an equilibrium solution that satisfies given opposing interests in an optimal trade off. ◇

3 In fact, the same dilemma is very general in that it also hinders nations/states in truly optimizing, by not being able to follow the natural optimal path from a physical/nature's (which includes physical, social, and general scientific laws) point of view due to conflicts of interests between parties, authorities, industries, people, etc..

Let us look at two examples demonstrating the two possible designs **[PD]** and **[MD]**, which can be obtained by the **(OMDIC)**-method as a result.

**Example 8.2.2** (Pure design solution). Consider the **(ODoECM)**-problem based on the following available material properties,

$$\begin{aligned} c_1 &= 1\,, \quad c_2 = 2\,,\\ \sigma_1 &= 1\,, \quad \sigma_2 = 2\,,\\ L^1 &= 1\,, \quad L^2 = 2\,,\\ A^1_{\mathrm{nfd}} &= 1\,, \quad A^2_{\mathrm{nfd}} = 2\,. \end{aligned} \tag{8.4}$$

These properties together with the general **Assumption (MDI)** in (8.3) turn into an interacting CHeS of competing designs governed by the following *interaction matrix*,

$$\hat{\mathrm{G}} := \begin{bmatrix} -2 & -\frac{5}{2} \\ -\frac{9}{2} & -\frac{33}{4} \end{bmatrix}, \tag{8.5}$$

with components defined by the values obtained via the objective function (8.2). The **Saddle Point Check** gives,

$$\begin{array}{cc|cc} -2 & -\frac{5}{2} & -\frac{5}{2} & -\frac{5}{2} \\ -\frac{9}{2} & -\frac{33}{4} & -\frac{33}{4} & \\ \hline -2 & -\frac{5}{2} & & \\ & -\frac{5}{2} & & \end{array}, \tag{8.6}$$

see Section 2.8. Computation (8.6) shows that the optimal design represents a *pure design* **[PD]** given by the material performance and price properties,

$$c_2 = 2\,, \quad \sigma_2 = 2$$

and the geometric design characteristics,

$$L^1 = 1\,, \quad A^1_{\mathrm{nfd}} = 1\,.$$

As a conclusion, under objective (8.2), the optimal material design consists of high performance and costly material choices in a small geometric setup. ◇

Interestingly, small changes in the available material characteristics can lead to an entirely different optimal material design.

**Example 8.2.3** (Mixed design solution). As in the previous Example 8.2.2, we study the **(ODoECM)**-problem, but for the second material, we consider different properties. That

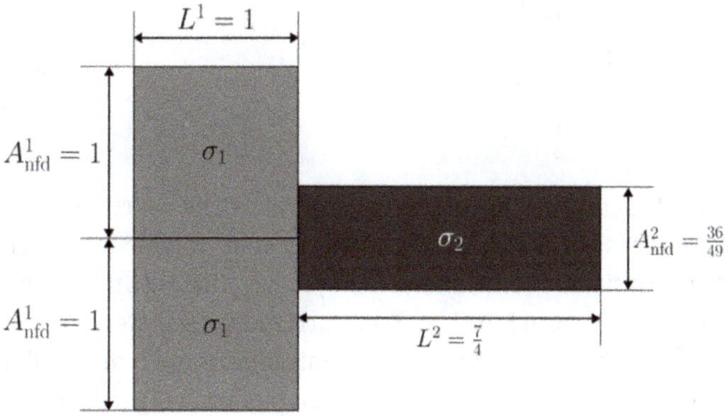


**Figure 8.7:** *Material Choices implying a Mixed Design:* Horizontal axis represents the electron conducting direction.

is, we have the following material characteristics available,

$$\begin{aligned} c_1 &= 1, & c_2 &= \frac{5}{4}, \\ \sigma_1 &= 1, & \sigma_2 &= 2, \\ L^1 &= 1, & L^2 &= \frac{7}{4}, \\ A^1_{\mathrm{nfd}} &= 1, & A^2_{\mathrm{nfd}} &= \frac{36}{49}. \end{aligned} \tag{8.7}$$

These parameters define conductive materials with designs depicted in Figure 8.7. Herewith, we can immediately determine the following *interaction matrix* by looking at it as an interacting CHeSs trading competing design, costs, and performance parameters,

$$\hat{\mathrm{G}} := \begin{bmatrix} -2 & -\frac{7}{4} \\ -\frac{13}{7} & -\frac{53}{28} \end{bmatrix}, \tag{8.8}$$

which can be computed with (8.2). As before, we make the **Saddle Point Check** to see, whether the solution is a *pure design*. We obtain the following,

$$\begin{array}{cc|cc} -2 & -\frac{49}{28} & -2 & \\ -\frac{52}{28} & -\frac{53}{28} & -\frac{53}{28} & -\boldsymbol{\frac{53}{28}} \\ \hline -\frac{52}{28} & -\frac{49}{28} & & \\ -\boldsymbol{\frac{52}{28}} & & & \end{array} . \tag{8.9}$$

The **(ODoECM)**-problem seems to be very close to a saddle point, but it is not because of the difference between $\frac{-52}{28}$ and $\frac{-53}{28}$. Hence, we compute the appropriate *mixed material design* by,

$$\begin{array}{cc|cc} 0 & 7 & -7 & \frac{1}{8} \\ 4 & 3 & 1 & \frac{7}{8} \\ \hline -4 & 4 & & \\ \frac{1}{2} & \frac{1}{2} & & \end{array}, \tag{8.10}$$

where we used the fact that we can multiply the interaction matrix by a factor such as 28, and so we can add a constant such as 56 to each component of the interaction matrix without violating the intrinsic rationale of the interaction system. As a result, the reasonable geometric action is given by $\mathbf{s}_1 = [0.5, 0.5]$, and the corresponding material action is given by $\mathbf{s}_2 = [1/8, 7/8]$. A possible realization of such a mixed material is depicted in Figure 8.8. A study of the MIG for various possible mixtures is plotted in Figure 8.9. ◇

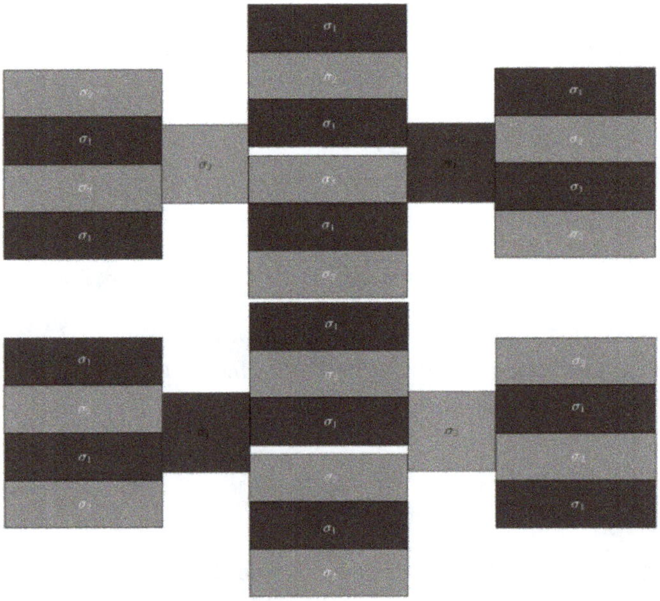

Figure 8.8: *Example of a 2D G-optimal Mixed Design based on the computed reasonable interaction (*8.10*):* Mixture based on applying conductivities $\sigma_1$ and $\sigma_2$ with ratio 1 : 1 and the geometric shapes $(A^1_{\text{nfd}} = 1, L^1 = 1)$ and $(A^2_{\text{nfd}} = \frac{36}{49}, L^2 = \frac{7}{4})$ with ratio 1 : 7. Note that the electron conduction is again in the horizontal direction.

**Remark 8.2.4** (Industrial relevance of Mixed Designs **[MD]**). At first, the necessity of a mixed design solution seems like a disadvantage from an industrial perspective, since the material designs available from the outset are generally given by an underlying production process being already in place. This is why a starting point with two possible geometries (belonging to two production processes for instance) such as $(A^1_{\text{ndf}} = 1, L^1 = 1)$ and $(A^2_{\text{ndf}} = \frac{36}{49}, L^2 = \frac{7}{4})$ is not so unlikely. However, it happens frequently that underlying production units for bringing materials into a certain shape/geometry can be applied to a wide range of different (conducting) materials. This immediately calls for an interest in the geometric mixing. Finally, the mixing of the materials depends on the possibility to purchase enough quantities under the prices accounted for in the material interaction system governed by $\hat{G}$ in Example 8.2.3. ◇

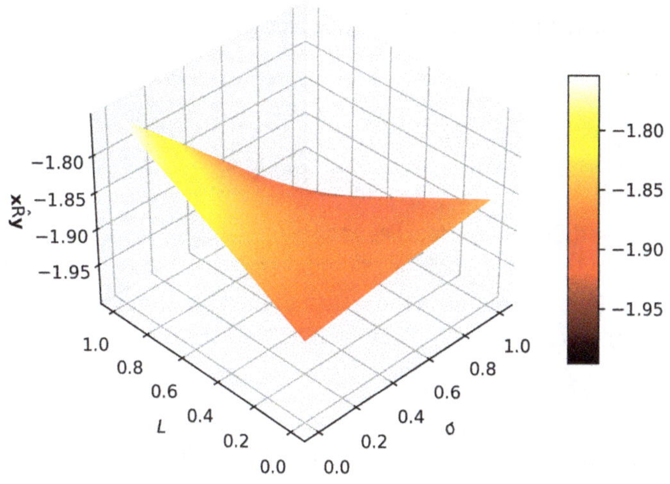


**Figure 8.9:** *Mean Interaction Gain (MIG):* The MIG for various choices of geometric $\mathbf{x} := [p_1, 1 - p_1]$ and material actions $\mathbf{y} = [q_1, 1 - q_1]'$ shows a saddle point, which represents the equilibrium between material and geometric properties.

## 8.3 Battery modeling

The general awareness of the need to reduce our carbon footprint requires us to rely on sustainable and nonfossil energy as much as possible. Immediate candidates for alternative energy resources are solar and wind energies, albeit they depend on increasingly uncertain weather and climate conditions. Similarly, electric vehicles seem currently to emerge as the most promising alternative to petrol-driven transport based on policy and the state of technological developments. To pave the way for sustainable and climate-preserving energy solutions, we need to develop efficient and reliable energy storage and management systems that would allow us to balance the gaps and surplus in our electricity networks. In this section, we focus on the more microscopic aspects of this endeavor. We aim to gain a systematic and fundamental understanding of how the geometry and material composition of battery electrodes affect their performance.

To this end, we rely on well-accepted transport and reaction models developed in the context of electrochemical systems [164] such as Li and Li-ion batteries [44]. It is a general fact that the performance of batteries requires intensive investment in basic science due to various challenges such as the reactivity of involved materials and the highly involved and barely understood electrode–electrolyte interactions [5]. Recent experimental evidence for one of the promising and recent battery materials such as $LiFePO_4$ shows the occurrence of a phase transformation during the Li-intercalation, i. e., a transformation into Li-rich and Li-poor phases. Hence, a phase field model was proposed in [68] to describe the interstitial diffusion in topotactic intercalation compounds. Hysteresis effects, encountered between dis- and charging voltage in electrodes containing multiple insertion particles, have been motivated from a thermodynamic point of view in [46]. Similarly, the Butler–Volmer reactions have also been revisited from a thermodynamic point of view in [8]. Moreover, a nonequilibrium thermodynamic investigation of interfacial transport and reaction processes for reactive charge transport systems can be found in [12].

### 8.3.1 Thermodynamic consistency: nonequilibrium formulation for electrolytes

Over the last decades, various different approaches have been proposed toward a more and more consistent nonequilibrium formulation of (multicomponent) electrolytes, see [10, 11, 45, 160], for instance. Here, we motivate the systematic and physically consistent nonequilibrium formulation of electrolytes relying on GENERIC, see Section 2.7 for its basic building blocks.

We consider the number densities $\tilde{n}_i(x,t)$, $x \in \mathcal{U} \subset \mathbb{R}^d$, $t \in \mathbb{R}_{\geq 0}$, which describe the density of charged species of type $i \in \{1,2,\ldots,N\}$ with charge number $z_i \in \mathbb{Z}$, except that for the density of the solvent, i. e., $i = N$, we have $z_N = 0$. Generally, due to the large particle numbers, the molar densities $n_i = \tilde{n}_i/N_A$, with the Avogadro constant $N_A$, are used. Hence, we can define the corresponding density of free charge in the electrolyte by $q^F = F\sum_{i=1}^N z_i n_i$, where $F = N_A e_0$ is the Faraday constant, and $e_0$ is the elementary charge. The velocity of the specific constituents is denoted by $\mathbf{v}_i$, and the electric potential associated with the free charge density is given by the Poisson equation,

$$-\operatorname{div}(\varepsilon_p \nabla\psi) = q^F \quad \text{in } \mathcal{U}. \tag{8.11}$$

We will further make use of the Green's function $G(\mathbf{x},\mathbf{x}')$ associated with (8.11) to rely on purely dynamic state variables. This allows us to write the electric potential $\psi$ in terms of the charge density $q^F$ by,

$$\psi = -\frac{1}{\varepsilon_p}\int_{\mathcal{U}} G(\mathbf{x},\mathbf{x}')q^F(\mathbf{x}',t)\,d\mathbf{x}', \tag{8.12}$$

where we neglect boundary contributions for simplicity. We will skip the explicit indication of the domain of integration if not required for clarity. Accordingly, we can introduce the mass density $\rho$ of the mixture together with its barycentric velocity $\mathbf{v}$,

$$\rho := \sum_{i=1}^N m_i n_i, \quad \mathbf{v} := \frac{1}{\rho}\sum_{i=1}^N \rho_i \mathbf{v}_i, \tag{8.13}$$

where $\rho_i := m_i n_i$ for molar masses $m_i$, $i = 1,2,\ldots,N$. The total molar number density is $n = \sum_{i=1}^N n_i$, such that $c_i = \frac{n_i}{n}$ is the mole fraction. Later on, we also use the variable $m := \sum_{i=1}^N m_i$.

Following the theory of electrochemical systems developed in [164], the following descriptions are required to reliably formulate transport in electrolytes:

**(i)** the movement of ionic species given, for instance, by the fluxes $\mathbf{J}_i$, $i \in \{1,2,\ldots,N\}$;

**(ii)** material balances of the form,

$$\frac{\partial n_i}{\partial t} = -\mathrm{div}(\mathbf{J}_i) + R_i\,, \tag{8.14}$$

for a possible source or sink term $R_i$ (reactions);

**(iii)** the current flow $\mathbf{i} = F\sum_i z_i \mathbf{N}_i$ for the mass fluxes $\mathbf{N}_i = m_i \mathbf{J}_i = \rho_i \mathbf{v}_i$, $i = 1, 2, \ldots, N$;[4]

**(iv)** a possible electro-neutrality assumption $F\sum_i z_i n_i = 0$, which according to [7] appears to be a valid leading-order approximation, but might be violated by higher-order corrections; and, finally,

**(v)** the momentum equation,

$$\rho(\mathbf{v}_t + (\mathbf{v}\cdot\nabla)\mathbf{v}) + \nabla p = \nu\Delta\mathbf{v} + \rho\mathbf{F}_{el}\,, \tag{8.15}$$

where $\mathbf{F}_{el} = \frac{F}{\rho}\sum_{i=1}^{N} z_i \rho_i \nabla\psi$ is the barycentric Coulomb force, which disappears in the case/assumption of electro-neutrality, and $\nu$ is the viscosity.

**Remark 8.3.1** (Current flow **i**). We note that the performance of electrochemical cells, i. e., a cell $\mathcal{U} \subset \mathbb{R}^d$ containing a particular electrolyte, is generally studied with respect to the current–voltage characteristic (or related state of charge plots) and the power density of the cell.

Before we look at the different widely used solution theories such as the regular, concentrated, and dilute solutions, we write down a general system of equations consistent with GENERIC and containing the above-mentioned solution theories as particular cases. We characterize the state of an electrolyte by the variables $\mathbf{z} := \{\rho, \{\rho_i\}_{i=1}^N, \mathbf{M}, u\}$, where $\rho := \sum_{i=1}^N \rho_i$ is the total molar mass density of the electrolyte with $\rho_N$ denoting the molar mass density of the solvent and $\rho_i$, $i \neq N$, denoting the molar mass densities of the charged species. The momentum of the electrolyte is denoted by $\mathbf{M}$, and the internal energy is denoted by $u$. These state variables satisfy the following GENERIC equations,

$$\left\{\begin{aligned}
&\rho_t + \mathrm{div}(\rho\mathbf{v}) = 0\,,\\
&\mathbf{M}_t + \mathrm{div}(\mathbf{M}\otimes\mathbf{v}) + \nabla p = -\mathrm{div}(\mathbb{T}) - \frac{q^F}{\varepsilon_p}\nabla\int_{\mathcal{U}} G(\mathbf{x},\mathbf{x}')q^F(\mathbf{x}',t)\,d\mathbf{x}'\,,\\
&\frac{\partial\rho_i}{\partial t} + \mathrm{div}(\rho_i\mathbf{v}) = \mathrm{div}(\mathbf{N}_i)\,, \quad \text{for } i = 1, 2, \ldots, N\,,\\
&u_t + (\nabla u + p\nabla)\mathbf{v} = \mathrm{div}\left(\underline{\lambda}_{31}'\nabla\frac{\mu^e}{T}\right) - \mathrm{div}(\lambda_{33}\nabla 1/T) - \mathbb{T} : (\nabla\mathbf{v})'\\
&\qquad + \underline{F}_{el}'\hat{\lambda}_{11}\nabla\frac{\mu^e}{T} + \underline{F}_{el}'\lambda_{13}\nabla\frac{1}{T}\,,
\end{aligned}\right. \tag{8.16}$$

4 For a detailed discussion about the definition of velocities and fluxes, we refer to the classical book [18].

where $\mathbb{T} = -[p + (2/3\eta - \nu)\mathrm{div}(\mathbf{v})]\mathbb{I} + 2\eta(\nabla\mathbf{v} + (\nabla\mathbf{v})')/2$ is Newton's viscous stress tensor [11], $\hat{\lambda}_{11} \in \mathbb{R}^{(N-1)\times(N-1)}$, $\underline{\lambda}_{13} \in \mathbb{R}^{1\times(N-1)}$, $\underline{\lambda}'_{31} \in \mathbb{R}^{(N-1)\times 1}$, $\lambda_{33} \in \mathbb{R}$, and $\lambda^d_{22} = \eta T(\delta_{ik}\delta_{jl} + \delta_{il}\delta_{jk}) + \kappa_2 T\delta_{ij}\delta_{kl}$, where $\delta_{mn}$ is the Kronecker delta function, are coefficients defining the positive, semidefinite Onsager matrix,

$$\Lambda := \begin{pmatrix} \hat{\lambda}_{11} & 0 & \underline{\lambda}_{13} \\ 0 & \lambda^d_{22} & 0 \\ \underline{\lambda}'_{31} & 0 & \lambda_{33} \end{pmatrix}, \tag{8.17}$$

as required later in the analysis of thermodynamic consistency. The parameters $\eta$ and $\kappa_2 = \kappa - 2/3\eta$ are the viscosity and second viscosity, respectively, where $\kappa$ is the dilatational viscosity. We denote the electrochemical potential by,

$$\mu^e_i = \mu_i + \frac{Fz_i}{R}\frac{n_i}{\varepsilon}\nabla \int_{\mathcal{U}} G(\mathbf{x}, \mathbf{x}')q^F(\mathbf{x}', t)\, d\mathbf{x}',$$

for $i = 1, 2, \ldots, N$. The mass flux of mobile species then reads,

$$\mathbf{N}_i := \sum_{j=1}^{N} \lambda^{ij}_{11}\left(\nabla\frac{\mu^e_j}{T} + \frac{\tilde{\lambda}^j_{13}}{T^2}\nabla T\right), \tag{8.18}$$

for $\hat{\lambda}_{11}\tilde{\underline{\lambda}}_{13} = \underline{\lambda}_{13}$, where the cross couplings $\tilde{\lambda}_{13}$ describe particle fluxes and heat fluxes induced by temperature gradients and hence account for the Soret and Dufour effects [40, 120], respectively.

As emphasized in [45], where the momentum is governed by the Euler equation, i. e., $\lambda^d_{22} = 0$, we generally write for thermodynamic consistency only $N - 1$ diffusion fluxes and $N - 1$ driving forces next to the continuity equation,

$$\mathbf{N}_i := \sum_{j=1}^{N-1} \lambda^{ij}_{11}\left(\nabla\frac{\mu^e_j - \mu^e_N}{T} + \frac{\tilde{\lambda}^j_{13} - \tilde{\lambda}^N_{13}}{T^2}\nabla T\right), \tag{8.19}$$

which is equivalent to (8.18) due to the property $\sum_{i=1}^N \lambda^{ij}_{11} = \sum_{j=1}^N \lambda^{ij}_{11} = 0$ as noted in [120]. The species specific mobility matrix $\hat{\lambda}_{11}$ is symmetric, positive semidefinite. A frequently employed particular case of (8.18) are the so-called Nernst–Planck fluxes as part of the dilute solution theory discussed in Section 8.3.3.2.

As noted in [120], the Onsager relations are guaranteed for the transport matrix $\hat{\lambda}_{11}$ after constructing $\lambda^{ij}_{11}$ in terms of the transport matrix,

$$D_{ij} = \frac{mR}{m_i m_j}\left(\frac{\lambda^{ii}_{11}}{\rho_i} - \frac{\lambda^{ij}_{11}}{\rho_j}\right), \tag{8.20}$$

which satisfies $D_{ii} = 0$, or, similarly,

$$\lambda_{11}^{ij} = \frac{m_i}{mR}\left(\sum_{k=1}^{N} m_k D_{ik}\frac{\rho_k}{\rho} - m_j D_{ij}\right)\rho_j, \tag{8.21}$$

where $m = \frac{1}{N}\sum_{i=1}^{N} m_i$. In the next section, we show that the coupled equations (8.16) are in fact General Equations for Nonequilibrium Reversible–Irreversible Couplings (GENERIC) [65, 121, 119].

### 8.3.2 System (8.16) satisfies the principles of GENERIC

We are going to verify that the state vector $\mathbf{z} := \{\rho, \{\rho_i\}_{i=1}^{N}, \mathbf{M}, u\}$, governed by system (8.16), satisfies for the total energy $E(\mathbf{z}) = \int_{\mathcal{U}} e(\mathbf{z})\, d\mathbf{x}$ and total entropy $S(\mathbf{z}) = \int_{\mathcal{U}} s(\mathbf{z})\, d\mathbf{x}$ the GENERIC equation,

$$\mathbf{z}_t = \mathcal{L}\frac{\delta E}{\delta \mathbf{z}} + \mathcal{M}\frac{\delta S}{\delta \mathbf{z}}, \tag{8.22}$$

together with the degeneracy requirements,

$$\mathcal{L}\frac{\delta S}{\delta \mathbf{z}} = 0 = \mathcal{M}\frac{\delta E}{\delta \mathbf{z}}. \tag{8.23}$$

Equations (8.22)–(8.23) have been proposed by Grmela and Öttinger [65, 121] in 1997 and seem currently to be the best available framework to systematically describe nonequilibrium systems.

First, we compute the variational derivatives,

$$\frac{\delta E}{\delta \mathbf{z}} = \begin{pmatrix} \frac{\delta E}{\delta \rho} \\ \sum_{i=1}^{N}\left(\frac{\delta E}{\delta \rho_i}\right)\underline{e}_i \\ \frac{\delta E}{\delta \mathbf{M}} \\ \frac{\delta E}{\delta u} \end{pmatrix} = \begin{pmatrix} -1/2\mathbf{v}^2 \\ \sum_{i=1}^{N}\left(\frac{Fz_i}{\varepsilon}\int_{\mathcal{U}} G(\mathbf{x},\mathbf{x}',t)q^F(\mathbf{x}',t)\, d\mathbf{x}'\right)\underline{e}_i \\ \mathbf{v} \\ 1 \end{pmatrix}, \tag{8.24}$$

and,

$$\frac{\delta S}{\delta \mathbf{z}} = \begin{pmatrix} \frac{\delta S}{\delta \rho} \\ \sum_{i=1}^{N}\left(\frac{\delta S}{\delta \rho_i}\right)\underline{e}_i \\ \frac{\delta S}{\delta \mathbf{M}} \\ \frac{\delta S}{\delta u} \end{pmatrix} = \begin{pmatrix} 0 \\ \sum_{i=1}^{N}(-\mu_i/T)\underline{e}_i \\ 0 \\ 1/T \end{pmatrix}, \tag{8.25}$$

for a chemical potential $\mu_i = \frac{\partial s}{\partial \rho_i} = \frac{1}{nm_i}\frac{\partial s}{\partial c_i}$.[5] With the canonical basis $\underline{e}_i \in \mathbb{R}^N$, $i = 1, 2, \ldots, N$, we can define the vectors $\sum_{i=1}^{N}(\frac{\delta E}{\delta \rho_i})\underline{e}_i = [\frac{\delta E}{\delta \rho_1}, \frac{\delta E}{\delta \rho_2}, \ldots, \frac{\delta E}{\delta \rho_N}]'$ and $\sum_{i=1}^{N}(\frac{\delta S}{\delta \rho_i})\underline{e}_i =$

5 In the literature on dilute solutions [7, 74, 78, 164, 132], this is generally $\frac{\partial s_{\mathrm{id}}(\rho_i,u)}{\partial \rho_i} = \ln c_i$.

$[\frac{\delta S}{\delta \rho_1}, \frac{\delta S}{\delta \rho_2}, \ldots, \frac{\delta S}{\delta \rho_N}]'$. To account for the reversible contributions, we introduce the Poisson matrix,

$$\mathcal{L}(\mathbf{x}) := \left(\begin{array}{c|c|c|c} 0 & 0 & -\mathrm{div}(\rho\,\cdot) & 0 \\ 0 & 0 & -\mathrm{div}(\underline{\rho}\,\cdot) & 0 \\ -\rho\nabla & -\underline{\rho}\nabla & -\mathrm{div}(\mathbf{M}\,\cdot) - \mathbf{M}\nabla & -[u\nabla + \nabla p] \\ 0 & 0 & [-\nabla u + p\,\mathrm{div}(\cdot)] & 0 \end{array}\right), \tag{8.26}$$

where we account for the dimension of mass of the Coulomb force in equation (8.15). Computing $L\frac{\delta E}{\delta z}$ leads to the following reversible terms in the state evolution equation (GENERIC),

$$\mathcal{L}\begin{pmatrix} -\frac{1}{2}\mathbf{v}^2 \\ \sum_{i=1}^{N}\left(\frac{Fz_i}{\varepsilon}\int_{\mathcal{U}} G(\mathbf{x},\mathbf{x}')c^F(\mathbf{x}',t)\,d\mathbf{x}'\right)\underline{e}_i \\ \mathbf{v} \\ 1 \end{pmatrix} = \begin{pmatrix} -\mathrm{div}(\rho\mathbf{v}) \\ -\mathrm{div}(\underline{\rho}\mathbf{v}) \\ \frac{q^F}{\varepsilon}\nabla\int_{\mathcal{U}} G(\mathbf{x},\mathbf{x}')q^F(\mathbf{x}',t)\,d\mathbf{x}' - \mathrm{div}(\mathbf{M}\otimes\mathbf{v}) - \nabla p \\ -[\nabla u + p\nabla]\mathbf{v} \end{pmatrix}. \tag{8.27}$$

The degeneracy requirement $\mathcal{L}\frac{\delta S}{\delta \mathbf{z}} = 0$ is satisfied due to the Gibbs–Duhem relation $d(\mu_i/T) = d(p/T) + ud(1/T)$ for compressible systems under local thermodynamic equilibrium.

Finally, the irreversible contributions are related to the $\mathcal{M}$-matrix.

The $\mathcal{M}$-matrix can be determined from the entropy production,

$$\sigma = \sum_k \mathbf{J}_k\mathbf{F}_k = \mathbf{J}_i\mathbf{F}_i + \mathbf{J}_v\mathbf{F}_v + \mathbf{J}_q\mathbf{F}_q, \tag{8.28}$$

which is the sum of the products of the total heat flux $\mathbf{J}_q$ and the gradient of the inverse temperature $\mathbf{F}_q := \nabla 1/T$, the species ion flux $\mathbf{J}_i$ and its driving force $\mathbf{F}_i := \nabla\frac{\mu_i}{T} - \frac{Fz_i}{\varepsilon RT}\nabla\int_{\mathcal{U}} G(\mathbf{x}',\mathbf{x})q^F(\mathbf{x}',t)\,d\mathbf{x}'$, as well as the "momentum flux" $\mathbf{J}_v$ and its driving force $\mathbf{F}_v = \frac{1}{T}\hat{\kappa}$, where $\hat{\kappa} = \{\frac{\partial v_i}{\partial x_j}\}_{i,j=1}^{d}$. This definition allows us to write for the Onsager coefficients the following linear flux-force relations,

$$\begin{aligned} \mathbf{J}_i &= -\lambda_{11}^{ij}\left(\nabla\frac{\mu_j}{T} - \frac{Fz_i}{\varepsilon RT}\nabla\int_{\mathcal{U}} G(\mathbf{x},\mathbf{x}')c^F(\mathbf{x}',t)\,d\mathbf{x}'\right) + \lambda_{13}^{i}\nabla\frac{1}{T}, \\ \mathbf{J}_v &= \lambda_{22}^{d}\frac{1}{T}\hat{\kappa}, \\ \mathbf{J}_q &= -\underline{\lambda}_{31}\left(\nabla\frac{\underline{\mu}}{T} - \underline{F}_{el}\right) + \lambda_{33}\nabla\frac{1}{T}, \end{aligned} \tag{8.29}$$

where $\underline{F}_{el} = \sum_{i=1}^{N-1}\left(\frac{Fz_i}{\varepsilon RT}\nabla\int_{\mathcal{U}} G(\mathbf{x},\mathbf{x}')q^F(\mathbf{x}',t)\,d\mathbf{x}'\right)\underline{e}_i \in \mathbb{R}^{1\times(N-1)}$ is the vector of Coulomb forces contributing to each charged species $i \in \{1,2,\ldots,N-1\}$ for the canonical basis elements $\underline{e}_i \in \mathbb{R}^{N-1}$. Hence, with the force field $\mathbf{F} := [\mathbf{F}_i, \mathbf{F}_v, \mathbf{F}_q]'$ and the Onsager matrix (8.17), we can provide the entropy production by $\sigma(\mathbf{x}) = \mathbf{F}'\Lambda\mathbf{F}$. If we choose a $C$-matrix according to the constraints imposed by GENERIC, $C\frac{\delta S}{\delta \mathbf{z}} = \mathbf{F}$ and $\frac{\delta E}{\delta \mathbf{z}}C' = 0$, we obtain the matrix,

$$C = \left(\begin{array}{c|c|c|c} 0 & \nabla\cdot & 0 & -\underline{F}_{el} \\ 0 & 0 & -\nabla\cdot & \hat{\kappa} \\ 0 & 0 & 0 & -\nabla\cdot \end{array}\right). \tag{8.30}$$

Herewith, the friction matrix reads,

$$\mathcal{M} = C'\Lambda C\,, \tag{8.31}$$

and hence, the irreversible contributions become,

$$\mathcal{M}\frac{\delta S}{\delta \mathbf{z}} = C'\Lambda C\begin{pmatrix} 0 \\ \{-\mu_i/T\}_{i=1}^{N} \\ 0 \\ 1/T \end{pmatrix} = \begin{pmatrix} 0 \\ \operatorname{div}\left(\hat{\lambda}_{11}\nabla\frac{\underline{\mu}^e}{T}\right) - \operatorname{div}\left(\underline{\lambda}_{12}\nabla\frac{1}{T}\right) \\ -\operatorname{div}(\mathbb{T}) \\ u_{\mathrm{ir}} \end{pmatrix}, \tag{8.32}$$

where $\underline{\mu}^e$ is the vector of electrochemical potential with components,

$$\mu_i^e = \mu_i + \frac{Fz_i}{R}\frac{n_i}{\varepsilon}\nabla\int_{\mathcal{U}} G(\mathbf{x},\mathbf{x}')c^F(\mathbf{x}',t)\,d\mathbf{x}'\,,$$

associated with each charge species $i \in \{1,2,\ldots,N-1\}$, and,

$$u_{\mathrm{ir}} = \underline{F}_{el}'\hat{\lambda}_{11}\nabla\frac{\underline{\mu}^e}{T} + \underline{F}_{el}'\underline{\lambda}_{13}\nabla 1/T - \mathbb{T} : (\nabla\mathbf{v})' + \operatorname{div}\left(\underline{\lambda}_{31}'\nabla\frac{\underline{\mu}^e}{T}\right) - \operatorname{div}(\lambda_{33}\nabla 1/T)\,. \tag{8.33}$$

Hence, putting these results together leads to the elementary system of equations describing the GENERIC state evolution, $\mathbf{z}_t = \mathcal{L}\frac{\delta E}{\delta \mathbf{z}} + \mathcal{M}\frac{\delta S}{\delta \mathbf{z}}$, i. e., (8.16).

### 8.3.3 Examples belonging to system (8.16)

With the rather general formulation presented in the previous two sections, let us demonstrate its generality by motivating special examples included in this formulation.

#### 8.3.3.1 Concentrated solutions (Maxwell–Stefan)

In the theory of electrochemical systems [19, 23, 35, 84, 87, 164], the concentrated solution theory plays an important role. This theory is governed by the following system of equations,

$$\begin{cases} \frac{\partial \rho_i}{\partial t} = -\mathrm{div}(\mathbf{N}_i + \rho_i \mathbf{v}), \\ \mathbf{N}_i = \sum_{j=1}^{N} \lambda_{11}^{ij} \mathbf{d}_j, \end{cases} \tag{8.34}$$

where $\mathbf{d}_i = \nabla \frac{\mu_i^e}{T} + \frac{\tilde{\lambda}_{13}^j}{T^2} \nabla T$ are generalized diffusional driving forces, and $\mathbf{v}$ is the barycentric velocity. If we do not assume that the mass and volume fraction are equal, then the term $(n_i \frac{V_i}{V} - w_i)\nabla p$ is frequently added to $\mathbf{d}_i$, where $V_i$ is the volume of species $i \in \{1, 2, \ldots, N\}$, such that $V = \sum_{i=1} V_i$ is the total volume. We note that an inversion of equations $(8.34)_2$ provides the Maxwell–Stefan form of the driving forces $\mathbf{d}_i$, i. e.,

$$\mathbf{d}_i = -\sum_{j=1, j\neq i}^{N} \frac{n_i n_j}{n^2 D_{ij}^{MS}} \left( \frac{\mathbf{N}_i}{\rho_i} - \frac{\mathbf{N}_j}{\rho_j} \right), \tag{8.35}$$

where $D_{ij}^{MS}$ denotes Maxwell–Stefan diffusional interactions.

#### 8.3.3.2 Dilute solution approximation

Dilute solutions describe the situation, where the molar densities $n_i \ll n_N$, for $i \in \{1, 2, \ldots, N-1\}$, are much smaller than the molar density of the solvent $n_N$. This assumption implies for the generalized fluxes (8.35), that only the terms where $i$ and $j = N$ appear, remain in the sum, i. e.,

$$\mathbf{d}_i := -\frac{n_i n_N}{n D_{iN}^{MS}} (\mathbf{v}_i - \mathbf{v}_N) \approx -c_i (\mathbf{v}_i - \mathbf{v}_N). \tag{8.36}$$

Hence the fluxes induced by temperature gradients (Dufour and Soret effects) are neglected. The evolution of the electrolyte is then governed by the species flux,

$$\mathbf{J}_i = c_i \mathbf{v}_i = -\mathbf{d}_i + c_i \mathbf{v}_N = -\nabla \frac{\mu_i^e}{T} + c_i \mathbf{v}_N. \tag{8.37}$$

The associated electrochemical potentials $\mu_i^e = \mu_i + \frac{F z_i}{R} \frac{n_i}{\varepsilon} \nabla \int_{\mathcal{U}} G(\mathbf{x}, \mathbf{x}') q^F(\mathbf{x}', t)\, d\mathbf{x}'$ are defined by the ideal gas entropy $s_{\mathrm{id}}(\rho_i) = c_i(\ln c_i - 1)$ such that $\mu_i = \ln c_i$, where $c_i = \frac{n_i}{n}$ are mole fractions.

**Binary symmetric electrolytes:** *The Navier–Stokes-Poisson–Nernst–Planck system.* A well-accepted and widely used model for binary symmetric dilute electrolytes, i. e., $z = z_1 = -z_2$ and $n_1, n_2 \ll n_3$, where $n_3$ is the number density of a solvent, is governed by the following dimensionless, incompressible, and nonlinearly coupled Navier–Stokes–Poisson–Nernst–Planck (NSPNP) system [7, 9, 74, 78, 132, 135, 151],

$$
\text{(NSPNP)}\quad\begin{cases} \mathbf{u}_t + \mathbf{u}\cdot\nabla\mathbf{u} = \Delta\mathbf{u} - \nabla p - \frac{(n^+-n^-)}{2s_r}\nabla\psi\,, \\ \operatorname{div}\mathbf{u} = 0\,, \\ n_t^\pm + \operatorname{div}(\mathbf{u}n^\pm) = \operatorname{div}(\nabla n^\pm \pm n^\pm\nabla\psi)\,, \\ -\operatorname{div}(\lambda\nabla\psi) = \frac{(n^+-n^-)}{2s_r}\nabla\psi\,, \end{cases} \tag{8.38}
$$

where $s_r$ is a reference salt concentration, $\psi = \frac{ez\tilde{\psi}}{FT}$ is the dimensionless electric potential, and $\lambda = \frac{\lambda_D}{L}$ is the dimensionless Debye length for a characteristic length $L$ and the Debye length $\lambda_D := \sqrt{\frac{\varepsilon_p RT}{2n_r F^2}}$. We note that the so-called drift-diffusion equations $(8.38)_3$–$(8.38)_4$ are also called the van Roosbroeck equations, in particular, in the context of semiconductors.

### 8.3.3.3 Regular solutions

According to Guggenheim [66], a so-called "(strictly) regular solution" refers to solutions that form an ideal mixture with a nonzero interaction energy $\omega_{ij}$ between species $i$ and $j$ in the mixture. The wording "strictly" is used to differ Guggenheim's definition from the original term "regular solution" going back to Hildebrand, who denotes hereby mixtures that experimentally show certain regularities. Hence we want to describe electrolytes where also interactions between the charged species of the electrolyte are taken into account. To this end, we consider multicomponent regular solutions characterized by the entropy density,

$$
s(\{\rho_i\}_{i=1}^N, u) = R\sum_{i=1}^N s_{\text{id}}(\rho_i, u) + \frac{1}{2}\sum_{i,j=1}^N \omega_{ij}c_ic_j - \sum_{i=1}^N \mu_i^R c_i\,, \tag{8.39}
$$

where $\mu_i^R$ are reference chemical potentials. The variables $c_i = n_i/n = \rho_i/(m_i n)$ are the mole fractions, and,

$$
s_{\text{id}}(\rho_i, u) = \frac{1}{2}c_i\ln\left(\frac{c_0 m_i^5 u^3}{3\rho_i^2/2\sum_{j=1}^N c_j}\right),
$$

is the Sackur–Tetrode-type ideal gas entropy density[6] for a constant $c_0 = c_0(\hbar, R)$, and $\omega_{ij}$ are the mean field interaction energies between species $i$ and $j$.

Hence system (8.16) describes a regular solution for the electrochemical potential $\mu_i^e = \mu_i + \frac{Fz_i}{R}\frac{n_i}{\varepsilon}\nabla\int_D G(\mathbf{x},\mathbf{x}')q^F(\mathbf{x}',t)\,d\mathbf{x}'$ in (8.18) (and in (8.19) for $N-1$ fluxes) by,

$$
\mu_i = \frac{1}{nm_i}\frac{\partial s}{\partial c_i} = \frac{1}{n}\left(RT\frac{1}{m_i}\frac{\partial s_{\text{id}}(\rho_i,u)}{\partial c_i} + \frac{1}{2}\sum_{l=1}^N\left(\frac{\omega_{il}}{m_i} + \frac{\omega_{li}}{m_i}\right)c_l - \frac{\mu_i^R}{m_i}\right). \tag{8.40}
$$

6 See [120], for instance. In the literature, we generally find the simplified expression $s_{\text{id}}(c_i) = c_i\ln c_i$.

As before, this immediately allows us to show that these equations are consistent with the nonequilibrium thermodynamic formulation of GENERIC, which relies on the entropy density (8.39) and the following energy density,

$$e(\rho, \{\rho_i\}_{i=1}^N, \mathbf{M}, u) = \frac{1}{2\rho}\mathbf{M}^2 + \frac{q^F(\mathbf{x}, t)}{2\varepsilon} \int G(\mathbf{x}, \mathbf{x}')q^F(\mathbf{x}', t)\, d\mathbf{x}' + u\,, \tag{8.41}$$

where $q^F = F \sum_{i=1}^N \frac{z_i}{m_i}\rho_i$ is the free charge density, and $F$ is the Faraday constant, see (8.11). The pressure in a multicomponent system is given by,

$$p\frac{\partial s}{\partial u} = s - \sum_{i=1}^N \rho_i \frac{\partial s}{\partial \rho_i} - u\frac{\partial s}{\partial u}\,, \tag{8.42}$$

which arises from the degeneracy condition imposed on the Poisson bracket [120].

### 8.3.4 Charge transport formulations for Li-batteries

#### 8.3.4.1 Electrochemical processes: Li-intercalation and phase transformation

The success of Li-ion batteries relies on a suitable composition of electrodes and electrolytes by allowing for a feasible mass production (and hence a reasonable purchasing price) and by providing convincing performance such as long enough *battery life* and large enough *storage capacity*. For instance, the frequently studied phase separating and competitive cathode material $LiFePO_4$ has been explored by John Goodenough and coauthors already in 1997, and the promising results supported ongoing research efforts in this direction. In fact, for his work on battery intercalation cathodes, he received the Nobel Prize in chemistry together with Stanley Whittingham and Akira Yoshino in 2019. Recalling the modeling and derivation of effective macroscopic intercalation electrodes based on homogenization/upscaling techniques introduced in Section 6.3, following the systematic derivation from [138], there are two kinds of moving charges, the intercalating Li-particles and the electrons entering via the current collector from the external circuitry/device back into the battery/cathode during discharging.

The Li-intercalation into an electrochemical solid is subject to constraints such as charge and mass conservation, see [60], for instance. *Interstitial diffusion* exploiting defects in lattice by occupying a generally empty site, *vacancy diffusion* driven by atomic vibrations due to thermal energy, or *impurity governed diffusion* originating from concentration gradients all represent general atomic transport or diffusion processes. Many available interstitial sites and their weaker attraction/bonding to neighboring atoms imply that vacancy diffusion is generally slower than interstitial diffusion. Successful battery designs and one of the key advantages of Li rely on its ability to reside in interstitial sites of *crystalline intercalation hosts* such as $FePO_4$. Naturally, these intercalation hosts are likely to change their structure under the gain or loss of intercalation mass/material, and in this case, they are referred to as *topotactic compounds*. Such compounds

(e. g., $LiFePO_4$) motivate to introduce a (phase) field variable indicating the location of Li-rich and Li-poor sites as well as the *interface* between them.

A systematic way for describing phase separation relies, for instance, on phase field formulations [8, 31, 52, 68, 122]. Such a formulation can be obtained by first introducing the underlying *homogeneous free energy*,

$$f_r(c_s) = \omega c_s(1 - c_s) + k_B T(c_s \log c_s + (1 - c_s)\log(1 - c_s)), \tag{8.43}$$

of a *regular solution*, which allows us to define the following thermodynamic *free energy functional*,

$$F_r(c_s) = n_v \int_{D_s} f_r(c_s) + \tilde{\lambda}(\nabla c_s)^2 \, d\mathbf{x}. \tag{8.44}$$

For a more thermodynamic motivated discussion in the context of the GENERIC framework, we refer the interested reader to the previous Sections 3.1 and 8.3.1. The effective interaction energy between Li-occupied/-interstitial and neighboring Li-free sites is denoted by $\omega$ in (8.43). In the case of $LiFePO_4$, the gradient penalty term in (8.44) is weighted by the parameter,

$$\tilde{\lambda} = a^2 \omega / 6, \tag{8.45}$$

for the nearest-neighbor *interaction energy* $\omega = 59$ [meV] and for the distance between neighboring sites $a = 2.5$[Å]. Moreover, $\tilde{\lambda}$ in (8.44) is proportional to the interfacial width. As motivated in [68], setting $n_v = 8.396 \times 10^{28}$ [1/m$^3$] allows us to identify the parameter $\lambda = n_v \tilde{\lambda} = 2.48 \times 10^{-11}$ [J/m].

Having now identified the relevant energetic contributions, we look at nonequilibrium aspects of Li-transport such as derived in [138], where a mass-conserving *gradient flow* leads to an evolution equation for the phase field $c_s$. To this end, we compute the variational derivative $\nabla_c^{H^1}$ of the free energy functional (8.44) and use it to iteratively minimize the free energy over time. Passing to the continuous limit in time gives us, up to boundary terms, the equation,

$$\partial_t c_s = -\mathrm{div}(\hat{\mathrm{M}} \nabla (\nabla_{c_s}^{H^1} F_r(c_s))), \tag{8.46}$$

which has to be understood in a weak sense, that is, after multiplication by test functions $v \in H^1(D_s)$. Note that the gradient $\nabla_c^{H^1}$ represents the Gâteaux derivative, such that for all $v \in H^1(D_s)$,

$$\begin{aligned} \frac{\delta^v F_r(c_s)}{\delta^v c_s} &= \lim_{\theta \to 0} \frac{F_r(c_s + \theta v) - F_r(c_s)}{\theta} \\ &= ((f_r'(c_s) - \lambda \Delta c_s), v) \\ &= (\nabla_{c_s}^{L^2} F_r(c_s), v). \end{aligned} \tag{8.47}$$

Riesz's theorem allows us to identify $\nabla^{L^2}_{c_s} F_r(c_s)$ in the last equality in (8.47) with respect to the $L^2$-inner product.

Applying Riesz's theorem in the context of the semi-inner product $(\hat{\mathrm{M}}\nabla u, \nabla v)_{H^1}$ for $u, v \in H^1(D_s)$ and a symmetric positive definite tensor $\hat{\mathrm{M}}$, we can define the variational derivative $\nabla^{H^1}_{c_s} F_r(c_s)$ for homogeneous Neumann boundary conditions by,

$$\begin{aligned}(\nabla^{H^1}_{c_s} F_r(c_s), v) &= (\hat{\mathrm{M}}\nabla\nabla^{L^2}_{c_s} F_r(c_s), \nabla v) \\ &= (-\mathrm{div}(\hat{\mathrm{M}}\nabla^{L^2}_{c_s} F_r(c_s)), v),\end{aligned} \tag{8.48}$$

for all $v \in H^1(D_s)$. Investigations (8.46)–(8.48) allow us to finally write the following phase field equation for intercalation compounds as a result of the above motivated gradient flow approach,

$$\partial_t c_s = \mathrm{div}(\hat{\mathrm{M}}\nabla(f_r'(c_s) - \lambda\Delta c_s)). \tag{8.49}$$

Since it is difficult to identify irreversible quantities from frictional processes on the microscale in a mean field formulation, the mobility tensor $\hat{\mathrm{M}}$ is generally chosen to be a constant $M_s$. Finally, we note that Han [68] suggests a concentration-dependent mobility $\hat{\mathrm{M}} = c_s(1 - c_s)M_0$, where $M_0$ is a constant.

**Microscopic composite cathode formulation.** A general composite cathode consists of solid intercalation particles $Y_s \subset Y := [0,1]^d$ for the space dimension $d \in \mathbb{N}$ and a polymer solution $Y_p \subset Y$. These two phases together define a characteristic reference cell $Y := Y_p \cup Y_s$, see Figure 8.10 (top), for a sample polymer-particle configuration of a composite cathode. Additionally, we account for thin carbon black pathways allowing for electron transport between the particles $Y_s$. We further state the basic microscopic equations that account for charge transport in the intercalation host of a composite as defined in the reference cell $Y$. This is generally one of the first steps in a systematic and reliable upscaling/homogenization process of a particular scientific problem of interest as described in detail in the preceding Sections 6.3 and 6.6.1 and demonstrated for phase changing intercalation hosts of Li-batteries in [138] and [111], for instance. The periodic extension/covering of a cathode domain $D_s$ with reference cells $Y$ defines the microscopic cathode domain $D_s^{\epsilon_p}$, where $\epsilon_p := \frac{l_p}{L}$ denotes the pore heterogeneity for a characteristic pore length and the macroscopic length scale of the cathode.

Herewith, the microscopic (i. e., on the pore scale) mean field transport in the intercalation host reads,

$$\begin{cases} \partial_t c_s^{\epsilon_p} = \mathrm{div}(M_s\nabla(f_r'(c_s^{\epsilon_p}) - \lambda\Delta c_s^{\epsilon_p})), & \text{in } D_s^{\epsilon_p}, \\ -\mathrm{div}(\sigma_s\nabla\phi_s^{\epsilon_p}) = 0, & \text{in } D_s^{\epsilon_p}, \end{cases} \tag{8.50}$$

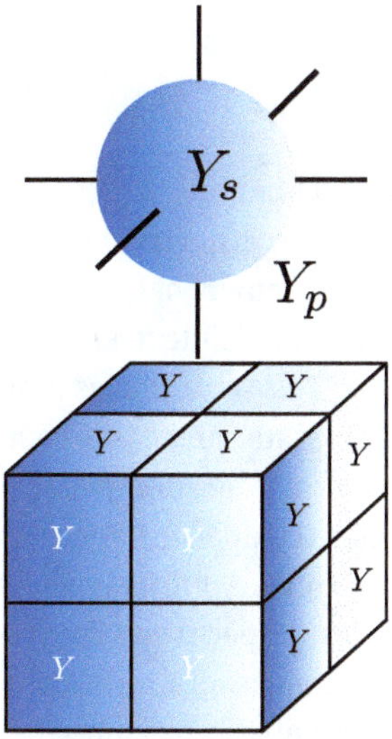


Figure 8.10: *Top:* Reference cell $Y$ consisting of a polymer solution $Y_p$ together with the solid intercalation compound $Y_s$, i. e., $Y = Y_p \cup Y_s$. The cell is electrically conducting due to the thin carbon black pathways connecting $Y_s$. *Bottom:* Reference cell $Y_D$, here composed of an elementary pore cell $Y$, for instance, such as depicted on *Top*. The cell $Y_D$ takes the influence of obstacles to momentum transport into account. The union of the $Y_p$ phases in $Y_D$ are then denoted by $Y_D^p$, and, similarly, $Y_D^s$ represents the union of the $Y_s$ phases in $Y_D$; see [138].

where $\sigma_s$ denotes the electric conductivity of the solid intercalation compound. Hence, the phase field equation $(8.50)_1$ accounts for the Li-transport, and $(8.50)_2$ for the electron transport.

We recall that as motivated in [138], next to the pore heterogeneity $\epsilon_p$ associated with the pore length scale $l_p$, there are other microscopic length scales such as the *Darcy scale*, whose characteristic length is $l_D > l_p$ in case that fluid flow is taken into account. Therefore, coupling of equations describing different scientific phenomena allows us to account for *multiple microscopic length scales* associated with each process involved.

The dynamic description (8.49) of the intercalation process obtained as a mass-conserving gradient flow associated with $f_r$ is referred to as the Cahn–Hilliard phase field equation. The fact that (8.50) is a thermodynamic, consistent description for the *dynamics of interfaces* as given in Section 3.1, where $(8.50)_1$ is shown to satisfy the equations of GENERIC, which currently represent the most general and systematic nonequilibrium framework available. Moreover, equation $(8.50)_1$ is a simple formulation for first-order phase transitions,[7] e. g., obtained by quickly quenching a stable single-phase solution, i. e., by quickly lowering the temperature, see [115], for instance.

7 *According to Ehrenfest* [48]*:* For temperature $T$, entropy $S$, pressure $p$, and volume $v$, and under constant Gibbs free energy $c(T,p) := G = U - TS + pv$, first order and second order phase transitions can be identified as kinks/discontinuities of first and second order derivatives of $c$, respectively, in a ($p$ versus $T$)-plot.

### 8.3.5 Fundamental and Self-consistent Battery Equations (FSBEs): binary and neutrally dilute electrolytes with active Li-hosts

To capture the essential impact of each elementary part of a Li-battery, we investigate Li-ion batteries with a single intercalation host surrounded by a binary polymer electrolyte and a Li-foil as anode, see Figure 8.12. Due to the high reactivity of Li-anodes, the electrolyte is considered to be nonliquid but modeled as neutrally dilute to keep the model complexity limited to the essential transport phenomena. This setup allows us to investigate the geometric and compositional influence of the separator and of the composite cathode on the battery performance. In Section 8.3.6, we formulate a complete set of equations and boundary conditions describing such two-phase composite batteries.

Note that we generally distinguish so-called primary and secondary lithium batteries, i. e., disposable/nonrechargeable and rechargeable batteries, respectively. The following conventions are used: (i) *Li-batteries* have generally lithium as an anode material, which, together with (an often liquid) electrolyte, leads to irreversible reactions that prevent recharging; (ii) *Li-ion batteries* use carbon materials as anodes such as graphite, lithium-titanate, or recently also silicon, which all allow Li-ions to intercalate.

Hence, our prototype model describes Li-batteries according to the above classification and is closely related to [44], albeit for a neutrally dilute electrolyte, e. g., justified for thin double layers [7] together with a mean field equation for interstitial diffusion taking a possible phase transformation into account [68]. We plan to investigate in a second step Li-ion batteries with anodes allowing for Li intercalation, so that we first can gain a proper understanding of cathodic intercalation by having a relatively simple anodic process. Finally, we also look at the influence of a carbon binder in so-called three-phase composite cathodes, see Figure 8.11.

### 8.3.6 Two-phase composite Li-battery

We first introduce the essential and well-accepted reactions that occur in Li-batteries as proposed in [44, 164]. Relevant values for the employed model parameters are given in Table 8.2.

**Interfacial reactions on $I_{ps}$ and $\Gamma_l$:** Generally, reactions take place on a *solid-electrolyte interface*, which we denote by $I_{ps} := \partial\mathcal{D}_p \cap \partial\mathcal{D}_s$. In Li-batteries, the reaction,

$$\mathrm{Li}^+ - \Theta_p \quad + \quad \Theta_s \quad + \quad e^- \quad \rightleftharpoons \quad \mathrm{Li} - \Theta_s \quad + \quad \Theta_p\,, \tag{8.51}$$

describes the Li insertion into the solid matrix. The variables $\Theta_p$ and $\Theta_s$ stand for the available sites in the polymer and the solid matrix (interstitial). We can determine the

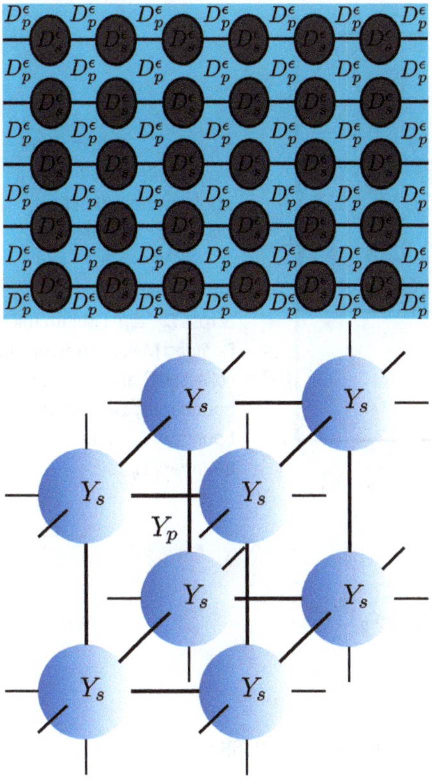


**Figure 8.11:** *Top:* A 2D periodic composite cathode $D = D_p^\epsilon \cup D_s^\epsilon$ of characteristic length $L$ with solid intercalation particles $D_s^\epsilon$ surrounded by a polymer electrolyte solution. The black lines are electric conducting pathways in the form carbon black. This reference cell motivates a layer-by-layer assembly. It represents a characteristic configuration of intercalation particles electrically connected in various ways via carbon black and surrounded by polymer electrolyte, see [138]. *Bottom:* A 3D sample reference cell of a possible cathode composed of $FePO_4$ particles connected via carbon black particles (thin lines). Obviously, more complex cells can be used.

| Parameter | Value | Ref. (system specific) |
|---|---|---|
| $i_0^l$ | $1.0 \cdot 10^4$ S/m | [44] |
| $c_s^{\max}$ | 29 k mol/m$^3$ | [44] |
| $c_p^{\max}$ | | |
| $\alpha_a, \alpha_c$ | 0.5 | [44] |
| $k_p$ | $10^{-10}$m$^4$/(mol · s) | [44] |
| $\nu_+, \nu_-$ | 1 | [44] |
| $\sigma_s$ | 1.0 | [44] |
| $D_s$ | $5.0 \cdot 10^{-13}$m$^2$/s | [44] |

**Table 8.2:** Choices for physical parameters.

current induced by this reaction by the well-accepted Butler–Volmer equations [164, p. 203],

$$\begin{aligned} i_{\mathrm{BV}}^{ps} &:= i_0^{ps}(c_p) R_{\mathrm{BV}}^{ps}(c_s, \eta_{ps}) \\ &:= i_0^{ps}(c_p)\left[c_s \exp\left(\frac{\alpha_a F}{R\theta}(\eta_{ps} - U_c)\right) - (c_{\max}^s - c_s)\exp\left(-\frac{\alpha_c F}{R\theta}(\eta_{ps} - U_c)\right)\right], \end{aligned} \tag{8.52}$$

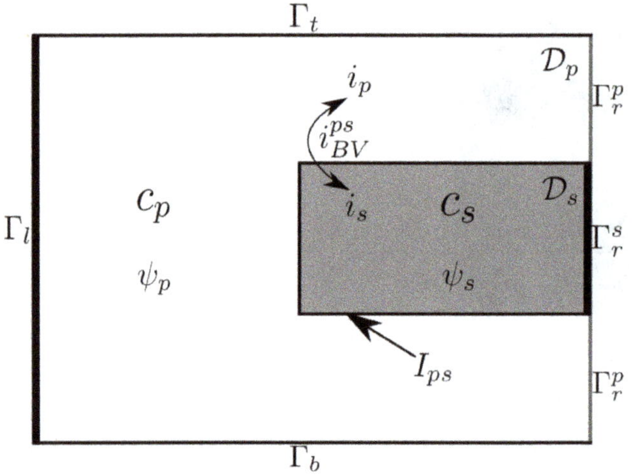


Figure 8.12: Schematic of a *composite cathode* with a single intercalation host, *solid-electrolyte interface* $I_{ps} := \partial\mathcal{D}_p \cap \partial\mathcal{D}_s$, and Butler–Volmer reaction $i_{BV}^{ps}$ across $I_{ps}$. *Left:* Lithium foil $\Gamma_l$. *Right:* Metal plate/current collector $\Gamma_r^s$ and polymer boundary $\Gamma_r^p$. *Top:* No-flux boundary $\Gamma_t$. *Bottom:* No-flux boundary $\Gamma_b$.

where $F$ is the Faraday constant,

$$i_0^{ps}(c_p) := Fk_p(c_{\max}^p - c_p)^{\alpha_c} c_p^{\alpha_a},$$

and,

$$U_c := U_p^\theta - U_{\text{ref}}^\theta + \frac{R\theta}{F}(\beta c_s + \zeta) = 2.17 + \frac{R\theta}{F}(-0.000558c_s + 8.1), \tag{8.53}$$

as chosen in [44] for instance, and $\eta_{ps} := \psi_s - \psi_p$ is the surface overpotential defined for $\phi_s$ and $\phi_p$, the electric potentials of the electrode (solid) and electrolyte, respectively. The parameters $c_{\max}^p$ and $c_{\max}^s$ are the maximum concentration in the polymer and the solid intercalation phase, respectively. Moreover, $\alpha_a$ and $\alpha_c$ are anodic and cathodic transfer coefficients, respectively. These are generally chosen to fit experimental data. We denote by $c_p$ and $c_s$ the Li-concentration in the polymer and the solid intercalation phase, respectively. To rewrite formulations in dimensionless form later on, we define the coupling parameter $\beta^{ps} := \frac{i_0^{ps} L_r s_+}{FD_r}$ of dimension $[L_r^{-1}]$, where $L_r$ is the reference length, and $s_+$ denotes the stochiometric coefficient of the positively charged ion.

At the anode/lithium foil $\Gamma_l$, we have a similar charge transfer reaction,

$$\text{Li} \quad + \quad \Theta_p \quad \rightleftharpoons \quad \text{Li} - \Theta_p \quad + \quad e^-, \tag{8.54}$$

which is governed by the Butler–Volmer equation,

$$i_a = i_{\text{BV}}^l := i_0^l R_{\text{BV}}^l(\eta_{lp}) := i_0^l \left[\exp\left(\frac{\alpha_{a_l} F\eta_{lp}}{R\theta}\right) - \exp\left(-\frac{\alpha_{c_l} F\eta_{lp}}{R\theta}\right)\right], \tag{8.55}$$

where $i_a := \frac{I_a}{|\Gamma_r^s|}$ is the superficial (/applied) current density of the cell, $\eta_{lp} := \psi_s - \psi_p - U_l$ is the local value of the surface overpotential, $U_l = 0$ is the theoretical open-circuit cell potential, and $i_0^l(c_p) := Fk_{a_l}^{\alpha_{c_l}} k_{c_l}^{\alpha_{a_l}} (c_{\max}^p - c_p)^{\alpha_{c_l}} c_p^{\alpha_{a_l}}$ for the anodic and cathodic reaction rate constants $k_{a_l}$ and $k_{c_l}$, respectively. Similarly, the coupling parameter of dimension $[FL^{-1}]$ associated with (8.55) is $\beta^l := \frac{i_0^l L_r s_+}{FD_r}$. The parameter $c_{\max}^p$ is the maximum Li concentration in the polymer phase.

**Charge transport in the polymer electrolyte $\mathcal{D}_p$:** Due to the Guggenheim relation $z_+\nu_+ + z_-\nu_- = 0$, where $\nu_+$ and $\nu_-$ are the numbers of cations and anions produced by the dissociation of one molecule of electrolyte, and the electroneutrality $z_+c_+ + z_-c_- = 0$ is valid, for instance, in systems with thin double layers, a binary electrolyte can be described by a single concentration variable $c_p = \frac{c_+}{\nu_+} = \frac{c_-}{\nu_-}$, which solves the equation,

$$\frac{\partial c_p}{\partial t} + \mathbf{u} \cdot \nabla c_p = \operatorname{div}(D_p \nabla c_p), \quad \text{in } \mathcal{D}_p, \tag{8.56}$$

where $D_p := \frac{z_+M_+D_- - z_-M_-D_+}{z_+M_+ - z_-M_-}$ with $M_i$, $i \in \{+,-\}$, being the mobilities associated with positively charged and negatively charge species. Since electroneutrality also removes the Coulomb driving force on the momentum equation described by the incompressible Navier–Stokes equations and since we do not impose any momentum driving force on the electrolyte, we can neglect the fluid velocity, i. e., $\mathbf{u} = 0$.

The above definition of the single concentration variable $c_p$ implies with the help of the Nernst–Planck equations [164, p. 277] the following equation for the associated electric potential,

$$-\operatorname{div}(c_p \nabla \psi_p) = -\mathcal{R} \Delta c_p, \quad \text{in } \mathcal{D}_p, \tag{8.57}$$

where $\mathcal{R} := \frac{(D_+ - D_-)}{F(z_+M_+ - z_-M_-)}$. Herewith, we can compute the local current density in the electrolyte phase $\mathcal{D}_p$ by,

$$\mathbf{i}_p = -z_+\nu_+F(z_+u_+ - z_-u_-)c_pF\nabla\psi_p + (D_+ - D_-)\nabla c_p, \quad \text{on } I_{ps}. \tag{8.58}$$

**Transport in the solid intercalation host $\mathcal{D}_s$:** The movement of Li inside the intercalation host $\mathcal{D}_s$ can be described by two fundamental transport processes:

**(A)** *intercalation by classical diffusion*, i. e.,

$$\frac{\partial c_s}{\partial t} - \operatorname{div}(D_s \nabla c_s) = 0, \quad \text{in } \mathcal{D}_s, \tag{8.59}$$

or **(B)** *intercalation by interstitial diffusion*, i. e.,

$$\partial_t c_s = \operatorname{div}(M_s \nabla (f_r'(c_s) - \lambda \Delta c_s)), \quad \text{in } \mathcal{D}_s, \tag{8.60}$$

where $f_r(c) := NR\theta[c \log c + (1-c)\log(1-c) + \omega c(1-c)]$ is the regular solution free energy density for the *exchange* energy $\omega := \frac{z}{R\theta}(\epsilon_{\alpha\beta} - (\epsilon_{\alpha\alpha} + \epsilon_{\beta\beta})/2)$ with the coordination number

$z$ (e. g., number of neighbors to a lattice site) and pairwise interaction energies $\epsilon_{\alpha\beta}$, $\epsilon_{\alpha\alpha}$, and $\epsilon_{\beta\beta}$, $N$ is the number of molecules per unit volume, $M_s = D_s c_s(1 - c_s)$ is the mobility with diffusion coefficient $D_s$, and $\lambda$ is a regularizing parameter (interfacial penalty) with units $[L_r^2]$, where $L_r$ is a reference length scale. The constant $R = k_B N_A$ denotes the ideal gas constant, where $k_B$ is the Boltzmann constant, and $N_A$ is the Avogadro constant.

Note that $\omega = \omega_c = 0$ denotes a critical exchange energy such that $f_r$ becomes an ideal solution, i. e., without any interaction energies. For $\omega > 0$, the mixture favors to be in a state where equal particles/species are next to each other, i. e., to minimize the free energy, the system is likely to phase separate. Finally, for $\omega < 0$, the system prefers the mixed state, i. e., a state where opposite species are next to each other. The free energy density $f_r$ shows the well-known double-well structure, which plays a crucial role in the description of systems with phase separation. For computational convenience, we will restrict ourselves to the following well-accepted double-well potential $f_r(c) := 1/4(1-c^2)^2$, such that $f_r'(c) = c^3 - c$. Moreover, we impose a constant mobility $M_s$ in this case. An increasingly relevant example is the description of lithium intercalation into crystalline host materials such as $FePO_4$, which is known to show phase separation under suitable operating conditions. Equation (8.60) has been proposed in [68] to model interstitial diffusion. On the contrary, $f_r$ is convex and represents a single well if $\theta \leq 1$.

These transport equations have to be complemented by the macroscopic equation describing electron conduction,

$$-\operatorname{div}(\sigma_s \nabla \psi_s) = 0 \quad \text{in } \mathcal{D}_s\,, \tag{8.61}$$

where $\sigma_s$ is the conductivity of the solid intercalation phase, and $\psi_s$ is the electric potential in $\mathcal{D}_s$. Equation (8.61) is widely used in the literature, e. g., [164].

To subsequently state complete dimensionless model equations, we introduce characteristic units such as a reference salt concentration $c_r$, thermal voltage $\frac{R\theta}{F}$ for temperature $\theta$, reference length $L_r$, e. g., Debye length $\lambda_D := \left(\frac{\varepsilon_p R\theta}{2z^2F^2c_r}\right)^2$, characteristic time $\tau_r$, e. g., diffusion time $\tau_p := \frac{L_r^2}{D_p}$ in the polymer phase, diffusion time $\tau_s := \frac{L_r^2}{D_s}$ or intercalation time $\tau_i := \frac{L_p^2}{D_s}$ for a characteristic diameter $L_p$ of the intercalation host, and a reference conductivity $\sigma_r$.[8] Since we generally expect the diffusion in the electrolyte phase to be smaller than the diffusion/interstitial diffusion in crystals, we choose as a reference time the larger diffusion time $\tau_d$. For a lithium reference density $c_r^p$ in the polymer phase $\mathcal{D}_p$ and for a lithium reference density $c_r^s$ in the solid phase $\mathcal{D}_s$, we define the dimensionless variables $\tilde{c}_p := \frac{c_p}{c_r^p}$, $\tilde{c}_s := \frac{c_s}{c_r^s}$, $\tilde{x} := \frac{x}{L_r}$, and $\tilde{t} := \frac{t}{\tau_d}$ such that the diffusion equation (8.56) reads in dimensionless form as follows,

$$\frac{\partial \tilde{c}_p}{\partial \tilde{t}} - \widetilde{\operatorname{div}}(\tilde{\nabla}\tilde{c}_p) = 0\,, \quad \text{in } \mathcal{D}_p\,. \tag{8.62}$$

8 The reference value $\sigma_r$ can be chosen according to the availability of experimental data, for instance.

Similarly, the dimensionless equation of classical diffusive transport in the intercalation material **(A)** as stated in (8.59) is,

$$a_1 \frac{\partial \tilde{c}_s}{\partial \tilde{t}} - \tilde{\operatorname{div}}(\tilde{\nabla} \tilde{c}_s) = 0\,, \quad \text{in } \mathcal{D}_s\,, \tag{8.63}$$

with the dimensionless parameter $\tilde{a}_1 := \frac{\tau_s}{\tau_p}$. The dimensionless form of the interstitial diffusion equation (8.60), i. e., formulation **(B)**, reads as follows,

$$\tilde{a}_1 \partial_t \tilde{c}_s = \tilde{\operatorname{div}}(\tilde{\nabla}(f_r'(\tilde{c}_s) - \tilde{\Delta}\tilde{c}_s))\,, \quad \text{in } \mathcal{D}_s\,. \tag{8.64}$$

Note that we use $\tilde{M} = \text{const.}$ in case of free energies defined by the classical double-well potential. It leaves to cast equations (8.57) and (8.61) for the electrostatic potentials in the polymer and electrolyte phases, respectively, into dimensionless form. Equation (8.57) with $\mathcal{R} = \frac{R\theta}{F}\tilde{\mathcal{R}}$ becomes,

$$-\tilde{\operatorname{div}}(\tilde{c}_p \tilde{\nabla} \tilde{\psi}_p) = -\tilde{\mathcal{R}}\tilde{\Delta}\tilde{c}_p\,, \quad \text{in } \mathcal{D}_p\,. \tag{8.65}$$

In the same way, equation (8.61) reads for $\sigma_s = \tilde{\sigma}_s \sigma_r$ in the dimensionless form as follows,

$$-\frac{\sigma_r R\theta}{F L_r^2} \tilde{\operatorname{div}}(\tilde{\sigma}_s \tilde{\nabla} \tilde{\psi}_s) = 0\,, \quad \text{in } \mathcal{D}_s\,, \tag{8.66}$$

Finally, we will subsequently skip the $\tilde{\cdot}$-notation for the dimensionless solutions of the equations, but not for the associated parameters.

Next, we state two fundamental formulations with a complete set of boundary conditions, *voltage-* and *current-driven* battery formulations, see Figure 8.13.

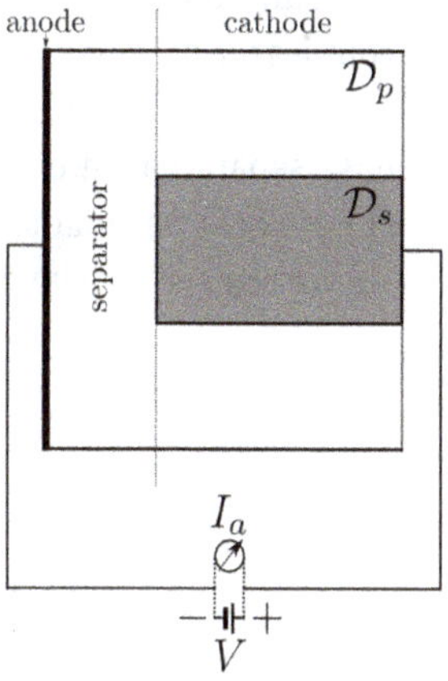


Figure 8.13: Schematic of a two-phase lithium battery with a Li-foil as anode, polymer electrolyte $\mathcal{D}_p$ as separator, and solid intercalation host $\mathcal{D}_s$ as lithium storage. We also indicate the basic two formulations that we state in the context of a neutrally dilute electrolyte, i. e., a current-driven battery with dis- or charging current $I_a$ and a voltage-driven battery with dis- or charging voltage $V$.

#### 8.3.6.1 Voltage-driven formulation

Under a voltage-driven battery we understand the situation, where the battery operates under a fixed applied voltage $V$ across the electrodes. We first consider **intercalation by classical diffusion (A)**. For suitable initial conditions, we consider the following equations,

$$\text{in } \mathcal{D}_p: \quad \begin{cases} \frac{\partial c_p}{\partial t} - \Delta c_p = 0\,, & \text{in } \mathcal{D}_p\,, \\ \nabla c_p \cdot \mathbf{n} = \frac{i_a}{FD_p} = \frac{D_r}{D_p}\beta^l R^l_{\text{BV}}\,, & \text{on } \Gamma_l\,, \\ \nabla c_p \cdot \mathbf{n} = 0\,, & \text{on } \Gamma^p_r \cup \Gamma_t \cup \Gamma_b\,, \\ \nabla c_p \cdot \mathbf{n} = \frac{D_r}{D_p}\beta^{ps} R^{ps}_{\text{BV}}\,, & \text{on } I_{ps}\,, \\ -\text{div}(c_p \nabla \psi_p) = -\tilde{\mathcal{R}}\Delta c_p\,, & \text{in } \mathcal{D}_p\,, \\ \psi_p = V\,, & \text{on } \Gamma_l\,, \\ \nabla\psi_p \cdot \mathbf{n} = \frac{\epsilon_s}{\epsilon_p}\nabla\psi_s \cdot \mathbf{n}\,, & \text{on } I_{ps}\,, \end{cases} \tag{8.67}$$

and,

$$\text{in } \mathcal{D}_s: \quad \begin{cases} \tilde{a}_1 \frac{\partial c_s}{\partial t} - \Delta c_s = 0\,, & \text{in } \mathcal{D}_s\,, \\ \nabla c_s \cdot \mathbf{n} = -\frac{D_r}{D_s}\beta^{ps} R^{ps}_{\text{BV}}\,, & \text{on } I_{ps}\,, \\ \nabla c_s \cdot \mathbf{n} = 0\,, & \text{on } \Gamma^p_r\,, \\ -\text{div}(\tilde{\sigma}_s \nabla\psi_s) = 0\,, & \text{in } \mathcal{D}_s\,, \\ \nabla\psi_s \cdot \mathbf{n} = \frac{\sigma_r}{\tilde{\sigma}_s}\beta_\psi R^s_{\text{BV}}\,, & \text{on } I_{ps}\,, \\ \psi_s = 0\,, & \text{on } \Gamma^s_r\,, \end{cases} \tag{8.68}$$

where $\beta_\psi := \frac{L_r F}{\sigma_r R\theta}$, and $\mathbf{n}$ is the unit normal pointing outward of the polymer electrolyte domain $\mathcal{D}_p$. Equation $(8.67)_5$ is a consequence of $c_p = \frac{c_+}{\nu_+} = \frac{c_-}{\nu_-}$. The variables $R^{ps}_{\text{BV}}$ and $R^l_{\text{BV}}$ are given by the Butler–Volmer reactions (8.52) and (8.55), respectively. Finally, we note that the last boundary condition $(8.67)_7$ for the electric potential accounts for the electric permittivity of the electrolyte and intercalation host and hence couples the process from the two domains $\mathcal{D}_p$ and $\mathcal{D}_s$.

To account for the interstitial diffusion and a possible phase separation in crystalline intercalation hosts, e. g., $FePO_4$ (see [44]), we employ the phase field formulation (8.60) to describe **intercalation by interstitial diffusion (B)**, i. e., we replace the classical diffusive transport (8.68) in formulation **(A)** with,

$$\begin{cases} \tilde{a}_1 \partial_t c_s = \text{div}(\nabla\mu_s)\,, & \text{in } \mathcal{D}_s\,, \\ \nabla\mu_s \cdot \mathbf{n} = -\beta^{ps} R^{ps}_{\text{BV}}\,, & \text{on } I_{ps}\,, \\ \nabla\mu_s \cdot \mathbf{n} = 0\,, & \text{on } \Gamma_r\,, \\ -\text{div}(\tilde{\sigma}_s \nabla\psi_s) = 0\,, & \text{in } \mathcal{D}_s\,, \\ \tilde{\sigma}_s \nabla\psi_s \cdot \mathbf{n} = \beta_\psi R^s_{\text{BV}} & \text{on } I_{ps}\,, \\ \psi_s = 0\,, & \text{on } \Gamma^s_r\,, \end{cases} \tag{8.69}$$

where $M_s$ = const. is the mobility taking irreversible processes into account, and $\mu_s = c^3 - c - \Delta c$.

#### 8.3.6.2 Current-driven formulation

As opposed to the voltage-driven case from Section 8.3.6.1, we state here the boundary conditions that account for a fixed external current density $i_a = I_a/|\Gamma_r^s|$.[9] We begin again with the **intercalation by classical diffusion (A)**,

$$\begin{cases} \frac{\partial c_p}{\partial t} - \Delta c_p = 0\,, & \text{in } \mathcal{D}_p\,, \\ \nabla c_p \cdot \mathbf{n} = -\frac{i_a}{D_p} = -\beta^l R_{\text{BV}}^l\,, & \text{on } \Gamma_l\,, \\ \nabla c_p \cdot \mathbf{n} = 0\,, & \text{on } \Gamma_r^p \cup \Gamma_t \cup \Gamma_b\,, \\ \nabla c_p \cdot \mathbf{n} = \beta^{ps} R_{\text{BV}}^{ps}\,, & \text{on } I_{ps}\,, \\ -\text{div}(c_p \nabla \psi_p) = -\tilde{\mathcal{R}} \Delta c_p\,, & \text{in } \mathcal{D}_p\,, \\ \psi_p = -\eta_{lp}\,, & \text{on } \Gamma_l\,, \\ \nabla \psi_p \cdot \mathbf{n} = 0\,, & \text{on } \Gamma_r^p \cup \Gamma_t \cup \Gamma_b\,, \\ \nabla \psi_p \cdot \mathbf{n} = \frac{\epsilon_s}{\epsilon_p} \nabla \psi_s \cdot \mathbf{n}\,, & \text{on } I_{ps}\,, \end{cases} \tag{8.70}$$

and,

$$\begin{cases} \tilde{a}_1 \frac{\partial c_s}{\partial t} - \Delta c_s = 0 & \text{in } \mathcal{D}_s\,, \\ \nabla c_s \cdot \mathbf{n} = -\beta^{ps} R_{\text{BV}}^{ps}\,, & \text{on } I_{ps}\,, \\ \nabla c_s \cdot \mathbf{n} = 0\,, & \text{on } \Gamma_r^s\,, \\ -\text{div}(\tilde{\sigma}_s \nabla \psi_s) = 0\,, & \text{in } \mathcal{D}_s\,, \\ \tilde{\sigma}_s \nabla \psi_s \cdot \mathbf{n} = \beta_\psi R_{\text{BV}}^s\,, & \text{on } I_{ps}\,, \\ \tilde{\sigma}_s \nabla \psi_s \cdot \mathbf{n} = -r_s i_a\,, & \text{on } \Gamma_r^s\,, \end{cases} \tag{8.71}$$

where $r_s := \frac{L}{\sigma_r} \frac{F}{R\theta}$, and $\eta_{lp} := \psi_l - \psi_p$ is the over potential between the anodic metal plate (/lithium foil) $\Gamma_l$ and polymer electrolyte phase $\mathcal{D}_p$. The potential $\psi_l$ of the lithium foil can be set to zero on $\Gamma_l$ (grounding), see [44], for instance. This leads to $\psi_p = \psi_l - \eta_{lp} = -\eta_{lp}$, where we obtain $\psi_p = -\eta_{lp}$ by solving the nonlinear problem,

$$i_a - i_{\text{BV}}^l(\eta_{lp}) = i_a - \beta^l R_{\text{BV}}^l(\eta_{lp}) = 0\,, \tag{8.72}$$

where $i_a := \frac{I_a}{|\Gamma_r^s|}$. For simplicity, we assume that the superficial current density $I_a$ is uniform along the lithium foil $\Gamma_l$ and the cathodic current collector $\Gamma_r^s$.

As for the voltage-driven formulation from Section 8.3.6.1, we obtain the **intercalation by interstitial diffusion (B)** by replacing (8.71) with,

9 Note that the sign of the current $I_a$ is associated with the (positive) direction of lithium flux in the electrolyte phase during discharge following [44].

$$\begin{cases} \tilde{a}_1 \partial_t c_s = \operatorname{div}(\nabla \mu_s), & \text{in } \mathcal{D}_s, \\ \nabla \mu_s \cdot \mathbf{n} = -\beta^{ps} R^{ps}_{\mathrm{BV}}, & \text{on } I_{ps}, \\ \nabla \mu_s \cdot \mathbf{n} = 0, & \text{on } \Gamma^s_r, \\ \nabla c_s \cdot \mathbf{n} = g(\mathbf{x}), & \text{on } \partial \mathcal{D}_s, \\ -\operatorname{div}(\tilde{\sigma}_s \nabla \psi_s) = 0, & \text{in } \mathcal{D}_s, \\ \tilde{\sigma}_s \nabla \psi_s \cdot \mathbf{n} = \beta_\psi R^{ps}_{\mathrm{BV}}, & \text{on } I_{ps}, \\ \tilde{\sigma}_s \nabla \psi_s \cdot \mathbf{n} = -r_s i_a, & \text{on } I_{sr}, \end{cases} \tag{8.73}$$

where the chemical potential $\mu_s = c^3 - c - \Delta c$ follows after choosing the classical double-well free energy density $f_r(c)$ in (8.60). We note that the function $g(x)$ allows us to describe a so-called “wetting” property, i. e., a contact angle property between the lithium rich and lithium poor phase and the boundary of the intercalation host. In the case of neutral “wetting”, i. e., a contact angle of 90 degrees, we can set $g(\mathbf{x}) = 0$.

### 8.3.7 Current-driven FSBE: linearization and 1D analytic solutions

Our main goal is the derivation of analytic solutions to the current-driven FSBE system (8.70)–(8.71). To simplify this task, we look at the following linearized equations reduced to 1D,

$$\begin{cases} \frac{\partial c_p}{\partial t} = \frac{\partial^2 c_p}{\partial x^2}, & \text{in } \mathcal{D}_p, \\ \frac{\partial c_p}{\partial x} = \frac{i_a}{FD_p} = \frac{D_r}{FD_p} \tilde{\beta}^l \tilde{R}^l_{\mathrm{BV}}, & \text{on } \Gamma_l, \\ \frac{\partial c_p}{\partial x} = \alpha_{ps} \frac{D_h}{D_p} \frac{\partial c^h_p}{\partial x}, & \text{on } I_{ps}, \\ \frac{\partial c_p}{\partial x} = (1-\alpha_{ps}) \frac{D_r}{D_p} \tilde{\beta}^{ps} \tilde{R}^{ps}_{\mathrm{BV}}, & \text{on } I_{ps}, \\ -\frac{\partial}{\partial x}\left(c_p \frac{\partial \psi_p}{\partial x}\right) = -\tilde{\mathcal{R}} \frac{\partial^2 c_p}{\partial x^2}, & \text{in } \mathcal{D}_p, \\ \psi_p = -\eta_{lp}, & \text{on } \Gamma_l, \\ \frac{\partial \psi_p}{\partial x} = \frac{\epsilon_s}{\epsilon_p} \frac{\partial \psi_s}{\partial x}, & \text{on } I_{ps}, \end{cases} \tag{8.74}$$

and,

$$\begin{cases} \tilde{a}_1 \frac{\partial c_s}{\partial t} - \frac{\partial^2 c_s}{\partial x^2} = 0, & \text{in } \mathcal{D}_s, \\ \frac{\partial c_s}{\partial x} = -\frac{D_r}{D_s} \tilde{\beta}^{ps} \tilde{R}^{ps}_{\mathrm{BV}}, & \text{on } I_{ps}, \\ \frac{\partial c_s}{\partial x} = 0, & \text{on } \Gamma^s_r, \\ -\frac{\partial}{\partial}\left(\tilde{\sigma}_s \frac{\partial \psi_s}{\partial x}\right) = 0, & \text{in } \mathcal{D}_s, \\ \tilde{\sigma}_s \frac{\partial \psi_s}{\partial x} = \tilde{\beta}_\psi \tilde{R}^s_{\mathrm{BV}}, & \text{on } I_{ps}, \\ \tilde{\sigma}_s \frac{\partial \psi_s}{\partial x} = -r_s i_a, & \text{on } \Gamma^s_r, \end{cases} \tag{8.75}$$

for Taylor approximations such as $\exp(x) = 1 + x$, for $x \ll 1$, that means,

$$\tilde{R}^l_{\mathrm{BV}}(c_s, \eta_{lp}) := (\alpha_{al} + \alpha_{cl}) \frac{F\eta_{lp}}{R\theta}, \tag{8.76}$$

$$\begin{aligned}\tilde{R}^{ps}_{\mathrm{BV}}(c_s, \eta_{ps}) := \Bigg[ c_s\left(1 + \frac{F(\alpha_a - \alpha_c)}{R\theta}(\eta_{ps} - U_c)\right) \\ - c^s_{\max}\left(1 - \frac{F\alpha_c}{R\theta}(\eta_{ps} - U_c)\right)\Bigg],\end{aligned} \tag{8.77}$$

$$\tilde{\beta}^l(c_p) := \tilde{\beta}^l_0 + \tilde{\beta}^l_1(c_p - \tilde{c}_p), \tag{8.78}$$

$$\tilde{\beta}^{ps}(c_p) := \tilde{\beta}^{ps}_0 + \tilde{\beta}^{ps}_1(c_p - \tilde{c}_p). \tag{8.79}$$

Furthermore, $U_c$ is defined in (8.53), and we set,

$$\begin{aligned}
r_s &:= \frac{L}{\sigma_r}\frac{F}{R\theta}, \\
\tilde{\beta}^l_0 &:= b_0(c^p_{\max} - \tilde{c}_p)^{\alpha_{cl}}\tilde{c}_p^{\alpha_{al}}, \\
\tilde{\beta}^l_1 &:= b_0(-\alpha_{cl}(c^p_{\max} - \tilde{c}_p)^{\alpha_{cl}-1}\tilde{c}_p^{\alpha_{al}} + \alpha_{al}(c^p_{\max} - \tilde{c}_p)^{\alpha_{cl}}\tilde{c}_p^{\alpha_{al}-1}), \\
\tilde{\beta}^{ps}_0 &:= r_pFk_p(c^p_{\max} - \tilde{c}_p)^{\alpha_c}\tilde{c}_p^{\alpha_a}, \\
\tilde{\beta}^{ps}_1 &:= r_pFk_p(-\alpha_c(c^p_{\max} - \tilde{c}_p)^{\alpha_c-1}\tilde{c}_p^{\alpha_a} + \alpha_a(c^p_{\max} - \tilde{c}_p)^{\alpha_c}\tilde{c}_p^{\alpha_a-1}), \\
b_0 &:= r_pFk_{al}^{\alpha_{cl}}k_{cl}^{al}k_{cl}^{\alpha_{al}}, \\
r_p &:= \frac{L_r s_+}{FD_r}.
\end{aligned} \tag{8.80}$$

We recall the overpotentials $\eta_{lp} := \psi_p - \psi_l$ and $\eta_{ps} := \psi_s - \psi_p$ and that $\psi_l = 0$ due to grounding, see [44], for instance. Finally, $\eta_{lp}$ is the solution of the linearized equation (8.72), that is,

$$i_a = \frac{I_a}{|\Gamma_r|} = \tilde{\beta}^l \tilde{R}^l_{\mathrm{BV}}(\eta_{lp}). \tag{8.81}$$

Hence, it follows that,

$$\eta_{lp} := \frac{i_a}{\tilde{\beta}^l(c_p)}\frac{R\theta}{F(\alpha_{al} + \alpha_{cl})}. \tag{8.82}$$

#### 8.3.7.1 Analytic solutions for $(8.74)_1$–$(8.74)_4$

A first observation is that the coupled system of equations, which is solved in the electrolyte phase $\mathcal{D}_p$, is *inhomogeneous*. This requires us to look for so-called *quasi-stationary solutions* $c^*_p$ in the process of deriving an analytic solution.

**1D Analytic Solutions to $(8.74)_1$–$(8.74)_4$**

Suppose that we are given a smooth enough initial condition $c_p^0(x)$ for the system $(8.74)_1$–$(8.74)_4$. Then, we can compute an analytic solution of the form,

$$c_p(x,t) = c_p^*(x,t) + \sum_{k=1}^{\infty} A_k c_p^k(x,t)\,, \tag{8.83}$$

where $c_p^* = A_0^* + A_1^* x$ with $A_1^* = \frac{i_a}{FD_p}$ and,

$$A_0^* = \frac{\frac{\frac{i_a}{FD_p}}{c_s(x_s)(1+\frac{F(\alpha_a-\alpha_c)}{R\theta}(\eta_{ps}-U_c))-c_{\max}^s(1-\frac{F\alpha_c}{R\theta}(\eta_{ps}-U_c))} - \tilde{\beta}_0^{ps}}{\tilde{\beta}_1^{ps}} + \tilde{c}_p - A_1^* x_s\,, \tag{8.84}$$

for,

$$\tilde{\beta}_0^{ps} = r_p F k_p \left(c_{\max}^p - \tilde{c}_p\right)^{\alpha_c} \tilde{c}_p^{\alpha_a}\,,$$
$$\tilde{\beta}_1^{ps} = r_p F k_p \left(-\alpha_c \left(c_{\max}^p - \tilde{c}_p\right)^{\alpha_c - 1} \tilde{c}_p^{\alpha_a} + \alpha_a \left(c_{\max}^p - \tilde{c}_p\right)^{\alpha_c} \tilde{c}_p^{\alpha_a - 1}\right),$$
$$U_c = 2.17 + \frac{R\theta}{F}(-0.000558 c_s + 8.1)\,,$$

and $\eta_{ps} = \psi_s - \psi_p$. Note that $\psi_s$ and $\psi_p$ are solutions of $(8.74)_5$–$(8.74)_7$ and $(8.75)_4$–$(8.75)_6$, respectively.[a] Moreover, we have that,

$$c_p^k(x,t) = \cos\left(\frac{k\pi}{x_s} x\right) \exp\left(-\frac{k^2\pi^2}{x_s^2} t\right), \tag{8.85}$$

and,

$$A_k = \frac{2}{x_s} \int_0^{x_s} \left(c_p^0(x) - c_p^*(x,0)\right) \cos\left(\frac{2k\pi}{2x_s} x\right) dx\,. \tag{8.86}$$

*a* This requires a coupled/iterative solution strategy.

**Derivation: (*1D Analytic Solutions*)**

**Step 1: *(Separation of Variables)*** We suppose that the Li-densities admit the following form,

$$c_p(x,t) = X_p(x) T_p(t)\,. \tag{8.87}$$

Inserting (8.87) into (8.74) gives,

$$\frac{\dot{T}_p(t)}{T_p(t)} = \frac{X_p''(x)}{X_p(x)}\,. \tag{8.88}$$

If we look at a certain $t = t_0$, then it holds for all $x$,

$$\frac{\dot{T}_p(t_0)}{T_p(t_0)} = \kappa = \frac{X_p''(x)}{X_p(x)}\,, \tag{8.89}$$

where $\kappa$ is a constant. In the same way, we conclude that for a certain $x = x_0$, we have that for all $t$,

$$\frac{\dot{T}_p(t_0)}{T_p(t_0)} = \kappa = \frac{X_p''(x)}{X_p(x)}\,. \tag{8.90}$$

Therefore, the ansatz (8.87) leads to the following *ordinary differential equations*,

$$\begin{cases} \dot{T}_p(t) = \kappa T_p(t)\,, \\ X_p''(x) = \kappa X_p(x)\,. \end{cases} \tag{8.91}$$

**Step 2:** ***(Stationary Solutions coping with inhomogeneities)*** Let us compute the *quasi-stationary solution,*[10] $c_p^*(x)$ which is the solution of the equation,

$$\dot{c}_p^* = \Delta c_p^* = 0\,. \tag{8.92}$$

Hence $c_p^* = A_0^* + A_1^* x$, where the constants $A_0^*$ and $A_1^*$ are determined by the inhomogeneous boundary conditions $(8.74)_3$, $(8.74)_2$, and $(8.74)_4$, that is,

$$\begin{cases} A_1^* = -\dfrac{D_h}{D_p}\alpha_{ps}\overline{A}_1^*\,, \\ A_1^* = \dfrac{D_r}{FD_p}\tilde{\beta}^l \tilde{R}_{\mathrm{BV}}^l \\ \quad = \dfrac{D_r}{FD_p}(\tilde{\beta}_0^l + \tilde{\beta}_1^l(A_0^* + A_1^* x - \tilde{c}_p))\left[(\alpha_{al} + \alpha_{cl})\dfrac{F\eta_{lp}}{R\theta}\right]\Big|_{x=0}\,, \\ A_1^* = \dfrac{D_r}{D_p}\tilde{\beta}_{\mathrm{BV}}^{ps}(1-\alpha_{ps}) \\ \quad = \dfrac{D_r}{D_p}(\tilde{\beta}_0^{ps} + \tilde{\beta}_1^{ps}(A_0^* + A_1^* x - \tilde{c}_p))\left[c_s\left(1 + \dfrac{F(\alpha_a - \alpha_c)}{R\theta}(\eta_{ps} - U_c)\right)\right. \\ \quad \left.- c_{\max}^s\left(1 - \dfrac{F\alpha_c}{R\theta}(\eta_{ps} - U_c)\right)\right]\Big|_{x=x_s}\,, \end{cases} \tag{8.93}$$

respectively, where $\overline{A}_1^*$ is the corresponding parameter of the stationary homogenized solution $\overline{c}_p(x,t)$ present in the composite cathode domain consisting of the solid intercalation host and the electrolyte.

Applying (8.82) in $(8.93)_2$ immediately gives,

$$A_1^* = \frac{i_a}{FD_p}\,. \tag{8.94}$$

**10** quasi-stationary in the sense of being stationary inside its domain but being driven by time-dependent boundary processes/conditions.

This also immediately defines the coefficient $\overline{A}_1^*$ of the homogenized equation for $\overline{c}_p$ in the polymer-$FePO_4$-composite $\mathcal{D}_s$, that is,

$$\overline{A}_1^* := -\frac{1}{\overline{D}a_{ps}}\frac{i_a}{F}\,. \tag{8.95}$$

In the same way, we can give an expression for $A_0^*$ by rewriting $(8.93)_3$,

$$\begin{aligned} A_0^* = \Bigg\{ \Bigg( \frac{i_a}{FD_p} / \Bigg[ c_s \Bigg( 1 + \frac{F(\alpha_a - \alpha_c)}{R\theta} (\eta_{ps} - U_c) \Bigg) \\ - c_{\max}^s \Bigg( 1 - \frac{F\alpha_c}{R\theta} (\eta_{ps} - U_c) \Bigg) \Bigg] - \tilde{\beta}_0^{ps} \Bigg) / \tilde{\beta}_1^{ps} + \tilde{c}_p - \frac{i_a}{FD_p} x_s \Bigg\} \,, \end{aligned} \tag{8.96}$$

which depends on $c_s$ due to the $c_s$-dependence of $U_c$. Moreover, note that both $\eta_{ps}$ and $c_s$ are time-dependent. Therefore, the stationary solution $c_p^*(x,t)$ receives a time-dependence driven by the inhomogeneous time-dependent boundary conditions.

**Step 3:** ***(Computing Homogeneous Solutions)*** We derive the equation for the homogeneous solution $C_p(x,t)$ with the help of the ansatz $c_p(x,t) = c_p^*(x,t) + C_p(x,t)$, which we insert into (8.74). As a result, we get,

$$\begin{cases} \dot{C}_p(x,t) = \frac{\partial^2}{\partial x^2} C_p(x,t)\,, & \text{in } \mathcal{D}_p\,, \\ \frac{\partial}{\partial x} C_p(x,t) = 0\,, & \text{on } I_{ps} \cup \Gamma_l\,, \\ C_p(x,0) = c_p^0(x) - c_p^*(x,0)\,, & \text{on } \mathcal{D}_p\,. \end{cases} \tag{8.97}$$

This equation again allows us to apply the method of separation of variables, i. e., for $C_p(x,t) = u(x)v(t)$, we obtain,

$$\begin{cases} \dot{v} & = \kappa v\,, \\ u'' & = \kappa u\,. \end{cases} \tag{8.98}$$

Allowing for periodic-type solutions,[11] i. e., $\kappa = -\omega^2 \le 0$, motivates us to look for,

$$u(x) = A\cos(\omega x) + B\sin(\omega x)\,. \tag{8.99}$$

Inserting (8.99) into $(8.97)_2$ gives,

$$\begin{aligned} -A\sin(\omega x)|_{x=0} + B\cos(\omega x)|_{x=0} &= 0\,, \\ -A\sin(\omega x)|_{x=x_s} + B\cos(\omega x)|_{x=x_s} &= 0\,. \end{aligned} \tag{8.100}$$

11 since the nonperiodic case, that is, $\kappa = \omega^2$, leads to $u(x) = A\cosh(\omega x) + B\sinh(\omega x)$ with trivial constants $A = B = 0$.

From $(8.100)_1$ it follows that $B = 0$, and from $(8.100)_2$ we have that,

$$-\sin(\omega x) = 0\,. \tag{8.101}$$

Hence, $\omega = k\frac{k\pi}{x_s}$, $k \in \mathbb{N}$, and therefore (8.99) becomes,

$$u_k(x) = A_k \cos(\omega_k x)\,, \quad \omega_k = \frac{k\pi}{x_s}\,, \quad k \in \mathbb{N}\,. \tag{8.102}$$

Herewith, $(8.98)_1$ reads $\dot{v} = -\omega_k^2 v$, so that,

$$v_k(t) = v_0 \exp(-\omega_k^2 t)\,, \quad \text{for } k \in \mathbb{N}\,. \tag{8.103}$$

For $k = 1, 2, \ldots$, this leads to the following fundamental solutions of the homogeneous equation (8.97),

$$c_p^k(x,t) = \cos\left(\frac{k\pi}{x_s}x\right)\exp\left(-\frac{k^2\pi^2}{x_s^2}t\right). \tag{8.104}$$

**Step 4:** ***(Initial Conditions)*** To satisfy the initial condition $(8.97)_3$, we superimpose the fundamental solutions (8.85). We obtain,

$$\sum_{k=1}^{\infty} A_k \cos\left(\frac{k\pi}{x_s}x\right) = c_p^0(x) - c_p^*(x,0)\,, \quad \text{in } \mathcal{D}_p\,. \tag{8.105}$$

Recalling the theory of Fourier series, we immediately see that $A_k$ are exactly the Fourier coefficients,

$$A_k = \frac{2}{x_s}\int_0^{x_s}\left(c_p^0(x) - c_p^*(x,0)\right)\cos\left(\frac{2k\pi}{2x_s}x\right)dx\,. \tag{8.106}$$

**Step 5:** ***(Superposition)*** Finally, we can write the solution as,

$$c_p(x,t) = c_p^*(x,t) + \sum_{k=1}^{\infty} A_k c_p^k(x,t)\,. \tag{8.107}$$

◇

### 8.3.8 Nonlinear and self-consistent Li-transport in concentrated polymer solutions

Let us first specify the basic physical setting for which we would like to investigate the concentrated solutions. Motivated by the model problem for diffuse charge dynamics of dilute solutions studied in [7], we look at a binary concentrated solution between two blocking parallel-plate electrodes, i. e., there are no Faradaic reactions on the *solid-electrolyte interface*. The two electrodes are separated by a distance $2L$, and an external

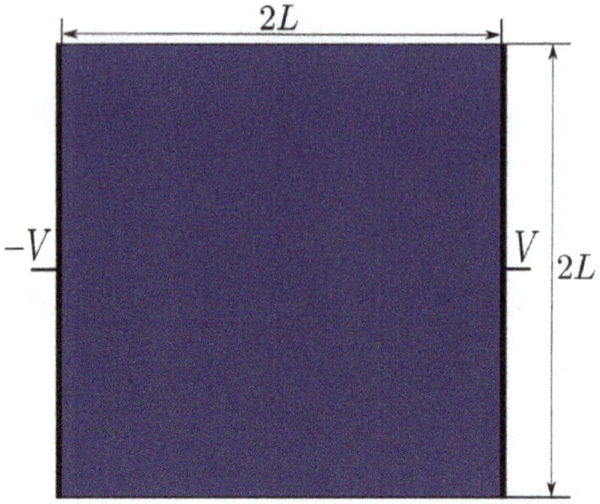


Figure 8.14: Physical setting: A concentrated electrolyte solution is enclosed by nonreactive electrodes separated by a distance $2L$ and over which a voltage of $2U$ is applied.

voltage of $2U$ is applied. Hence, the electrolyte is supposed to occupy a domain $\mathcal{U} := [-L, L]^d \subset \mathbb{R}^d$ of dimension $1 \le d \le 3$, see Figure 8.14.

Let $\mathcal{U} \subset \mathbb{R}^d$ denote the domain occupied by the polymer electrolyte, where $d \in \mathbb{N}_{>0}$ is the dimension of space. We consider the polymer solution to be a binary salt LiX, which dissociates into $Li^+$ and $X^-$. Motivated by the well-established concentrated solution theory [44, 164], we describe the Li-transport by the equation,

$$n_t = \operatorname{div}(D\nabla n) - \frac{\mathbf{i}_p \cdot \nabla t_+^0(c)}{z_+ \nu_+ F}, \quad \text{in}\, \mathcal{U}, \tag{8.108}$$

where $D = \mathcal{D}(1 - \frac{d \ln n_0}{d \ln n})$ with $\mathcal{D}$ being the diffusion coefficient measured with respect to the gradient of concentration [164, eqs. (12.12) and (12.13)]. Equation (8.108) relies on the flux,

$$\mathbf{j}_+ := \frac{\mathbf{N}_+}{\nu_+} = -D\nabla n + \frac{\mathbf{i}_p t_+^0}{z_+ \nu_+ F}, \tag{8.109}$$

since $\operatorname{div} \mathbf{i}_p = 0$, and the transference number,

$$t_+^0(n) := a + bn, \tag{8.110}$$

for $a = 0.0107907$ and $b = 1.48837 \cdot 10^{-4}$. In the case of a solution of uniform composition, we assign to the transference number $t_i$ the fraction of the current carried by the $i$th species [160]. Conservation of total mass is then guaranteed for the following corresponding flow of negatively charged species,

$$\mathbf{j}_- = \frac{\mathbf{N}_-}{\nu_-} = -D\nabla n + \frac{\mathbf{i}_p t_-^0}{z_- F}, \tag{8.111}$$

where $t_-^0(n) = 1 - t_+^0(n)$. In the concentrated solution theory, the current density in the polymer phase is defined by,

$$\mathbf{i}_p = -\sigma_p(n)\nabla\phi_p - \frac{\sigma_p(c)RT}{F}\left(1 + \frac{\partial \ln f_\pm}{\partial \ln c}\nabla n\right)\left(\frac{s_+}{n\nu_+} + \frac{t_+^0(n)}{z_+\nu_+}\right)\nabla \ln n\,, \tag{8.112}$$

where we make use of the conductivity $\sigma_p(n) := F^2(z_+^2 M_+\nu_+ + z_-^2 M_-\nu_-)n$ of a dilute binary electrolyte [164, eq. (11.7)], for simplicity, for the mobilities $M_+$ and $M_-$ of the positive and negative ions, respectively. The parameter $f_\pm$ accounts for the mean molar activity of an electrolyte, and its variation with respect to the salt density $c$ is only included, if data are available [44]. Generally, we measure $\phi_p$ with a reference electrode [44], but to avoid the experimental challenges of using reference electrodes, we determine $\phi_p$ for a given electric permittivity $\varepsilon_p$ of the polymer electrolyte as in the dilute solution theory by,

$$-\mathrm{div}(\varepsilon_p\nabla\phi_p) = Fz_n n\,, \tag{8.113}$$

where $z_c := (z_+\nu_+ + z_-\nu_-)$, and the charge density on the right-hand side is a result of electroneutrality ($z_+n_+ + z_-n_- = 0$) of the bulk solution and the concentration $n = \frac{n_+}{\nu_+} = \frac{n_-}{\nu_-}$ of a binary electrolyte.

For the subsequent investigation of this self-consistent form of the polymer electrolyte formulation proposed and discussed in [44, 164], we rewrite equations (8.108), (8.112), and (8.113) as the nonlinear concentrated solution problem on the domain $\mathcal{U} \subset \mathbb{R}^d$ enclosed by the *ideally polarizable/blocking* parallel-plate electrodes,

$$\begin{cases} \frac{\partial n}{\partial t} = \mathrm{div}(D\nabla n + M_p n^2\nabla\phi_p) + Q\left(1 + \frac{d\ln f_\pm}{d\ln n}\right)(\nabla n)^2 + Nn^3\,, & \text{in}\,\mathcal{U}\times(0,T)\,,\\ (D\nabla n + M_p n^2\nabla\phi_p)\cdot\mathbf{n} = 0\,, & \text{on}\,\partial\mathcal{U}\times(0,T)\,,\\ -\mathrm{div}(\varepsilon_p\nabla\phi_p) = z_n n\,, & \text{in}\,\mathcal{U}\times(0,T)\,,\\ \phi_p - \lambda_s\nabla\phi_p\cdot\mathbf{n} = V\,, & \text{on}\,\partial\mathcal{U}\times(0,T)\,, \end{cases} \tag{8.114}$$

where $f_\pm$ is a mean molar activity coefficient, and the remaining parameters are defined as follows,

$$\begin{aligned} 2M_p &= \frac{b}{z_+\nu_+}F(z_+^2M_+\nu_+ + z_-^2M_-\nu_-)\,, \quad N = \frac{b(z_+\nu_+ + z_-\nu_-)F^2}{z_+\nu_+\varepsilon_p}\,,\\ Q &= Q_1 + Q_2c\,, \quad Q_0 = \frac{bRT}{z_+\nu_+}(z_+^2M_+\nu_+ + z_-^2M_-\nu_-)\,,\\ Q_1 &= Q_0\left(\frac{s_+}{n\nu_+} + \frac{a}{z_+\nu_+}\right), \quad Q_2 = Q_0\frac{b}{z_+\nu_+}\,. \end{aligned} \tag{8.115}$$

In ideal solutions, $f_\pm = 1$. We note that the boundary condition $(8.114)_4$ describes a so-called Stern layer, which accounts for an intrinsic surface capacitance of the electrode surface. Finally, problem (8.114) has to be complemented with appropriate initial conditions.

**Dimensionless formulation of (8.114):** For the characteristic parameters such as the length $L$, reference concentration $n_r$, reference time $\tau$, and thermal voltage $U_\theta := \frac{F}{RT}$, we introduce the dimensionless variables,

$$\tilde{\mathbf{x}} = \frac{\mathbf{x}}{L}, \quad \tilde{n} = \frac{n}{n_r}, \quad \tilde{t} = \frac{t}{\tau}, \quad \tilde{\phi}_p = U_\theta \phi_p . \tag{8.116}$$

This immediately leads to the dimensionless transport equation,

$$\begin{cases} \tilde{n}_{\tilde{t}} - \tilde{\Delta}\tilde{n} = \mathcal{M}\tilde{\operatorname{div}}(\tilde{n}^2\tilde{\nabla}\tilde{\phi}_p) + \mathcal{Q}_1(\tilde{\nabla}\tilde{n})^2 + \mathcal{Q}_2(\tilde{\nabla}\tilde{n})^2\tilde{n} + \mathcal{N}\tilde{n}^3 , \\ -\tilde{\operatorname{div}}(\lambda^2\tilde{\nabla}\tilde{\phi}_p) = \tilde{z}_n\tilde{n} , \end{cases} \tag{8.117}$$

for $\tilde{z}_n := \frac{z_n}{2}$, the diffusion time $\tau = \tau_D = \frac{L^2}{D}$, and the dimensionless parameters,

$$\begin{aligned} \lambda &= \frac{\lambda_d}{L}, \quad \mathcal{M} := \frac{M_p}{D} n_r \frac{RT}{F}, \quad \mathcal{Q}_1 := \frac{Q_1}{D} n_r , \\ \mathcal{Q}_2 &:= \frac{Q_2}{D} n_r^2, \quad \mathcal{N} := 2M_p \frac{N}{D} (\mathrm{L} n_r)^2 , \end{aligned} \tag{8.118}$$

where $\lambda_D^2 := \frac{\varepsilon_p RT}{2n_r F^2}$ is the dimensionless Debye length. We used the Einstein relation $D = MRT$ to ensure that parameters (8.118) are dimensionless. For notational convenience, we will further work with the dimensionless system and skip the tilde from the dimensionless variables. Choosing $L$ such that $1/\lambda^2 = \frac{L^2}{\lambda_D^2} = 10$, we have the inequalities,

$$10^{-4} \le \mathcal{Q}_2 \le \mathcal{M} \le \mathcal{Q}_1 \le \mathcal{N} \le 0.3 , \tag{8.119}$$

for the following characteristic sizes of the dimensionless numbers (8.118) under $n_r = 10^2$: $\mathcal{M} = 1.488 \cdot 10^{-4} \cdot n_r \sim 10^{-2}$, $\mathcal{Q}_1 = 2.976 \cdot 10^{-4} \cdot (1 + 0.0108) \cdot n_r \sim 10^{-2}$, $\mathcal{Q}_2 = (1.488 \cdot 10^{-4})^2 \cdot 2n_r^2 \sim 10^{-4}$, $\mathcal{N} = \frac{L^2}{\lambda_D^2} 2.96 \cdot 10^{-4} \cdot n_r \sim \frac{L^2}{\lambda_D^2} \cdot 10^{-2}$, see Table 8.3.

| Parameter | Value(s) | Ref. |
|---|---|---|
| $D$ | $7.5 \cdot 10^{-12} \frac{m^2}{s}$ | [44, p. 8] |
| $t_+^0(n) := a + bn$ | $a := 0.0107907, b := 1.48837 \cdot 10^{-4} \frac{\mathrm{m}^3}{\mathrm{mol}}$ | [44, p. 8] |
| $M = M_+ = M_-$ | $M = DRT$ | |
| $z = z_+ = z_- = 1$ | $z = 1$ | |
| $v = v_+ = v_-$ | $v = 1$ | |
| $n$ | $n = 1$ | |
| $\varepsilon_p$ | $9.8 \cdot 10^3 \frac{\mathrm{F}}{\mathrm{m}} \le \varepsilon_p \le 2.93 \cdot 10^5 \frac{\mathrm{F}}{\mathrm{m}}$ | [172, p. 17] |

Table 8.3: Characteristic parameters for polymer insertion cells. We set $\varepsilon_p = 1.0 \cdot 10^4 \frac{\mathrm{F}}{\mathrm{m}}$.

**Reduced formulation based on small reference concentrations $n_r \sim 10^{-3}$:** In the case of a small reference concentration $\left(n_r = \mathcal{O}(10^{-3})\right)$, we have that $\mathcal{Q}_2 = 10^{-17} \ll \mathcal{Q}_1 = 10^{-7}$.

This motivates the consideration of a reduced problem, where we neglect the nonlinear term scaled with $\mathcal{Q}_2$,

$$\begin{cases} n_t - \Delta n = f_n(n), & \text{in } \mathcal{U} \times (0, T), \\ -\operatorname{div}(\lambda^2 \nabla \phi_p) = \tilde{z}_n n, & \text{in } \mathcal{U} \times (0, T), \end{cases} \tag{8.120}$$

where the highly nonlinear right-hand side $f_n(n)$ is defined for a general function $v$ by,

$$f_v(n) := \mathcal{M} \operatorname{div}(v^2 \nabla \phi_p) + \mathcal{Q}_1 \nabla v \cdot \nabla n + \mathcal{N} v^2 n. \tag{8.121}$$

We will rigorously investigate the existence of solutions to the nonlinear system (8.120) based on a fixed point argument and the concepts of weak and strong solutions in Section 9.

# Part II: Rigorous mathematical methods: CHeSs relevant applications

# 9 Electrochemical CHeSs: from well-posedness to battery failure/blow-up

## 9.1 Well-posedness of a binary electrolyte for v = 0

We investigate under what conditions system (8.120) has a local solution in time and look at the existence of blow-up solutions. To this end, we consider the following homogeneous Dirichlet initial value problem,

$$\begin{cases} n_t - \Delta n = f_n(n)\,, & \text{in } \mathcal{U} \times (0,T)\,, \\ n = 0\,, & \text{on } \partial\mathcal{U} \times (0,T)\,, \\ n = n_0\,, & \text{in } \mathcal{U} \times \{t=0\}\,, \\ -\operatorname{div}(\lambda^2 \nabla \phi_p) = \tilde{z}_n n\,, & \text{in } \mathcal{U} \times (0,T)\,, \\ \phi_p = 0\,, & \text{on } \partial\mathcal{U} \times (0,T)\,, \end{cases} \tag{9.1}$$

where the nonlinear driving force $f_n$ is defined in (8.121), and $\mathcal{U} \subset \mathbb{R}^3$ is an open bounded set. We subsequently apply Banach's fixed point theorem. To this end, we first introduce the following two standard classes of solutions for parabolic equations.

**Weak solutions**

**Definition 9.1.1.** The pair of functions $(n, \phi_p) \in L^2(0,T;H_0^1(\mathcal{U})) \times L^2(0,T;H_0^1(\mathcal{U}))$ with $\dot{n} \in L^2(0,T;H^{-1}(\mathcal{U}))$ and $n \geq 0$ that solves for all $(\varphi_1, \varphi_2) \in [H_0^1(\mathcal{U})]^2$ the system,

$$\begin{cases} (n_t, \varphi_1) + (\nabla n, \nabla \varphi_1) = (f, \varphi_1)\,, \\ \left(\lambda^2 \nabla \phi_p, \nabla \varphi_2\right) = \tilde{z}_n (n, \varphi_2)\,, \end{cases} \tag{9.2}$$

where $f \in L^2(0,T;L^2(\mathcal{U}))$ and $n(\mathbf{x},0) = n_0 \in L^2(\mathcal{U})$, is called a *weak solution* of (9.1), if $f = f_n(n)$. We will frequently refer to such a weak solution $n$ as being of class *(W)*.

We also need slightly higher regular solutions for the subsequent fixed point argument.

**Strong solutions**

**Definition 9.1.2.** A pair of weak solutions $(n, \phi_p) \in L^2(0,T;H_0^1(\mathcal{U})) \times H_0^1(\mathcal{U})$ with $\dot{n} \in L^2(0,T;H^{-1}(\mathcal{U}))$ of (9.1), initial condition $n_0 \in H_0^1(\mathcal{U})$, and the right-hand side $f \in L^2(0,T;L^2(\mathcal{U}))$ that satisfies,

$$\begin{aligned} n \in L^2\big(0,T;H^2(\mathcal{U})\big) \cap L^\infty\big(0,T;H_0^1(\mathcal{U})\big), \quad \dot{n} \in L^2\big(0,T;L^2(\mathcal{U})\big), \\ \text{and} \quad \phi_p \in L^2\big(0,T;H^2(\mathcal{U})\big) \cap L^\infty\big(0,T;H_0^1(\mathcal{U})\big) \end{aligned} \tag{9.3}$$

is called a *strong solution* of (9.1) for $f = f_n(n)$. We will frequently refer to a strong solution $n$ as being of class *(S)*.

Before we define the fixed point map, which represents the key tool for establishing existence of solutions, we prove that solutions of (9.2) are nonnegative.

https://doi.org/10.1515/9783110579543-009

**Nonnegative solutions**

**Lemma 9.1.3.** *Weak solutions n solving (9.2) are nonnegative a.e..*

*Proof.* First, note that we can write $n = C^+ - C^-$, where $C^+ := \sup\{n, 0\}$ and $C^- := \sup\{-n, 0\}$. Herewith, we introduce the auxiliary problem,

$$\begin{cases} n_t - \Delta n = \mathcal{M} \operatorname{div}(nC^+ \nabla \phi_p) + \mathcal{Q}_1 \nabla n \cdot \nabla C^+ + \mathcal{N} n^2 C^+ , \\ -\operatorname{div}\left(\lambda^2 \nabla \phi_p\right) = \tilde{z}_n n , \end{cases} \tag{9.4}$$

which we test with $C^-$,

$$-\frac{1}{2}\frac{d}{dt}\|C^-\|^2 - \|\nabla C^-\|^2 = -\mathcal{M}(nC^+ \nabla\phi_p, \nabla C^-) + \mathcal{Q}_1(\nabla n \nabla C^+, C^-) \\ + \mathcal{N}(n^2 C^+, C^-) = 0 . \tag{9.5}$$

This implies that $C^- = 0$ a. e., and hence $n \geq 0$ a. e. for solutions of the auxiliary problem (9.5) and also for solutions of (9.2), since (9.5) turns into (9.2) for $n \geq 0$, which proves the lemma. □

**Local well-posedness**

**Theorem 9.1.4.** *For small enough $T > 0$, there exists a local unique weak solution in the sense of Definition* 9.1.1.

*Proof.* For $X := C(0, T; L^2(\mathcal{U})) \cap L^2(0, T; H_0^1(\mathcal{U}))$ with the associated norm,

$$\|\cdot\|_X^2 := \sup_{t\in(0,T)} \|\cdot\|^2(t) + \int_0^T \|\nabla \cdot\|^2(t)\, dt ,$$

we introduce for all $v$ of class (S) the following fixed point map $\mathcal{F}_v : X \to X, n \mapsto w = \mathcal{F}_v(n)$, which is defined for $n \in X$ as the solution of the following auxiliary problem,

$$\mathcal{F}_v : \quad \begin{cases} w_t - \Delta w = f_v(n) , & \text{in } \mathcal{U} \times (0, T) , \\ w = 0 , & \text{on } \partial\mathcal{U} \times (0, T) , \\ w = w_0 , & \text{in } \mathcal{U} \times \{t = 0\} . \end{cases} \tag{9.6}$$

As for weak solutions (W), we accordingly introduce the canonical norm,

$$\|\cdot\|_Y^2 := \sup_{t\in(0,T)} \|\nabla \cdot\|^2(t) + \int_0^T \|\nabla \cdot\|^2(t)\, dt , \tag{9.7}$$

for strong solutions (S), which are a subset of $Y := L^\infty(0, T; H_0^1(\mathcal{U})) \cap L^2(0, T; H^2(\mathcal{U}))$.

Before we establish the existence of a fixed point based on Banach's fixed point theorem in the space $X$, we establish necessary properties of the right-hand side $f_v(n)$.

**Property I of $f_v$**

**Lemma 9.1.5.** *For all v of class* (S) *and n of class* (W), *we have that* $f_v(n) \in L^2\left(0, T; L^2(\mathcal{U})\right)$.

**Remark 9.1.6** (Solvability and strong solutions to $\mathcal{F}_v$). We note that Lemma 9.1.5 immediately guarantees the existence of a weak solution $w$ to (9.6) of class *(W)* and the existence of a strong solution $w$ of class *(S)*.

*Proof.* We are going to show that,

$$\sup_{\varphi\in L^2(\mathcal{U}),\, \|\varphi\|\le 1} \frac{|(f_v(n), \varphi)|}{\|\varphi\|} \le C(v, \phi_p), \tag{9.8}$$

where the bound depends on $v$ and $\phi_p$ whose norms are bounded due to the given regularity $v$ in (S). First, we have the following estimate for all $\varphi \in L^2(\mathcal{U})$,

$$\begin{aligned} |(f_v(n), \varphi)| &\le \mathcal{M}(|(2\nabla v v \nabla \phi_p, \varphi)| + |(v^2 \Delta \phi_p, \varphi)|) + \mathcal{Q}_1|(\nabla v \nabla n, \varphi)| + \mathcal{N}|(v^2 n, \varphi)| \\ &\le \left( C\|\phi_p\|_{H^2}^2 \|v\|_{H^1}^2 + \frac{1}{2}\|v\|_{H^2}^2 \right)\|\varphi\| . \end{aligned} \tag{9.9}$$

The first term on the right-hand side in (9.9) is bounded by $C(v, \phi_p)$, since $v \in L^\infty(0, T; H^1(\mathcal{U}))$ as being part of (S).

Similarly, we have

$$|(v^2 \Delta \phi_p, \varphi)| \le C\|v\|_{L^6}^2 \|n\|_{L^6} \|\varphi\| \le C\|v\|_{H^1}^2 \|\nabla n\| \|\varphi\| , \tag{9.10}$$

so that also the second term on the right-hand side in (9.9) is bounded by $C(v, \phi_p)$. In the same way, we can proceed with the remaining terms. □

**Property II of $f_v$**

**Lemma 9.1.7.** *For all v of class* (S), *all $\varphi$ of class* (W), *and $n_1 - n_2$ of class* (W), *the right-hand side $f_v$ is Lipschitz continuous, i. e.,*

$$|(f_v(n_1) - f_v(n_2), \varphi)| \le C\left( \epsilon, \sup_{t\in(0,T)} \|v\|_{H^1}^2 \right) \|v\|_Y^2 \|n_1 - n_2\|_X^2 + \epsilon\|\nabla\varphi\|^2 , \tag{9.11}$$

*where $\epsilon > 0$ can be chosen arbitrarily small.*

*Proof.* For all $\varphi$ of class (W), we can estimate the left-hand side of (9.11) with $\epsilon$-Young and Hölder inequalities,

$$\begin{aligned} |(f_v(n_1) - f_v(n_2), \varphi)| \le\, & \mathcal{M}|(v^2 \nabla \phi_p^{12}, \nabla\varphi)| \\ & + \mathcal{Q}_q|(\nabla v \nabla (n_1 - n_2), \varphi)| \end{aligned}$$

$$
\begin{aligned}
&+\mathcal{N}\left|(v^2(n_1-n_2),\varphi)\right| \\
\leq\, & C(\epsilon)\|v\|_{L^6}^4\|n_1-n_2\|^2+\epsilon\|\nabla\varphi\|^2 \\
&+C(\epsilon)\|v\|_{H^2}^2\left\|\nabla(n_1-n_2)\right\|^2+\epsilon\|\nabla\varphi\|^2 \\
&+C(\epsilon)\|v\|_{H^1}^2\|n_1-n_2\|^2+\epsilon\|\nabla\varphi\|^2\,, \\
\leq\, & C\Big(\epsilon,\sup_{t\in(0,T)}\|v\|_{H^1}^2\Big)\|v\|_Y^2\|n_1-n_2\|_X^2+\epsilon\|\nabla\varphi\|^2\,,
\end{aligned} \tag{9.12}
$$

where $\phi_p^{12}$ is the weak solution of the following equation,

$$
-\lambda^2\Delta\phi_p^{12}=\tilde{z}_n(n_1-n_2)\,. \tag{9.13}
$$

This proves the claim. □

For the proof of the next result, we recall the definition of the spaces $X := C(0,T;L^2(\mathcal{U}))\cap L^2(0,T;H_0^1(\mathcal{U}))$ and $Y := L^\infty(0,T;H_0^1(\mathcal{U}))\cap L^2(0,T;H^2(\mathcal{U}))$ with their associated norms

$$
\begin{cases}
\|\cdot\|_X^2 := \sup_{t\in(0,T)}\|\cdot\|^2(t)+\int_0^T\|\nabla\cdot\|^2(t)\,dt\,, \\
\|\cdot\|_Y^2 := \sup_{t\in(0,T)}\|\nabla\cdot\|^2(t)+\int_0^T\|\nabla\cdot\|^2(t)\,dt\,,
\end{cases} \tag{9.14}
$$

which belong to weak (W) and strong solutions (S), respectively.

**Contraction property**

**Lemma 9.1.8.** *The fixed point map $w=\mathcal{F}_v(n)$ defined by (9.6) for all $v$ of class (S) is a contraction, i. e., for all $n_1,n_2\in X$, it holds that,*

$$
\|w_1-w_2\|_X\leq q\|n_1-n_2\|_X\,, \tag{9.15}
$$

*where $0<q<1$. Naturally, the contraction property also holds for $\mathcal{F}_n(n)$, if we start with an initial guess $n$ of class (S).*

*Proof.* First, we subtract equation (9.6), whose solution $w_2$ is governed by the right-hand side $f_v(n_2)$, from the equation (9.6) with solution $w_1$ governed by $f_v(n_1)$ and test the resulting equation for the difference $w_1-w_2$ with $w_1-w_2$ itself, so that we obtain,

$$
\begin{aligned}
\frac{1}{2}\frac{d}{dt}\|w_1-w_2\|^2+\left\|\nabla(w_1-w_2)\right\|^2 &\leq C\Big(\epsilon,\sup_{t\in(0,T)}\|v\|_{H^1}^2\Big)\|v\|_Y^2\|n_1-n_2\|_X^2 \\
&\quad+\epsilon\left\|\nabla(w_1-w_2)\right\|^2\,,
\end{aligned} \tag{9.16}
$$

where $\phi_p^{12}$ is the weak solution of (9.13), and where we applied Property II (Lemma 9.1.7) of $f_v$ with $\varphi=(w_1-w_2)$. Integrating over time leads for $0\leq s\leq T$ to,

$$\|w_1 - w_2\|^2(s) + \int_0^s \|\nabla(w_1 - w_2)\|^2(t)\,dt \le C(\epsilon, v)\int_0^T \|n_1 - n_2\|^2_{H^1}(t)\,dt$$
$$\le C(\epsilon, v)T\|n_1 - n_2\|^2_X\,. \tag{9.17}$$

Hence, maximizing the left-hand side with respect to $s$ leads for the norm $\|\cdot\|_X$ to the required contraction property,

$$\|w_1 - w_2\|^2_X \le C(\epsilon, v)T\|n_1 - n_2\|^2_X\,, \tag{9.18}$$

with a contraction parameter $q := \sqrt{C(\epsilon, v)T} < 1$ for $T$ small enough. Since solutions $w = \mathcal{F}_v(n)$ of the fixed point map (9.6) are both of class (S)⊂(W) and (W), for $n$ of class (W), according to Remark 9.1.6, the contraction property remains valid for $v = n$, if we start the fixed point iteration with an initial guess $c$ of class (S), so that the norms containing $v = n$ in (9.16) are under control. □

Finally, the uniqueness follows from inequality (9.17) by Gronwall's inequality. □

### 9.1.1 Conditions for blow-up in finite time

We are going to show that there exist solutions $n(\mathbf{x}, t)$ to the initial value problem,

$$\begin{cases} n_t - \Delta n = \mathcal{M}\,\mathrm{div}(n^2\nabla\phi_p) + \mathcal{Q}_1(\nabla n)^2 + \mathcal{N}n^3\,, & \text{in } \mathcal{U}\times(0,T)\,,\\ n = 0\,, & \text{on } \partial\mathcal{U}\times(0,T)\,,\\ n(\mathbf{x},0) = g(\mathbf{x})\,, & \text{in } \mathcal{U}\times\{0\}\,,\\ -\lambda^2\,\mathrm{div}(\nabla\phi_p) = n\,, & \text{in } \mathcal{U}\times(0,T)\,,\\ \phi_p = 0\,, & \text{on } \partial\mathcal{U}\times(0,T) \end{cases} \tag{9.19}$$

that blow up in finite time in case of large enough initial conditions $g$, i. e., if $g$ satisfies,

$$\eta^2(0) = \left(\int_{\mathcal{U}} g v_1\,d\mathbf{x}\right)^2 > \lambda_1\,, \tag{9.20}$$

where $v_1$ and $\lambda_1 > 0$ are the eigenfunction and the associated principal eigenvalue of the following eigenvalue problem considered in the $H^1_0$-sense,

$$\begin{cases} -\Delta v_1 = \lambda_1 v_1\,, & \text{in } \mathcal{U}\,,\\ v_1 = 0\,, & \text{on } \partial\mathcal{U}\,. \end{cases} \tag{9.21}$$

Additionally, we may assume that,

$$v_1 > 0\,, \quad \text{and} \quad \int_{\mathcal{U}} v_1 \, d\mathbf{x} = 1\,. \tag{9.22}$$

These considerations allow us to demonstrate a finite time blow-up of the function,

$$\eta(t) := \int_{\mathcal{U}} n v_1 \, d\mathbf{x}\,, \tag{9.23}$$

where $n$ is a smooth solution of (9.19) with initial condition $g \geq 0, g \neq$ const., such that $n > 0$ within $\mathcal{U}$.

**Blowup**

**Lemma 9.1.9.** *Let $n$ be a smooth solution of problem (9.19). Then there exists a time,*

$$T^* := -\frac{1}{2\lambda_1} \ln\left(1 - \frac{\lambda_1}{\eta^2(0)}\right) > 0\,,$$

*such that,*

$$\lim_{t \to T} \eta(t) = \infty\,, \quad \textit{for some} \quad 0 < T \leq T^*\,, \tag{9.24}$$

*where $\lambda_1 > 0$ is the principle eigenvalue in (9.21).*

*Proof.* Starting with definition (9.23) and applying $(9.19)_1$, we obtain the equation,

$$\eta_t = (n_t, v_1) = -(\nabla n, \nabla v_1) - \mathcal{M}(n^2 \nabla \phi_p, \nabla v_1) + \mathcal{Q}_1((\nabla n)^2, v_1) + \mathcal{N}(n^3, v_1)\,. \tag{9.25}$$

We can also estimate as follows,

$$\eta = (n v_1^{1/3}, v_1^{2/3}) \leq \|n v_1^{1/3}\|_{L^3} \|v_1^{2/3}\|_{L^{3/2}} = \|n v_1^{1/3}\|_{L^3}\,, \tag{9.26}$$

so that,

$$\mathcal{N}\eta^3 \leq \mathcal{N}(n^3, v_1)\,. \tag{9.27}$$

Testing the eigenvalue problem (9.21) with $n$ and integrating by parts immediately give,

$$(\nabla v_1, \nabla n) = \lambda_1 (v_1, n) = \lambda_1 \eta\,. \tag{9.28}$$

Thanks to the nonnegativity of $v_1$, we get the following inequality for the $\mathcal{Q}_1$-term,

$$\mathcal{Q}_1((\nabla n)^2, v_1) \geq 0\,. \tag{9.29}$$

To study our last term $-\mathcal{M}(n^2 \nabla \phi_p, \nabla v_1)$, we test $(9.19)_4$ with $-\lambda_1 n^2 v_1$ and obtain,

$$-\lambda_1 (n^3, v_1) = \lambda^2 \lambda_1 ((n^2 \nabla \phi_p, \nabla v_1) + (v_1 \nabla \phi_p, \nabla (n^2)))\,,$$

so that,

$$-\mathcal{M}(n^2\nabla\phi_p, \nabla v_1) = \frac{\mathcal{M}}{\lambda^2}((n^3, v_1) + (\nabla n^2 \nabla\phi_p, v_1)) \geq \frac{\mathcal{M}}{\lambda^2}\eta^3, \tag{9.30}$$

where the last inequality is a consequence of (9.27) and the following Claim I, which we will prove at the end.

**Claim I.** *For the pair* $(n, \phi_p)$ *solving* (9.19) *and for the solution* $v_1$ *of the eigenvalue problem* (9.21), *we have,*

$$(n\nabla n\nabla\phi_p, v_1) \geq 0\,.$$

Putting things together leads to,

$$\eta_t \geq -\lambda_1\eta + \left(\frac{\mathcal{M}}{\lambda^2} + \mathcal{N}\right)\eta^3\,. \tag{9.31}$$

Setting $\rho(t) := \exp(\lambda_1 t)\eta(t)$ gives,

$$\rho_t = \exp(\lambda_1 t)\eta_t + \lambda_1 \exp(\lambda_1 t)\eta \geq \exp(\lambda_1 t)\left(\frac{\mathcal{M}}{\lambda^2} + \mathcal{N}\right)\eta^3 = \exp(-2\lambda_1 t)\rho^3(t)\,, \tag{9.32}$$

for $0 \leq t \leq T$. Integrating the resulting inequality,

$$-\frac{d}{dt}\left(\frac{1}{2}\rho^{-2}\right) = \rho^{-3}\rho_t \geq \exp(-2\lambda_1 t)\,, \tag{9.33}$$

over time leads to,

$$-\rho^{-2}(t) \geq \frac{-\lambda_1 + \rho^2(0)(1 - \exp(-2\lambda_1 t))}{\rho^2(0)\lambda_1}\,. \tag{9.34}$$

After rewriting the last expression, we obtain the inequality,

$$\rho(t) \geq \sqrt{\frac{\rho^2(0)\lambda_1}{\lambda_1 - \rho^2(0)(1 - \exp(-2\lambda_1 t))}}\,, \tag{9.35}$$

provided that the denominator $\lambda_1 - \rho^2(0)(1 - \exp(-2\lambda_1 t))$ is not zero.

Hence, due to assumption (9.20) of large initial conditions, i. e.,

$$\eta^2(0) = \rho^2(0) > \lambda_1\,,$$

we have that $\rho(t) \to \infty$ as $t \to T^*$, where $T^*$ is such that $\lambda_1 = \rho^2(0)(1 - \exp(-2\lambda_1 T_*))$, and therefore,

$$T^* := -\frac{1}{2\lambda_1}\ln\left(1 - \frac{\lambda_1}{\rho^2(0)}\right), \tag{9.36}$$

which proves the lemma after justifying Claim I. □

*Proof of Claim* I. Since we consider the eigenvalue problem (9.21) in the separable Hilbert space $H_0^1(\mathcal{U})$, there exists a basis $\{v_l\}_{1\leq k\leq\infty}$ consisting of eigenfunctions of the compact self-adjoint operator $-\Delta$ in $H_0^1(\mathcal{U})$. Since the solution $\phi_p$ to the Poisson equation $(9.19)_4$ is an element of the Hilbert space $H_0^1(\mathcal{U})$, we can represent it by $\phi_p = \sum_{k=1}^{\infty} \lambda_k^p v_k$. Herewith and due to orthogonality, we have that,

$$
\begin{aligned}
(n\nabla n\nabla\phi_p, v_1) = \lambda^2\big(n\nabla(-\Delta\phi_p)\nabla\phi_p, v_1\big) &= \lambda^2 \sum_{k=1}^{\infty} (\lambda_k^p)^2 \big(n(\nabla v_k)^2, v_1\big) \\
&\geq \lambda^2 (\lambda_1^p)^2 \big(n(v_1)^2, v_1\big) \geq 0\,,
\end{aligned}
\tag{9.37}
$$

which proves the claim. □

The above rigorous derivations are a valuable fundamental basis for applications to estimate and predict battery failure as summarized in the following method.

---

**!** **Estimation of battery life(/blow-up) and its dependencies**
Lemma 9.1.9 represents a systematic tool for the rigorous and promising practical investigation and prediction of battery failures together with an upper bound $T^* > 0$ on its lifetime, which depends on the amount of initial Li-concentration $n(\mathbf{x}, 0)$ in the electrolyte, see equation (9.20).

---

# 10 Rigorous upscaling of CHeSs: the two-scale convergence method

## 10.1 Introduction

In this section, we present a rigorous method that allows us to pass to the limit in problems, where we encounter strong oscillations, primarily induced by heterogeneities. Before we begin with introducing the rigorous and constructive two-scale convergence method going back to Nguetseng [113] and Allaire [3], we motivate the underlying mechanisms leading to a lack of weak convergence either due to missing regularity or due to oscillations in the subsequent and classical examples found in [37], for instance.

**Example 10.1.1.** (Weak convergence in $L^1$) Consider the sequence,

$$a_n(x) := \begin{cases} n, & 0 \leq x \leq 1/n\,, \\ 0 & \text{else}\,. \end{cases} \tag{10.1}$$

A direct computation shows that $a_n \in L^1((-1,1))$, since,

$$\|a_n\|_{L^1((-1,1))} = \int_{-1}^{1} a_n(x)\,dx = \int_{0}^{1/n} n\,dx = n[x]_0^{1/n} = n(1/n - 0) = 1\,.$$

After multiplying the sequence $a_n$ by continuous compactly supported functions $v \in C_0^0((-1,1))$, the mean value theorem gives,

$$\int_{-1}^{1} a_n(x)v(x)\,dx = (a_n, v)_{L^1((-1,1)),L^\infty((-1,1))} = n\int_{0}^{1/n} v\,dx = nv(\alpha)(1/n - 0)\,, \tag{10.2}$$

where $\alpha \in (0, 1/n)$. Hence, after passing to the limits as $n \to \infty$, we obtain the limit,

$$(a_n, v)_{L^1((-1,1)),L^\infty((-1,1))} \to v(0)\,.$$

This means that the sequence $a_n$ converges to the Dirac function $\delta_0(x)$, which is zero for $x \neq 0$ and one for $x = 0$. Note that the Dirac function satisfies the criteria for a measure and it is not in $L^1((-1,1))$. This example motivates the concept of weak$^*$ compactness in the space of Radon measures $M(D)$, see [37, Proposition 1.48, p. 24], for instance. ◇

The effects of oscillations are investigated in the following example.

**Example 10.1.2** (Weak convergence of oscillating sequences). Consider the periodic function $u(y) = \sin(2\pi y)$ for $y \in \mathbb{R}$ and with setting $y = x/\epsilon$ define $u_\epsilon(x) = v(x/\epsilon)$. Let us restrict ourselves to the interval $I = (\alpha, \beta)$ for $\alpha, \beta \in \mathbb{R}$.

An immediate question is what can we say about the convergence of $u_\epsilon(x)$ as $\epsilon \to 0$?

https://doi.org/10.1515/9783110579543-010

If we study the range of $u_\epsilon$ for different $\epsilon$, then it becomes clear that $u_\epsilon$ cannot converge a. e. in $I$.

Is there hope for some kind of weak convergence?

The answer is positive, if we recall [37, Lemma 1.46, p. 22], for instance, which guarantees that,

$$u_\epsilon \rightharpoonup 0 \quad \text{in } L^2(I)\,. \tag{10.3}$$

This requires us to verify properties (a) and (b) of [37, Lemma 1.46, p. 22]:
(a) The boundedness of $u_\epsilon$ in $L^2(I)$ is clear.
(b) For any interval $I_1 = (\alpha_1, \beta_1) \subset (\alpha, \beta)$, we have,

$$\int_{I_1} \sin(2\pi x/\epsilon)\, dx = -\frac{\epsilon}{2\pi}\left[\cos(2\pi x/\epsilon)\right]_{\alpha_1}^{\beta_1} \to 0 \quad \text{as } \epsilon \to 0\,, \tag{10.4}$$

which proves (10.3). ◇

These examples will motivate that it is very useful in practice to have a good and convenient notion of convergence for problems, where we have to pass to the limit as $\epsilon \to 0$ and $\epsilon$ represents a small problem-specific parameter $\epsilon > 0$.

## 10.2 The classical two-scale convergence method

The aim of this section is to present a rigorous alternative to the formal asymptotic two-scale expansion method introduced in Section 6.3. The advantage of this rigorous method is that we can derive the homogenized/upscaled equations based on the rigorous analytical concepts such as a priori estimates, variational/weak formulations, and the existence and uniqueness theory generally relying on initial formal, often physically motivated derivations, which are generally referred to as *a priori estimates*.

A powerful framework for such rigorous investigations represents the so-called two-scale convergence method, which goes back to Nguetseng [113] and Allaire [3]. Their work rigorously established the following central concept.

**Two-scale convergence**

**Definition 10.2.1.** Let $\mathcal{D} \subset \mathbb{R}^d$ be an open subset, and let $d \in \mathbb{N}$ be the dimension of space. Moreover, let $Y \subset [0,1]^d$ denote the closed unit cube such that $C_\sharp^\infty(Y)$ is the set of infinitely differentiable and periodic functions on $Y$. We say that a sequence $u^\epsilon(\mathbf{x}) \in L^2(\mathcal{D})$, where $\epsilon > 0$ is sequence of real numbers going to zero, is *two-scale convergent* to a limit $u^0(\mathbf{x}, \mathbf{y}) \in L^2(\mathcal{D} \times Y)$, if for all $\psi(\mathbf{x}, \mathbf{y}) \in C^\infty(\mathcal{D}; C_\sharp^\infty(Y))$,

$$\int_{\mathcal{D}} u^\epsilon(\mathbf{x})\psi(\mathbf{x}, \mathbf{x}/\epsilon)\, d\mathbf{x} \quad \to \quad \int_{\mathcal{D}\times Y} u^0(\mathbf{x}, \mathbf{y})\psi(\mathbf{x}, \mathbf{y})\, d\mathbf{x}d\mathbf{y}\,. \tag{10.5}$$

For property (10.5), we will further apply the shorthand notation,

$$u^\epsilon(\mathbf{x}) \stackrel{2}{\rightarrow} u(\mathbf{x},\mathbf{y}), \tag{10.6}$$

as $\epsilon \to 0$.

Allaire [3] also motivates the following type of test functions.

**Admissible test function**

**Definition 10.2.2.** We call a $Y$-periodic function $\psi(\mathbf{x},\mathbf{y})$ an *admissible test function*, if it holds that,

$$\lim_{\epsilon\to 0} \int_{\mathcal{D}} \psi^2\left(\mathbf{x}, \frac{\mathbf{x}}{\epsilon}\right) d\mathbf{x} = \int_{\mathcal{D}} \int_{Y} \psi^2(\mathbf{x},\mathbf{y})\, d\mathbf{x}d\mathbf{y}. \tag{10.7}$$

**Remark 10.2.3** (Admissibility condition). As pointed already out in [3], we note that a function $\psi(\mathbf{x},\mathbf{y}) \in L^2(\mathcal{D}\times Y)$ does not satisfy the admissibility condition (10.7). The minimal regularity requirements on $\psi(\mathbf{x},\mathbf{y})$ ensuring the admissibility property (10.7) are still not clear. ◇

Definition 10.2.1 allows us to collect the key results of the two-scale convergence theory as developed in [3, Theorem 1.2 and Proposition 1.14] for applications in the following theorem.

**Compactness**

**Theorem 10.2.4.** **1.** *From each bounded sequence $u^\epsilon \in L^2(\mathcal{D})$, i. e.,*

$$\left\|u^\epsilon\right\|_{L^2(\mathcal{D})} < C, \tag{10.8}$$

*with $C > 0$ independent of $\epsilon > 0$, we can extract a subsequence that two-scale converges to a limit $u^0(\mathbf{x},\mathbf{y}) \in L^2(\mathcal{D}\times Y)$.*

**2.** *Let $u^\epsilon \in H^1(\mathcal{D})$ be a bounded sequence, that is,*

$$\|u^\epsilon(\mathbf{x})\|_{L^2(\mathcal{D})} + \|\nabla_{\mathbf{x}} u^\epsilon(\mathbf{x})\|_{L^2(\mathcal{D})} < C, \tag{10.9}$$

*with $C > 0$ independent of $\epsilon > 0$. Then, from $u^\epsilon(\mathbf{x})$ we can extract a subsequence (still denoted by $u^\epsilon$) with the properties,*

$$\begin{aligned} u^\epsilon &\stackrel{2}{\rightarrow} u(\mathbf{x}), \quad \text{as } \epsilon \to 0, \\ \nabla_{\mathbf{x}} u^\epsilon &\stackrel{2}{\rightarrow} \nabla_{\mathbf{x}} u(\mathbf{x}) + \nabla_{\mathbf{y}} u^1(\mathbf{x},\mathbf{y}), \quad \text{as } \epsilon \to 0, \end{aligned} \tag{10.10}$$

*where $u(\mathbf{x}) \in H^1(\mathcal{D})$, and $u^1(\mathbf{x},\mathbf{y})$ is a function of the macroscale variable $\mathbf{x} \in \mathcal{D}$ and the microscale variable $\mathbf{y} \in Y$ such that $u^1(\mathbf{x},\mathbf{y}) \in L^2(\mathcal{D}; H^1_\sharp(Y))$.*

**3.** *Let $u^\epsilon$ and $\epsilon\nabla_{\mathbf{x}} u^\epsilon$ be two bounded sequences in $L^2(\mathcal{D})$, that is,*

$$\left\|u^\epsilon\right\|_{L^2(\mathcal{D})} \leq C, \quad \epsilon\left\|\nabla_{\mathbf{x}} u^\epsilon\right\|_{L^2(\mathcal{D})} \leq C, \tag{10.11}$$

*with $C > 0$ independent of $\epsilon > 0$. Then, there exist a subsequence $u^\epsilon$ and a function $u^0(\mathbf{x},\mathbf{y}) \in L^2(\mathcal{D}; H^1_\sharp(Y))$ such that,*

$$u^\epsilon(\mathbf{x}) \xrightarrow{2} u^0(\mathbf{x},\mathbf{y}), \quad \text{as } \epsilon \to 0, \qquad \epsilon\nabla_{\mathbf{x}} u^\epsilon(\mathbf{x}) \xrightarrow{2} \nabla_{\mathbf{y}} u^0(\mathbf{x},\mathbf{y}), \quad \text{as } \epsilon \to 0. \tag{10.12}$$

**4.** *Let $u^\epsilon$ be a divergence-free bounded sequence in $[L^2(\mathcal{D})]^d$ such that,*

$$u^\epsilon(\mathbf{x}) \xrightarrow{2} u^0(\mathbf{x},\mathbf{y}) \in \left[L^2(\mathcal{D}\times Y)\right]^d. \tag{10.13}$$

*Then, the two-scale limit $u^0$ satisfies,*

$$\mathrm{div}_{\mathbf{y}}\, u^0(\mathbf{x},\mathbf{y}) = 0, \qquad \int_Y \mathrm{div}_{\mathbf{y}}\, u^0(\mathbf{x},\mathbf{y})\, d\mathbf{y} = 0. \tag{10.14}$$

This result can be easily generalized to bounded sequences in $L^p(\mathcal{D})$ for $1 < p \leq \infty$. Hence, we recall [3, Corollary 1.15] here.

**Compactness in $L^p$**

**Corollary 10.2.5.** *Let $u^\epsilon$ be a bounded sequence in $L^p(\mathcal{D})$ with $1 < p \leq \infty$. Then, there exist a function $u^0(\mathbf{x},\mathbf{y}) \in L^2(\mathcal{D}\times Y)$ and a subsequence $u^\epsilon$ such that for all $\psi(\mathbf{x},\mathbf{y}) \in C_0^\infty(\mathcal{D}; C_\sharp^\infty(Y))$, we have that,*

$$u^\epsilon(\mathbf{x}) \xrightarrow{2} u^0(\mathbf{x},\mathbf{y}). \tag{10.15}$$

### 10.2.1 Two-scale convergence of conductivity problems

The two-scale convergence method allows us to rigorously derive the effective macroscopic conductivity equation (6.9) derived in Section 6.3 with the asymptotic two-scale expansion. As we will further see, the two-scale convergence method leads to a two-scale limiting problem and consists of the upscaled equation plus the cell problem.

Let $\mathcal{D}$ denote a composite material that consists of different conducting phases defined with respect to a periodic reference cell $Y$. This leads to the conductivity tensor $\hat{\Sigma}^\epsilon(\mathbf{x}) = \hat{\Sigma}(\mathbf{x},\mathbf{x}/\epsilon)$, which is $\mathbf{y} = \frac{\mathbf{x}}{\epsilon}$-periodic and varies fast on the microscale $\mathbf{y} \in Y$ but slowly on the macroscale $\mathbf{x} \in \mathcal{D}$. Moreover, we assume that $\hat{\Sigma}^\epsilon$ is strongly elliptic, see (1.3). As already mentioned in [3], strong ellipticity (1.3) does not guarantee that $\hat{\Sigma}^\epsilon(\mathbf{x}) = \hat{\Sigma}(\mathbf{x},\mathbf{x}/\epsilon)$ is measurable, and also the convergence as $\epsilon \to 0$ to $\int_Y \hat{\Sigma}(\mathbf{x},\mathbf{x}/\epsilon)\, d\mathbf{y}$ cannot be concluded. Instead of making the abstract admissibility assumption (10.7) for the conductivity tensor $\hat{\Sigma}(\mathbf{x},\mathbf{x}/\epsilon)$, we make the more restrictive but also more intuitive,

$$\text{(Admissibility Assumption)} \quad \hat{\Sigma}(\mathbf{x},\mathbf{x}/\epsilon) \in \left[C(\mathcal{D}; L_\sharp^\infty(Y))\right]^{d\times d}. \tag{10.16}$$

Based on this preparations, the conductivity formulation for a composite material reads as follows,

$$\textbf{(Microscopic problem)}\quad \begin{cases} -\operatorname{div}(\hat{\Sigma}(\mathbf{x},\mathbf{x}/\epsilon)\nabla\phi^\epsilon) = 0\,, & \text{in } \mathcal{D}\,, \\ \phi^\epsilon = V\,, & \text{on } \partial\mathcal{D}\,, \end{cases} \tag{10.17}$$

where $V > 0$ is a voltage applied on the boundary $\partial\mathcal{D}$. We recall that $\epsilon = \frac{\ell}{L} \ll 1$ generally denotes the so-called heterogeneity parameter for an intrinsic microscopic length $\ell$ (e. g., the characteristic pore diameter of a porous medium) and a macroscopic length $L$ (e. g., the actual size of the medium of interest). Recall that the trace space $H^{\frac{1}{2}}(\partial\mathcal{D})$ can be identified with the help of the trace map $\gamma$ as the range $H^{\frac{1}{2}}(\partial\mathcal{D}) := \gamma(H^1(\mathcal{D}))$, see [37, Definition 3.29, p. 51], for instance. We immediately have the following well-posedness of problem (10.17).

**Existence**

**Theorem 10.2.6.** *Let $\mathcal{D} \subset \mathbb{R}^d$, $d \in \mathbb{N}$, be open and bounded set with Lipschitz-continuous boundary $\partial\mathcal{D}$. Moreover, let $\hat{\Sigma}(\mathbf{x},\mathbf{x}/\epsilon)$ be a strongly elliptic conductivity tensor, and let $V \in H^{1/2}(\partial\mathcal{D})$. The microscopic problem* (10.17) *has a unique solution $\phi^\epsilon \in H^1(\mathcal{D})$, which satisfies the following a priori bound,*

$$\|\phi^\epsilon\|_{H^1(\mathcal{D})} \le C(\mathcal{D})\|V\|_{H^{1/2}(\partial\mathcal{D})}\,. \tag{10.18}$$

*Proof.* Thanks to the trace theorem (see, e. g., [37, Theorem 3.28, p. 49] together with [37, Proposition 3.31, p. 51] and [37, Proposition 3.32, p. 51]), we have that for $V \in H^{1/2}(\mathcal{D})$, there exists $v \in H^1(\mathcal{D})$ such that $\gamma(v) = V$ and,

$$\|v\|_{H^1(\mathcal{D})} \le C(\mathcal{D})\|V\|_{H^{1/2}(\partial\mathcal{D})}\,. \tag{10.19}$$

By [37, Proposition 3.42, p. 54] we can conclude that $\operatorname{div}(\hat{\Sigma}(\mathbf{x},\mathbf{x}/\epsilon)\nabla v) \in H^{-1}(\mathcal{D})$. We denote by $u \in H_0^1(\mathcal{D})$ the weak solution of the following homogeneous Dirichlet problem,

$$\begin{cases} -\operatorname{div}(\hat{\Sigma}(\mathbf{x},\mathbf{x}/\epsilon)\nabla u) = \operatorname{div}(\hat{\Sigma}(\mathbf{x},\mathbf{x}/\epsilon)\nabla v) & \text{in } \mathcal{D}\,, \\ u = 0 & \text{on } \partial\mathcal{D}\,, \end{cases} \tag{10.20}$$

which exists (see, e. g., [37, Theorem 4.16, p. 72]) together with the estimate,

$$\|\nabla u\|_{L^2(\mathcal{D})} \le \frac{1}{\alpha_1}\|\operatorname{div}(\hat{\Sigma}(\mathbf{x},\mathbf{x}/\epsilon)\nabla v)\|_{H^{-1}(\mathcal{D})}\,. \tag{10.21}$$

After setting $g = u+v$, using the linearity of the trace operator $\gamma$ and [37, Proposition 3.34, p. 52], we have that $\gamma(g) = V$. It is straightforward to show that $g$ is a distributional solution, that is, $-\operatorname{div}(\hat{\Sigma}(\mathbf{x},\mathbf{x}/\epsilon)\nabla g) = 0$ for test functions $\varphi \in C_0^\infty(\mathcal{D})$ with $\gamma(g) = V \in H^{1/2}(\partial\mathcal{D})$. It remains to establish inequality (10.18). With the help of the trace theorem, Poincaré's inequality, and estimate (10.19), we obtain,

$$\begin{aligned} \|g\|_{H^1(\mathcal{D})} &\le \|g - v\|_{H^1(\mathcal{D})} + \|v\|_{H^1(\mathcal{D})} \le \|u\|_{L^2(\mathcal{D})} + \|\nabla u\|_{L^2(\mathcal{D})} + \|v\|_{L^2(\mathcal{D})} \\ &\le (1 + C(\mathcal{D}))\|\nabla u\|_{L^2(\mathcal{D})} + C(\mathcal{D})\|V\|_{H^{1/2}(\partial\mathcal{D})} \\ &\le (1 - C(\mathcal{D}))/\alpha_1\|\operatorname{div}(\hat{\Sigma}(\mathbf{x},\mathbf{x}/\epsilon)\nabla v)\|_{H^{-1}(\mathcal{D})} + C(\mathcal{D})\|V\|_{H^{1/2}(\partial\mathcal{D})}\,, \end{aligned} \tag{10.22}$$

where we used (10.21) in the last inequality, and $C(\mathcal{D})$ is a generic constant depending on the domain $\mathcal{D}$. Moreover, we have that,

$$\|\operatorname{div}(\hat{\Sigma}(\mathbf{x},\mathbf{x}/\epsilon)\nabla v)\|_{H^{-1}(\mathcal{D})} \leq C(\mathcal{D})\|V\|_{H^{1/2}(\partial\mathcal{D})}, \tag{10.23}$$

since for all $\varphi \in H_0^1(\mathcal{D})$, we get,

$$\begin{aligned}|(\operatorname{div}(\hat{\Sigma}(\mathbf{x},\mathbf{x}/\epsilon)\nabla v),\varphi)| &= \left|\int_{\mathcal{D}} \hat{\Sigma}(\mathbf{x},\mathbf{x}/\epsilon)\nabla v\nabla\varphi\, d\mathbf{x}\right| \\ &\leq C(\mathcal{D})\|V\|_{H^{1/2}(\partial\mathcal{D})}\|\nabla\varphi\|_{L^2(\mathcal{D})}. \end{aligned}\tag{10.24}$$

This proves the statement. □

The two-scale convergence method introduced in [113, 3] then allows us to recover the upscaling result for the conductivity problem from Section 6.3 in the form of the following theorem.

**Rigorous 2-scale convergence result**

**Theorem 10.2.7.** *Solutions $\phi^\epsilon$ of the microscopic problem* (10.17) *satisfy the following convergence properties,*

$$\begin{aligned}\phi^\epsilon(\mathbf{x}) &\overset{H^1(\mathcal{D})}{\rightharpoonup} \phi(\mathbf{x}),\\ \nabla\phi^\epsilon(\mathbf{x}) &\overset{2}{\rightharpoonup} \nabla\phi(\mathbf{x}) + \nabla_{\mathbf{y}}\phi^1(\mathbf{x},\mathbf{y}),\end{aligned}\tag{10.25}$$

*where $\phi^1$ is Y-periodic, and the limiting functions $(\phi(\mathbf{x}),\phi^1(\mathbf{x},\mathbf{y})) \in H^1(\mathcal{D})\times L^2(\mathcal{D};H^1_\#(Y)/\mathbb{R})$ solve the coupled system of limiting equations,*

$$\begin{cases}-\operatorname{div}\big(\hat{\Sigma}(\mathbf{x},\mathbf{y})\big(\nabla\phi(\mathbf{x}) + \nabla_{\mathbf{y}}\phi^1(\mathbf{x},\mathbf{y})\big)\big) = 0, & \text{in } \mathcal{D}\times Y,\\ -\operatorname{div}\big(\int_Y \hat{\Sigma}(\mathbf{x},\mathbf{y})\big(\nabla\phi(\mathbf{x}) + \nabla_{\mathbf{y}}\phi^1(\mathbf{x},\mathbf{y})\big)\, d\mathbf{y}\big) = 0, & \text{in } \mathcal{D},\\ \phi(\mathbf{x}) = V, & \text{on } \partial\mathcal{D}.\end{cases}\tag{10.26}$$

*Finally, using the usual relation,*

$$\phi^1(\mathbf{x},\mathbf{y}) = \sum_{i=1}^{d} \xi^i(\mathbf{y})\frac{\partial\phi}{\partial x_i}(\mathbf{x}), \tag{10.27}$$

*we can rewrite system* (10.26) *in the classical form,*

$$\begin{aligned}-\operatorname{div}\big(\hat{\Sigma}^0(\mathbf{x})\nabla\phi(\mathbf{x})\big) &= 0, \quad \text{in } \mathcal{D},\\ \phi(\mathbf{x}) &= V, \quad \text{on } \partial\mathcal{D},\end{aligned}\tag{10.28}$$

*where the effective conductivity tensor $\hat{\Sigma}^0(\mathbf{x}) = \{\sigma^0_{ij}(\mathbf{x})\}_{i,j=1}^d$ is given by,*

$$\sigma^0_{ij}(\mathbf{x}) := \int_Y \hat{\Sigma}(\mathbf{x},\mathbf{y})\big(\nabla_{\mathbf{y}}\xi^i(\mathbf{x},\mathbf{y}) + \mathbf{e}_i\big)\big(\nabla_{\mathbf{y}}\xi^j(\mathbf{x},\mathbf{y}) + \mathbf{e}_j\big)\, d\mathbf{y}, \tag{10.29}$$

*and the correctors $\xi^i$, $1 \leq i \leq d$, are solutions of the following cell problem,*

$$\begin{cases}-\operatorname{div}_{\mathbf{y}}\big(\hat{\Sigma}(\mathbf{x},\mathbf{y})\big(\nabla_{\mathbf{y}}\xi^i(\mathbf{x},\mathbf{y}) + \mathbf{e}_i\big)\big) = 0, & \text{in } Y,\\ \xi^i \text{ is } Y\text{-periodic.}\end{cases}\tag{10.30}$$

**Remark 10.2.8** (Effective conductivity tensor). The definition of the effective conductivity tensor (10.29) is of the form (6.26) derived in Section 6.4. ◇

*Proof.* Estimate (10.18) in Theorem 10.2.6 guarantees the existence of a subsequence $\phi^\epsilon(\mathbf{x})$ and a limit $\phi(\mathbf{x})$ such that $(10.25)_1$ holds. Moreover, the $H^1(\mathcal{D})$-boundedness allows us to apply Theorem 10.2.4.2, i. e., there exist a subsequence $\phi^\epsilon(\mathbf{x})$ and a function $\phi^1(\mathbf{x},\mathbf{y}) \in L^2(\mathcal{D}; H^1_\sharp(Y))$ such that,

$$\nabla_{\mathbf{x}}\phi^\epsilon(\mathbf{x}) \ \overset{2}{\rightharpoonup}\ \nabla_{\mathbf{x}}\phi(\mathbf{x}) + \nabla_{\mathbf{y}}\phi^1(\mathbf{x},\mathbf{y})\,.$$

These limits suggest the leading-order behavior $\phi^\epsilon(\mathbf{x}) = \phi(\mathbf{x}) + \epsilon\phi^1(\mathbf{x},\mathbf{y})$. Hence, we multiply the microscopic formulation (10.17) by the test function $\varphi(\mathbf{x}) + \epsilon\varphi^1(\mathbf{x},\mathbf{x}/\epsilon)$ for $\varphi(\mathbf{x}) \in C_0^\infty(\mathcal{D})$ and $\varphi^1(\mathbf{x},\mathbf{y}) \in C_0^\infty(\mathcal{D}; C_\sharp^\infty(Y))$, that is,

$$(\hat{\Sigma}(\mathbf{x},\mathbf{x}/\epsilon)\nabla\phi^\epsilon(\mathbf{x}), (\nabla\varphi(\mathbf{x}) + \nabla_{\mathbf{y}}\varphi^1(\mathbf{x},\mathbf{x}/\epsilon) + \epsilon\nabla_{\mathbf{x}}\varphi^1(\mathbf{x},\mathbf{x}/\epsilon)))_{\mathcal{D}} = 0\,. \tag{10.31}$$

Thanks to the regularity assumption (10.16), the function $\hat{\Sigma}'(\mathbf{x},\mathbf{x}/\epsilon)(\nabla\varphi(\mathbf{x}) + \nabla_{\mathbf{y}}\varphi^1(\mathbf{x},\mathbf{x}/\epsilon) + \epsilon\nabla_{\mathbf{x}}\varphi^1(\mathbf{x},\mathbf{x}/\epsilon))$ two-scale converges strongly to its limit and hence can be considered as a test function in (10.31). This allows us to pass to the two-scale limit in (10.31),

$$(\hat{\Sigma}(\mathbf{x},\mathbf{x}/\epsilon)(\nabla\phi(\mathbf{x}) + \nabla_{\mathbf{y}}\phi^1(\mathbf{x},\mathbf{y})), (\nabla\varphi(\mathbf{x}) + \nabla_{\mathbf{y}}\varphi^1(\mathbf{x},\mathbf{y})))_{\mathcal{D}\times Y} = 0\,. \tag{10.32}$$

This is also true for all $(\varphi,\varphi^1) \in H^1(\mathcal{D}) \times L^2(\mathcal{D}; H^1_\sharp(Y))$ due to denseness. Moreover, integrating by parts in (10.32) leads to (10.26). We immediately have the existence of a unique solution $(\phi,\phi^1) \in H^1(\mathcal{D}) \times L^2(\mathcal{D}; H^1_\sharp(Y))$ by Lax–Milgram's theorem with respect to the Hilbert space $H^1(\mathcal{D}) \times L^2(\mathcal{D}; H^1_\sharp(Y))$ endowed with the norm $\|\nabla\phi(\mathbf{x})\|_{L^2(\mathcal{D})} + \|\nabla_{\mathbf{y}}\phi^1(\mathbf{x},\mathbf{y})\|_{L^2(\mathcal{D}\times Y)}$. It only remains to insert the ansatz (10.27) into the two-scale limit (10.26), and we arrive at the upscaled/homogenized system (10.28)–(10.30). □

# 11 CHeSs governed by interparticle forces

## 11.1 Introduction

In this chapter, we take a more microscopic point of view on interfacial descriptions by rather accounting for physical forces that keep different species apart than introducing an ideal sharp interface together with corresponding balance/mass exchange requirements. In this context, a related and established approach represents the so-called *Density Functional Theory (DFT)*, whose original motivation is rather the reliable understanding of microscopic interaction forces between species such as water molecules and a solid wall than looking at very complex and highly heterogeneous systems involving various different species and material phases such as *reactive flows* in *porous media*.

## 11.2 Upscaling of a hard-sphere particle system

The subsequent discussion reflects the work presented in [137]. Let us look at transport in heterogeneous media governed by experimentally observed, physical interaction energies/potentials instead of sharp interfacial boundary conditions, generally taken as an effective substitute for such physical interaction forces. Examples of such potentials are *van der Waals*, *Lennard-Jones*, and DLVO-type interactions, which are observed in composites and complex materials such as heterogeneous and multiphase materials, e. g., transport in batteries [164] or biology [83], desalination devices [79], fuel cells [49], ionic solutions [94], and interfacial dynamics [145]. In fact, the here exploited concept of interaction potentials provides a promising and physically more consistent alternative for upscaling/homogenization by not starting with a microscopic sharp interface formulation (such as in the so-called perforated domain methodology [3, 73, 136, 162, 177]) depending on effective macroscopic boundary conditions. Instead, a reliable description of interaction potentials (often called *potentials of mean force*) appearing between the species constituting the heterogeneous/composite medium of interest are taken into account and systematically and reliably scaled in the upscaling/homogenization limit. This setup has the practical advantage that we can take the characteristic randomness of materials into account in the upscaling/averaging/homogenization process. The subsequently presented upscaling framework provides promising new possibilities for modeling and describing physical, chemical, and biological CHeSs by reliably and systematically accounting for repulsion, attraction, and penetration forces of different species and media as well as their associated different scales.

We describe this promising approach by looking at a prototype diffusion around randomly placed obstacles, which represent a random porous medium. This is a generalization of the single cylindrical channel case investigated in [26] toward general diffusion processes in porous materials governed by local hard-sphere interactions between pore walls and diffusing species $s$ of density $c^\epsilon$, i. e.,

https://doi.org/10.1515/9783110579543-011

$$-\mathrm{div}(\hat{\tilde{\mathrm{D}}}\nabla\tilde{c}^{\epsilon} - \tilde{c}^{\epsilon}\hat{\tilde{\mathrm{M}}}\nabla\tilde{\psi}^{\theta,\epsilon}_{\mathrm{HS}}) = \tilde{f}\,, \quad \text{in } D := [0, L]^{d}\,. \tag{11.1}$$

The function $\tilde{f}$ accounts for possible external sources/sinks, and $\epsilon := \frac{\ell}{L}$ is the heterogeneity of the porous medium defined by the characteristic/average distance $\ell$ between rigid spherical hard-sphere particles $p$ through which the species $s$ cannot enter. Moreover, $L$ is the length scale of the porous medium composed of the solid obstacles $p$ and the species $s$ diffusing through a supporting medium/material, and $d \in \mathbb{N}_{>0}$ is the dimension of space.

As generally required in upscaling/averaging/homogenization, we identify the so-called characteristic reference cells/volume elements. It is exactly here, where we can account for the material randomness by randomly distributing the obstacles' barycenter $\mathbf{y}_b^{\epsilon}$ according to the characteristic porosity, i. e., the volume fraction of the pore space to the volume of the whole porous medium, see Figure 11.1.

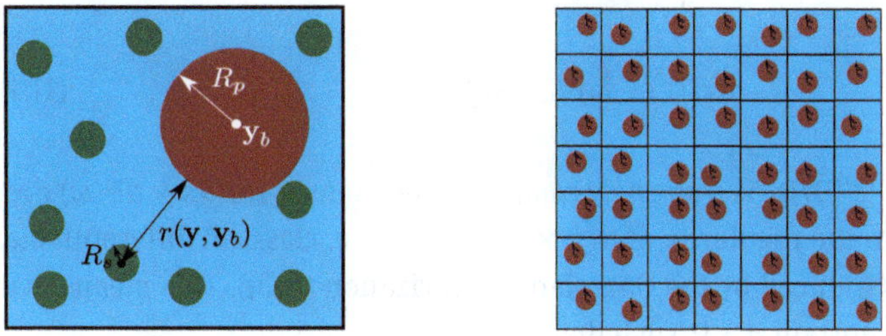


Figure 11.1: *Left:* Reference cell $Y$ containing a single solid hard-sphere obstacle $p$ (red) with barycenter $\mathbf{y}_b$ and radius $R_p$ and species $s$ (green) with radius $R_s$ and density $c^{\epsilon}$. $r(\mathbf{y}, \mathbf{y}_b)$ is the distance of species' barycenter to the obstacle's barycenter. *Right:* Random medium formed by uniformly distributed hard-sphere particles $p$.

The diffusion and mobility enter as tensors $\hat{D}$ and $\hat{M}$ in (11.1) to allow for possible anisotropic characteristics, if available. The standard hard-sphere potential avoids the parameter $\theta$, which weights the strength of repulsion in (11.3), i. e.,

$$\tilde{\psi}^{\epsilon}_{\mathrm{HS}} := \begin{cases} 0\,, & \text{if } r_{ps} \geq \sigma\,, \\ \infty\,, & \text{if } r_{ps} < \sigma\,. \end{cases} \tag{11.2}$$

For a Gaussian regularization of the Dirac delta function $\delta(x)$, $x \in \mathbb{R}$, that is, $\delta_{\theta}(x) := \frac{1}{\theta\sqrt{\pi}} \mathrm{e}^{-\frac{x^2}{\theta^2}}$, and the hard-sphere radius $\sigma := R_p + R_s$ being the sum of the solid particle radius $R_p$ and species radius $R_s$, we can write the regularized repulsion entering (11.1) as follows,

$$\tilde{\psi}^{\theta,\epsilon}_{\mathrm{HS}} := \begin{cases} \delta_{\theta}(r^{\epsilon}_{ps} - \sigma)\,, & \text{if } r^{\epsilon}_{ps} \geq \sigma\,, \\ \delta_{\theta}(0)\,, & \text{if } r^{\epsilon}_{ps} < \sigma\,. \end{cases} \tag{11.3}$$

In fact, we can obtain wall repulsions of *Lennard-Jones* type by selecting a physically appropriate size of the parameter $\theta$ in (11.3).

It remains to identify the distance between the obstacle's barycenter and the barycenter of species $s$ by the variable $r^{\epsilon}_{ps}(\mathbf{x}) := r(\mathbf{x}/\epsilon, \mathbf{y}^{\epsilon}_{b})$ for a point $\mathbf{x} \in D$ in the domain and the barycenter $\mathbf{y}^{\epsilon}_{b}$ of the nearest solid particle $p$ given by $\mathbf{y}^{\epsilon}_{b}(\mathbf{x};\omega) := \mathbf{y}_{b}(\tau_{\mathbf{x}/\epsilon}\omega)$. On a standard probability space $(\Omega, \mathcal{F}, \mathbb{P})$, the map $\tau_{\mathbf{x}} : \Omega \to \Omega$ denotes a *d-dimensional dynamical system* characterized by the following properties:

(DS1) $\tau_{\mathbf{x}}$ is *measure preserving*;
(DS2) $\tau_{\mathbf{x}}$ is *invertible*;
(DS3) $\tau_{\mathbf{x}}$ satisfies the *group property* $\tau_{\mathbf{x}+\mathbf{y}} = \tau_{\mathbf{x}} \circ \tau_{\mathbf{y}}$ with $\tau_{0}$ denoting the identity map;
(DS4) the set $\{(\mathbf{x};\omega) \in D \times \Omega \mid \tau_{\mathbf{x}}\omega \in \mathcal{F}\}$ is $d\mathbf{x} \times d\mathbb{P}$*-measurable*;

In a first simplifying step, we assume that the barycenters $\mathbf{y}^{\epsilon}_{b}(\mathbf{x};\omega) \sim \mathcal{U}(\mathcal{Y}_{\mathbf{x}})$ are uniformly distributed in the domain $\mathbf{x} \in D \subset \mathbb{R}^{d}$. We can rewrite (11.1) in dimensionless form as,

$$-\operatorname{div}(\hat{\mathrm{D}}^{\epsilon}(\mathbf{x};\omega)\nabla u^{\epsilon}) = f\,, \tag{11.4}$$

where $\hat{\mathrm{D}}^{\epsilon}(\mathbf{x};\omega) := \mathrm{e}^{\psi^{\theta,\epsilon}_{\mathrm{HS}}(\mathbf{x};\omega)}\hat{\tilde{\mathrm{D}}}/\mathrm{D}_0$ is a *strongly random elliptic* tensor for $d\mathbf{x} \times d\mathbb{P}(\omega)$-a. e. $(\mathbf{x},\omega) \in D \times \Omega$ thanks to the regularity of $\psi^{\theta,\epsilon}_{\mathrm{HS}}(\mathbf{x};\omega)$ (i. e., $\mathbb{P}$-a. e. classical differentiable, and its boundedness is induced by the Gaussian regularization (11.3)). This means that there exists a positive constant $\eta > 0$, such that,

$$\eta|\mathbf{z}|^{2} \le \sum \mathrm{D}^{\epsilon}_{ij}(\mathbf{x};\omega)z_i z_j \le \eta^{-1}|\mathbf{z}|^{2}\,, \quad \forall \mathbf{z} \in \mathbb{R}^{d}\,, \text{ and for } \mathcal{L} \times \mathbb{P}\text{-a.e. } (\mathbf{x},\omega) \in D \times \Omega\,. \tag{11.5}$$

Note that this equation relies on the Stokes–Einstein relation $\hat{\tilde{\mathrm{D}}} := \frac{\tilde{\mathrm{M}}}{k_B T}$ and the Slotboom-type variable $u^{\epsilon} := \mathrm{e}^{-\psi^{\theta,\epsilon}_{\mathrm{HS}}}c^{\epsilon}$ for the dimensionless variables $\psi^{\theta,\epsilon}_{\mathrm{HS}} := \frac{\tilde{\psi}^{\theta,\epsilon}_{\mathrm{HS}}}{k_B T}$, $c^{\epsilon} := \tilde{c}^{\epsilon}/c_0$, and $f := \tilde{f}\frac{L^2}{\mathrm{D}_0 c_0}$. Moreover, $\mathrm{D}_0$ is a characteristic diffusion parameter, $T$ is the temperature, $c_0$ is a reference concentration, and $k_B$ is the Boltzmann constant.

We have depicted computational results obtained with equation (11.4) in Figure 11.2 for *lightly heterogeneous media/CHeSs*, that is, media with heterogeneity $\epsilon > 0$ closer to 1 than to 0. In the literature about subsurface flows and elasticity, the term *weakly heterogeneous* is used to denote variations in the diffusion/permeability that are lower than one order of magnitude [63]. On the contrary, we refer to media with heterogeneity $\epsilon \ll 1$ as *strongly heterogeneous materials*. Problem descriptions for media with $\epsilon \ll 1$ are high-dimensional, and hence, upscaled/effective macroscopic formulations represent a more systematic approach to gain an efficient, lower-dimensional, and reliable computational problem description, see Section 11.2.1.

Our main upscaling result relies on basic definitions concerning the *stochastic two-scale convergence* in the mean [24, 162, 177], which we recall here for readability.

Figure 11.2: *Left:* Uniformly distributed spherical obstacles with Gaussian-regularized hard-sphere repulsion. *Right:* Sample average $\langle c^\epsilon \rangle_N(\mathbf{x}) := \frac{1}{N}\sum_{n=1}^{N} c^\epsilon(\mathbf{x};\omega_n)$ for $N = 1000$ realizations. (Figures are from [137]).

**Admissibility**

**Definition 11.2.1.** We call a function $u(\mathbf{x};\omega) \in L^2(D\times\Omega)$ *admissible*, if the function $u(\mathbf{x};\tau_{\mathbf{x}}\omega)$ is an element of $L^2(D\times\Omega)$, where $\tau_{\mathbf{x}}$ is a $d$-dimensional dynamical system (DS1)–(DS4).

**Stochastic two-scale convergence in the mean**

**Definition 11.2.2.** We call a sequence $(u^\epsilon) \subset L^2(D\times\Omega)$ *stochastically two-scale convergent in the mean* to $u \in L^2(D\times\Omega)$, if for every admissible $v \in L^2(D\times\Omega)$,

$$\lim_{\epsilon\to 0} \int_{D\times\Omega} u^\epsilon(\mathbf{x};\omega)v(\mathbf{x},\tau_{\mathbf{x}/\epsilon}\omega)\,d\mathbf{x}d\mathbb{P}(\omega) = \int_{D\times\Omega} u(\mathbf{x};\omega)v(\mathbf{x},\omega)\,d\mathbf{x}d\mathbb{P}(\omega)\,. \tag{11.6}$$

In this context, the Weyl decomposition $L^2_{\mathbb{P}}(\Omega;\mathbb{R}^d) = V^2_{\mathrm{pot}}(\Omega)\oplus L^2_{\mathrm{sol}}(\Omega) = V^2_{\mathrm{sol}}(\Omega)\oplus L^2_{\mathrm{pot}}(\Omega)$ (see, e. g., [2, 162] and [177, p. 228]) serves useful and relies on the following function spaces,

$$\begin{aligned}
L^2_{\mathrm{pot}}(\Omega) &:= \left\{f \in L^2_{\mathbb{P}}(\Omega;\mathbb{R}^d)\,\middle|\,\exists F \in H^1(\Omega)\text{ s.t. }\nabla_\omega F = f\right\}\,,\\
L^2_{\mathrm{sol}}(\Omega) &:= \left\{f \in L^2_{\mathbb{P}}(\Omega;\mathbb{R}^d)\,\middle|\,\mathrm{div}_\omega f = 0\right\}\,,\\
V^2_{\mathrm{pot}}(\Omega) &:= \left\{f \in L^2_{\mathrm{pot}}(\Omega;\mathbb{R}^d)\,\middle|\,\mathrm{E}[f] = 0\right\}\,,\\
V^2_{\mathrm{sol}}(\Omega) &:= \left\{f \in L^2_{\mathrm{sol}}(\Omega;\mathbb{R}^d)\,\middle|\,\mathrm{E}[f] = 0\right\}\,.
\end{aligned} \tag{11.7}$$

### 11.2.1 Main results: effective macroscopic transport in strongly heterogeneous materials

First, we establish the existence of solutions to problem (11.4) in the form of the following proposition and recall the corresponding proof in Section 11.2.2.1 from [137].

**Solvability of (11.4)**

**Proposition 11.2.3.** *Let $f(\mathbf{x}) \in H^{-1}(D)$, and let $\psi^{\epsilon}(\mathbf{x};\omega) := \psi_{HS}^{\theta,\epsilon} \in L^2_{\mathbb{P}}(\Omega; W^{2,d/2}(D))$ be the regularized hard-sphere potential (11.3). We impose $u^{\epsilon} = 0$ $\mathbb{P}$-a. e. on the boundary $\partial D$. Under these conditions, there exists a unique weak solution $u^{\epsilon} \in L^2_{\mathbb{P}}(\Omega; H^1_0(D))$ that satisfies,*

$$
\begin{aligned}
\left(\hat{\mathrm{D}}^{\epsilon}(\mathbf{x};\omega)\nabla u^{\epsilon}(\mathbf{x};\omega), \nabla v\right)_{D\times\Omega} &:= \int_{\Omega}\int_{D} \hat{\mathrm{D}}^{\epsilon}(\mathbf{x};\omega)\nabla u^{\epsilon}(\mathbf{x};\omega)\nabla v \, d\mathbf{x}d\mathbb{P} \\
&= \int_{D} f(\mathbf{x}) v \, d\mathbf{x}d\mathbb{P} =: (f, v)_{D\times\Omega}\,, 
\end{aligned}
\tag{11.8}
$$

*for all $v(\mathbf{x};\omega) \in L^2_{\mathbb{P}}(\Omega; H^1_0(D))$.*

The framework of $d$-dimensional dynamical systems seems to go back to [24, 162, 177]. It allows us to pass to the limit as $\epsilon \to 0$ in (11.8), as stated in the following theorem established in [137].

**Effective hard-sphere Smoluchowski equation**

**Theorem 11.2.4.** *Let $D \subset \mathbb{R}^d$ be a porous medium formed by uniformly distributed rigid spherical obstacles, that is, the barycenters satisfy,*

$$
\mathbf{y}_b^{\epsilon}(\mathbf{x};\omega) := \mathbf{y}_b(\tau_{\mathbf{x}/\epsilon}\omega) \sim \mathcal{U}(Y_{\mathbf{x}})\,, \tag{11.9}
$$

*for a d-dimensional dynamical system $\tau_{\mathbf{x}} : \Omega \to \Omega$ on the standard probability space $(\Omega, \mathcal{F}, \mathbb{P})$, where $Y_{\mathbf{x}}$ represents the reference cell associated with the macroscopic point $\mathbf{x} \in D$. The definition of the stochastic tensor $\hat{\mathrm{D}}^{\epsilon}(\mathbf{x};\omega)$ in (11.4) guarantees the random ellipticity (11.5) and the regularity property $\mathrm{D}_{ij} \in L^2_{\mathbb{P}}(\Omega; [C(D; L^{\infty}_{\#}(Y))]^{d^2})$.*

*Then the unique solution $u^{\epsilon}(\mathbf{x};\omega) \in H^1(D; L^2_{\mathbb{P}}(\Omega))$ of (11.4) stochastically two-scale converges in the mean to $u \in H^1(D; L^2_{sol}(\Omega))$ as $\epsilon \to 0$ such that with the auxiliary function $\xi(\mathbf{x};\omega) \in L^2(D; V^2_{pot}(\Omega))$, the pair $(u, \xi)$ solves uniquely the problem,*

$$
\begin{cases}
-\mathrm{div}_{\omega}\left(\hat{\mathrm{D}}(\mathbf{x};\omega)\left[\nabla_{\mathbf{x}} u(\mathbf{x};\omega) + \xi(\mathbf{x};\omega)\right]\right) = 0\,, & \text{in } D\times\Omega\,, \\
-\mathrm{div}_{\mathbf{x}}\, \mathrm{P}_{L^2_{sol}}\left(\hat{\mathrm{D}}(\mathbf{x};\omega)\left[\nabla_{\mathbf{x}} u(\mathbf{x};\omega) + \xi(\mathbf{x};\omega)\right]\right) = f(\mathbf{x})\,, & \text{in } D\,,
\end{cases}
\tag{11.10}
$$

*in the weak sense, where the projection $\mathrm{P}_{L^2_{sol}} : L^2 \to L^2_{sol}$ is defined for $v(\mathbf{x};\omega) \in L^2(D\times\Omega)$ as follows,*

$$
\mathrm{P}_{L^2_{sol}}\, v(\mathbf{x};\omega) := \lim_{r\to\infty} \frac{1}{(2r)^d} \int_{[-r,r]^d} v(\mathbf{x};\tau_{\mathbf{y}}\omega)\, d\mathbf{y} = \frac{1}{L^d} \int_{[0,L]^d} v(\mathbf{x};\tau_{\mathbf{y}}\omega)\, d\mathbf{y}\,,
$$

*where the last equality holds in the so-called "periodic case", not to be confused with "periodic homogenization".*

### 11.2.2 Proof of the main results

We follow the rigorous derivations from [137].

#### 11.2.2.1 Proof of Proposition 11.2.3: solvability of the microscopic stochastic problem

We apply the Lax–Milgram theorem to verify the coercivity of the bilinear form in the Hilbert space $H := L^2_{\mathbb{P}}(\Omega; H^1_0(D))$, i. e.,

$$b(u,v) := (\hat{\mathrm{D}}^{\epsilon}(\mathbf{x};\omega)\nabla u(\mathbf{x};\omega), \nabla v(\mathbf{x};\omega))_{\Omega\times D} := \int_{\Omega}\int_{D} \hat{\mathrm{D}}^{\epsilon}\nabla u \nabla v \, d\mathbf{x} d\mathbb{P}\,. \tag{11.11}$$

The strong random ellipticity (11.5) of the bilinear form (11.11) immediately implies $b(u,u) \geq \eta \|\nabla u(\cdot;\cdot)\|^2_{L^2_{\mathbb{P}}(\Omega;L^2(D))}$ for all $u \in H$. Hence, we have the existence and uniqueness of a weak solution. □

#### 11.2.2.2 Proof of Theorem 11.2.4: homogenization of the stochastic Smoluchowski equation (11.4)

We restate the proof from [137], which adapts ideas from [3] and [24, 177] to our specific "diffusion coefficient" $\hat{\mathrm{D}}^{\epsilon}(\mathbf{x};\omega) = \mathrm{e}^{\psi^{\theta,\epsilon}_{\mathrm{HS}}(\tau_{\mathbf{x}/\epsilon}\omega)}\hat{\mathrm{D}}$.

The verification of the admissibility property for $\hat{\mathrm{D}}^{\epsilon}$ according to Definition 11.2.1 is the main step in this proof. With the local microscopic probability spaces $(\mathcal{Y}_{\mathbf{z}}, \mathcal{F}_{\mathcal{Y}_{\mathbf{z}}}, \mathbb{P}_{\mathcal{Y}_{\mathbf{z}}})$ for $\mathbf{z} \in \mathcal{Z}^d_{\epsilon} := \epsilon\mathbb{Z}^d \cap D$ (see [2]), we can define the macroscopic probability space $(\mathcal{D}_{\epsilon}, \mathcal{F}_{\mathcal{D}_{\epsilon}}, \mathbb{P}_{\mathcal{D}_{\epsilon}})$ by

$$\begin{aligned}
\mathcal{D}_{\epsilon} &:= \prod_{\mathbf{z}\in\mathcal{Z}^d_{\epsilon}} \mathcal{Y}_{\mathbf{z}}\,,\\
\mathcal{D}_{\epsilon} &:= \prod_{\mathbf{z}\in\mathcal{Z}^d_{\epsilon}} \mathcal{Y}_{\mathbf{z}}\,,\\
\mathcal{F}_{\mathcal{D}_{\epsilon}} &:= \prod_{\mathbf{z}\in\mathcal{Z}^d_{\epsilon}} \mathcal{F}_{\mathcal{Y}_{\mathbf{z}}}\,, \text{ and,}\\
\mathbb{P}_{\mathcal{D}_{\epsilon}} &:= \prod_{\mathbf{z}\in\mathcal{Z}^d_{\epsilon}} \mathbb{P}_{\mathcal{Y}_{\mathbf{z}}}\,.
\end{aligned} \tag{11.12}$$

After introducing the Lebesgue measure $\mathcal{L}$ and the associated Borel $\sigma$-algebra $\mathcal{B}(\mathbb{T}^d)$ on the torus $\mathbb{T}^d$, we define the following product space as our multiscale probability space,

$$(\Omega_{\epsilon}, \mathcal{F}_{\epsilon}, \mathbb{P}_{\epsilon}) := (\mathcal{D}_{\epsilon}, \mathcal{F}_{\mathcal{D}_{\epsilon}}, \mu_{\mathcal{D}_{\epsilon}}) \otimes (\mathbb{T}^d, \mathcal{B}(\mathbb{T}^d), \mathcal{L})\,, \tag{11.13}$$

in which we account for translations by the unit cell $\mathbb{T}^d := ]0,1[^d$ with opposite faces identified. According to (11.13), elements $\omega \in \Omega_{\epsilon}$ decompose into a *macroelement* $\hat{\mathbf{y}} := \{\mathbf{y}_{\mathbf{z}}\}_{\mathbf{z}\in\mathcal{Z}^d_{\epsilon}} \in \mathcal{D}_{\epsilon}$ and a *microelement* $\mathrm{t} \in \mathbb{T}^d$, that is, $\omega = (\hat{\mathbf{y}}, \mathrm{t})$.

With projections $\mathrm{P}_1 : \mathbb{R}^d \to \mathbb{Z}^d$ and $\mathrm{P}_2 : \mathbb{R}^d \to \mathbb{T}^d$ defined by $\mathrm{P}_1\mathbf{x} := [\mathbf{x}]$ and $\mathrm{P}_2\mathbf{x} := \mathbf{x} - [\mathbf{x}]$, respectively, we can define the spatial decomposition,

$$\mathbf{x} = \mathrm{P}_1\mathbf{x} + \mathrm{P}_2\mathbf{x}\,, \tag{11.14}$$

where $[x]$ denotes the nearest smaller integer of $x \in \mathbb{R}$ and is canonically extended for vector-valued arguments $\mathbf{x} \in \mathbb{R}^d$.

Set $\mathbf{J}_\epsilon := \sum_{k=1}^{d}[1/\epsilon]\mathbf{e}_k$ and let $\mathbf{e}_k, 1 \le k \le d$, denote the canonical basis in $\mathbb{R}^d$. Defining the dynamical systems $\tau^{\mathcal{D}_\epsilon}$ acting on $\mathcal{D}_\epsilon$ and $\tau^{\mathbb{T}^d}$ acting on $\mathbb{T}^d$ by

$$\begin{aligned} \tau_{\mathbf{x}}^{\mathcal{D}_\epsilon}(\hat{\mathbf{y}}) &:= \left(\{\mathbf{y}_{(\mathbf{z}+\mathrm{P}_1(\mathbf{x}+\mathrm{t}))\,\mathrm{mod}\,\mathbf{J}_\epsilon}\}_{\mathbf{z}\in\mathcal{Z}_\epsilon^d}\right), \text{ and} \\ \tau_{\mathbf{x}}^{\mathbb{T}^d}(\mathrm{t}) &:= \mathrm{P}_2(\mathbf{x}+\mathrm{t}), \end{aligned} \tag{11.15}$$

respectively, allows us to identify the two-scale dynamical system $\{\tau_{\mathbf{x}}(\omega)\}_{\mathbf{x}}$, $\mathbf{x} \in \mathbb{R}^d$, operating on $\Omega_\epsilon$ by,

$$\tau_{\mathbf{x}}(\omega) = \tau_{\mathbf{x}}(\hat{\mathbf{y}}, \mathrm{t}) = \left(\tau^{\mathcal{D}_\epsilon}_{\mathrm{P}_1(\mathbf{x}+\mathrm{t})}(\hat{\mathbf{y}}), \tau_{\mathbf{x}}^{\mathbb{T}^d}(\mathrm{t})\right). \tag{11.16}$$

Herewith, we immediately have that our heterogeneous medium showing uniformly distributed spherical obstacles with barycenters (11.9) represents a measure-preserving and ergodic two-scale dynamical system ,see [2, Propositions 6 and 7]. $\mathbb{P}$-a. e. classical differentiability (see (11.3)) guarantees that $\psi_{\mathrm{HS}}^{\theta,\epsilon} \in C(\overline{D}; L^\infty_{\mathbb{P}}(\Omega))$, which establishes *admissibility*, see [24, Proposition 3.2].

Finally, the uniform a priori bound,

$$\eta\left\|\nabla u^\epsilon(\cdot;\omega)\right\|_{L^2(D)} \le C(D)\|f\|_{L^2(D)}, \qquad \text{for } \mathbb{P}\text{-a.e. } \omega \in \Omega \tag{11.17}$$

follows after testing equation (11.4) with $u^\epsilon \in L^2_{\mathbb{P}}(\Omega; H^1_0(D))$ and the subsequent use of the strong stochastic ellipticity (11.5) and Poincaré's inequality.

For $X = H^1_0(D) \subset H^1(D)$, [24, Theorem 3.7] guarantees the existence of $u(\mathbf{x};\omega) \in H^1_0(D; L^2_{\mathbb{P}}(\Omega))$, $\xi(\mathbf{x};\omega) \in L^2(D; V^2_{\mathrm{pot}}(\Omega))$ and of a subsequence $u^\epsilon$, such that,

$$\begin{aligned} \mathrm{curl}_\omega\, \xi(\mathbf{x};\omega) &= 0, \\ \mathrm{P}_{V^2_{\mathrm{sol}}}\, \xi(\mathbf{x};\omega) &= 0, \\ u^\epsilon(\mathbf{x};\omega) &\overset{\mathrm{s2im}}{\rightharpoonup} u(\mathbf{x};\omega), \\ \nabla_{\mathbf{x}} u^\epsilon(\mathbf{x};\omega) &\overset{\mathrm{s2im}}{\rightharpoonup} \nabla_{\mathbf{x}} u(\mathbf{x};\omega) + \xi(\mathbf{x};\omega). \end{aligned} \tag{11.18}$$

With [24, Proof of Theorem 4.1.1], we immediately obtain that $u$ and $\xi$ are unique solutions of (11.10). □

## 11.3 Locating obstacles in transport problems: forward and inverse problems

Consider the general problem of *inferring* the location of a single obstacle, e. g., inside a square domain, based on density and flux measurements of a particular species (so-

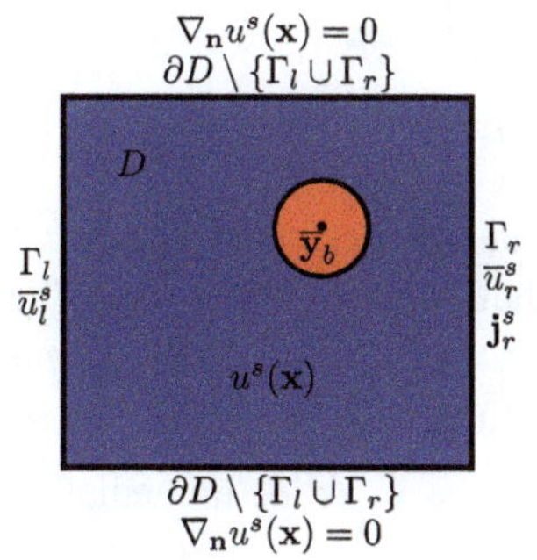


Figure 11.3: *Forward problem:* Computing the density profile $u^s(\mathbf{x})$ and hereby the wanted flux $\mathbf{j}_r^s := \mathrm{e}^{\psi_{\mathrm{HS}}^\theta(r(\mathbf{x},\overline{\mathbf{y}}_b))} \nabla u^s(\mathbf{x})|_{\Gamma_r}$ for given boundary data on $\Gamma_l$, $\Gamma_r$, and $\partial D \setminus \{\Gamma_l \cup \Gamma_r\}$ and the barycenter $\overline{\mathbf{y}}_b$ of the circular obstacle.

lute) undergoing a diffusive transport in a solvent along the horizontal direction from the left- to right-hand side of the square domain, see Figure 11.3. This immediately motivates inference processes as a (measurement) *data-driven approach*. The subsequent systematic framework extends the basic physical setting of the *transport problem* (11.1) from Section 11.2.

### 11.3.1 Forward problem $\mathcal{F}$

We begin with the better accessible forward problem $\mathcal{F} : Y_{\mathrm{ad}} \times L^2(\Gamma_l) \times L^2(\Gamma_r) \to L^2(\Gamma_r)$, i. e., $(\overline{\mathbf{y}}_b, \overline{u}_l^s, \overline{u}_r^s) \mapsto \mathbf{j}_r^s$, where $\overline{\mathbf{y}}_b \in Y_{\mathrm{ad}} \subset \mathbb{R}^d$ is a given obstacle barycenter for a specific admissible set $Y_{\mathrm{ad}}$ defined later on, $\overline{u}_l^s$ is Dirichlet data on the left boundary $\Gamma_l \subset \partial D$, and $\overline{u}_r^s$ is Dirichlet data on the right boundary $\Gamma_r \subset \partial D$ for $s = 1, 2, 3, \dots, S$ with $S \in \mathbb{N}$. The image,

$$\mathbf{j}_r^s := \mathrm{e}^{\psi_{\mathrm{HS}}^\theta(r(\mathbf{x},\overline{\mathbf{y}}_b))} \nabla u^s(\mathbf{x})\big|_{\Gamma_r}, \tag{11.19}$$

denotes the flux through the right boundary $\Gamma_r$ induced by the imposed concentration difference across the left and right boundaries. The forward map requires us to find solutions to the following intermediate problems $\mathcal{F}_s$ indexed by a parameter $s = 1, 2, \dots, S$,

$$\mathcal{F}_s : \begin{cases} -\mathrm{div}(\mathrm{e}^{\psi_{\mathrm{HS}}^\theta(r(\mathbf{x},\overline{\mathbf{y}}_b))} \nabla u^s(\mathbf{x})) = 0\,, & \text{in } D\,, \\ u^s = \overline{u}_l^s := \exp(-\psi_{\mathrm{HS}}^\theta)\overline{c}_l^s\,, & \text{on } \Gamma_l\,, \\ u^s = \overline{u}_r^s := \exp(-\psi_{\mathrm{HS}}^\theta)\overline{c}_r^s\,, & \text{on } \Gamma_r\,, \\ \mathrm{e}^{\psi_{\mathrm{HS}}^\theta(r(\mathbf{x},\overline{\mathbf{y}}_b))} \nabla_{\mathbf{n}} u^s(\mathbf{x}) = 0\,, & \text{on } \partial D \setminus \{\Gamma_r \cup \Gamma_l\}\,. \end{cases} \tag{11.20}$$

The physical species density is then the variable $c^s := \exp\left(\psi_{\mathrm{HS}}^\theta(r(\mathbf{x}, \overline{\mathbf{y}}_b))\right) u^s$ defined via the intermediate solution $u^s$ of (11.20). The hard-sphere potential $\psi_{\mathrm{HS}}^\theta$ involves the Gaussian regularization $\delta_\theta(r) := \frac{1}{\theta\sqrt{\pi}} \mathrm{e}^{-\frac{r^2}{\theta^2}}$ governed by the parameter $\theta$, that is,

$$\psi^{\theta}_{\text{HS}}(r(\mathbf{x},\overline{\mathbf{y}}_b)) := \begin{cases} \delta_\theta(r-\sigma)\,, & \text{if } r >= \sigma\,, \\ \delta_\theta(0)\,, & \text{if } r < \sigma\,. \end{cases} \tag{11.21}$$

The variable $r(\mathbf{x},\overline{\mathbf{y}}_b)$ describes the distance between the barycenter of a transported species $\mathbf{x} \in D$ and the obstacle's barycenter $\overline{\mathbf{y}}_b \in Y_{\text{ad}} \subset D$, that is, $r(\mathbf{x},\overline{\mathbf{y}}_b) := |\mathbf{x} - \overline{\mathbf{y}}_b|$. The parameter $\sigma$ is called the hard-sphere radius and represents the sum of the radii $R_p$ and $R_\zeta$ of a circular obstacle $p$ and species $\zeta$, respectively. For simplicity, we set $R_p = R_s = 0$. The definition of the diffusion/permeability coefficient $\mathrm{d}_{ij} := \mathrm{e}^{\psi^{\theta}_{\text{HS}}(\mathbf{x})}\delta_{ij}$ immediately implies the strong ellipticity,

$$\eta|\mathbf{z}|^2 \le \sum_{ij} \mathrm{d}_{ij}(\mathbf{x})z_i z_j \le \eta^{-1}|\mathbf{z}|^2 \quad \forall \mathbf{z} \in \mathbb{R}^d\,,\ \text{and}\ \forall \mathbf{x} \in D\,, \tag{11.22}$$

where $\delta_{ij}$ is the Kronecker delta. Property (11.22), together with the Lax–Milgram theorem, finally gives the well-posedness of the forward problem (11.20), see [137]. Hence, problem (11.20) constitutes a physically meaningful forward problem $\mathcal{F}_s$.

A variational characterization of the forward problem (11.20) will be useful for the study of the inverse problem later on. Testing equation (11.20) with its solution $u^s \in H^1(D)$, i. e., multiplying it by $u^s$ and integrating over the domain $D$ lead to the following variational energy,

$$\begin{aligned} E(\overline{\mathbf{y}}_b, u^s, \overline{u}^s_l, \overline{u}^s_r) &:= (\mathrm{e}^{\psi^{\theta}_{\text{HS}}(r(\mathbf{x},\mathbf{y}_b))}\nabla u^s, \nabla u^s)_D \\ &\quad + \int\limits_{\partial D\setminus\{\Gamma_r \cup \Gamma_l\}} (\mathrm{e}^{\psi^{\theta}_{\text{HS}}(r(\mathbf{x},\overline{\mathbf{y}}_b))}\nabla u^s \cdot \mathbf{n})u^s\, do = 0\,, \end{aligned} \tag{11.23}$$

which naturally includes the vanishing flux boundary condition. We can also include the Dirichlet boundary conditions by looking for solutions $u^s$ in a space of homogeneous Dirichlet boundary conditions, for instance, i. e., we should have $u^s - \overline{u}^s \in H^1_0(D)$, where $\overline{u}^s|_{\Gamma_l} = \overline{u}^s_l$, $\overline{u}^s|_{\Gamma_r} = \overline{u}^s_r$, and $\overline{u}^s \in H^1_0(D)$ is defined on the whole domain $D$, for simplicity, together with the no-flux condition on $\partial D \setminus \{\Gamma_l \cup \Gamma_r\}$.

Our main interest however is the identification of the obstacle's location $\overline{\mathbf{y}}_b$ for given measured data $(\overline{c}^s, \overline{c}^s_l, \overline{c}^s_r)$, $s = 1, 2, 3, \dots, S$, with $S \in \mathbb{N}$.

### 11.3.2 Inverse problem $\mathcal{I}$

Our next goal is to invert the forward map $\mathcal{F} : Y_{\text{ad}} \times L^2(\Gamma_l) \times L^2(\Gamma_r) \to L^2(\Gamma_r)$ to the inverse map $\mathcal{I} : L^2(\Gamma_r) \times L^2(\Gamma_l) \times L^2(\Gamma_r) \to Y_{\text{ad}}$, that is, $(\{\overline{\mathbf{j}}^s_r\}_{s=1}^S, \{\overline{c}^s_l\}_{s=1}^S, \{\overline{c}^s_r\}_{s=1}^S) \mapsto \mathbf{y}_b$, which determines the *obstacle's barycenter* $\mathbf{y}_b = \mathcal{I}(\{\overline{\mathbf{j}}^s_r\}_{s=1}^S, \{\overline{c}^s_l\}_{s=1}^S, \{\overline{c}^s_r\}_{s=1}^S) \in Y_{\text{ad}} \subset \mathbb{R}^d$ from given data $(\{\overline{\mathbf{j}}^s_r\}_{s=1}^S, \{\overline{c}^s_l\}_{s=1}^S, \{\overline{c}^s_r\}_{s=1}^S) \in L^2(\Gamma_r) \times L^2(\Gamma_l) \times L^2(\Gamma_r)$. The first step is to introduce a *cost/objective function* that penalizes unfavorable barycenter locations,

$$\tilde{C}(\mathbf{y}_b) := \frac{1}{2}\sum_{s=1}^{S} \|\overline{\mathbf{j}}_r^s - \mathcal{F}(\mathbf{y}_b, \overline{c}_l^s, \overline{c}_r^s)\|_{L^2(\Gamma_r)}^2 . \tag{11.24}$$

As a consequence, the barycenter $\mathbf{y}_b \in \mathbb{R}^d$, $1 \le d \le 3$, that minimizes the cost function (11.24), is then expected to be the best estimate of $\mathbf{y}_b$ for the given data $(\overline{\mathbf{j}}_r^s, \overline{c}_l^s, \overline{c}_r^s) \in L^2(\Gamma_r) \times L^2(\Gamma_l) \times L^2(\Gamma_r)$, $s = 1, 2, \dots, S$. Since inverse problems do not continuously depend on the data $(\{\overline{\mathbf{j}}_r^s\}_{s=1}^S, \{\overline{c}_l^s\}_{s=1}^S, \{\overline{c}_r^s\}_{s=1}^S) \in L^2(\Gamma_r) \times L^2(\Gamma_l) \times L^2(\Gamma_r)$ in general, Tikhonov [165] introduced a regularization that allows us to resolve this issue albeit with the cost of a new parameter $\alpha$, see (11.25). Hence, we can guarantee the existence of a minimizer $\mathbf{y}_b \in \mathbb{R}^d$ by regularizing the cost function (11.24) by adding an additional term multiplied by a small parameter $\alpha > 0$,

$$C(\{\mathbf{j}_r^s(\mathbf{y}_b)\}_{s=1}^S, \mathbf{y}_b) := \frac{1}{2}\sum_{s=1}^{S} \|\overline{\mathbf{j}}_r^s - \mathbf{j}_r^s(\mathbf{y}_b, \overline{c}_l^s, \overline{c}_r^s)\|_{L^2(\Gamma_r)}^2 + \frac{\alpha}{2}|\mathbf{y}_b|^2 , \tag{11.25}$$

which is subject to the side constraint (imposed by $\mathbf{j}_r^s = \mathrm{e}^{\psi_{\mathrm{HS}}^\theta(r(\mathbf{x},\mathbf{y}_b))}\nabla_n u^s|_{\Gamma_r} = \mathcal{F}(\mathbf{y}_b, \overline{c}_l^s, \overline{c}_r^s)$),

$$\begin{cases} -\mathrm{div}(\mathrm{e}^{\psi_{\mathrm{HS}}^\theta(r(\mathbf{x},\mathbf{y}_b))}\nabla u^s(\mathbf{x})) = 0\,, & \text{in } D\,, \\ u^s = \overline{u}_l^s := \exp(-\psi_{\mathrm{HS}}^\theta)\overline{c}_l^s\,, & \text{on } \Gamma_l\,, \\ u^s = \overline{u}_r^s := \exp(-\psi_{\mathrm{HS}}^\theta)\overline{c}_r^s\,, & \text{on } \Gamma_r\,, \\ \mathrm{e}^{\psi_{\mathrm{HS}}^\theta(r(\mathbf{x},\mathbf{y}_b))}\nabla_{\mathbf{n}} u^s(\mathbf{x}) = 0\,, & \text{on } \partial D \setminus \{\Gamma_r \cup \Gamma_l\}\,, \end{cases} \tag{11.26}$$

whose intermediate solution defines the species density $c^s = \exp(\psi_{\mathrm{HS}}^\theta)u^s$.

### 11.3.3 Data inferred barycenter: optimality constraints

We want to find a reliable analytic formulation that guarantees the systematic computation of the optimal barycenter location $\mathbf{y}_b \in \mathbb{R}^d$ for given data $\overline{c}^s, \overline{c}_l^s, \overline{c}_r^s$ for $s = 1, 2, \dots, S$. To this end, we first formally introduce for $c^s = \exp(\psi_{\mathrm{HS}}^\theta)u^s$ the Lagrangian,

$$\begin{aligned} L(\{\mathbf{j}_r^s(\mathbf{y}_b)\}_{s=1}^S, \mathbf{y}_b, \{\lambda^s\}_{s=1}^S) &:= C(\{\mathbf{j}_r^s(\mathbf{y}_b)\}_{s=1}^S, \mathbf{y}_b) \\ &\quad - \sum_{s=1}^{S} (-\mathrm{div}(\mathrm{e}^{\psi_{\mathrm{HS}}^\theta(r(\mathbf{x},\mathbf{y}_b))}\nabla u^s), \nabla\Lambda^s)_D\,, \end{aligned} \tag{11.27}$$

where the function $\lambda^s = \exp(\psi_{\mathrm{HS}}^\theta)\Lambda^s$ plays the role of a Lagrange multiplier. With the help of (11.27), we would like to address the following mathematical questions:

(i) *Does an optimal barycenter* $\mathbf{y}_b$ *of the obstacle's location together with the associated optimal states* $\{\mathbf{j}_r^s(\mathbf{y}_b, \overline{c}_l^s, \overline{c}_r^s)\}_{s=1}^S$ *exist?*

(ii) *Are there optimality constraints for optimal solutions* $(\mathbf{y}_b, \{\mathbf{j}_r^s\}_{s=1}^S)$ *to be satisfied?*

(iii) *How can we construct a numerical method that would allow us to reliably compute optimal solutions* $(\mathbf{y}_b, \{\mathbf{j}_r^s\}_{s=1}^S)$*?*

To study optimality, we first introduce the basic mathematical setting. We treat the barycenter $\mathbf{y}_b \in \mathbb{R}^d$ as a control variable, which is confined to the following *admissible set*,

$$Y_{\text{ad}} := \{\mathbf{y} \in \mathbb{R} \,|\, \mathbf{y} \in [0,1]^2\}. \tag{11.28}$$

Accordingly, to every $\mathbf{y}_b \in Y_{\text{ad}}$, there exists a unique weak solution $\mathbf{j}_r^s \in L^2(\Gamma_r)$ subject to the Dirichlet boundary conditions $\overline{c}_l^s$ and $\overline{c}_r^s$ for each $1 \le s \le S$. This solution is called *the state associated with the control* $\mathbf{y}_b \in Y_{\text{ad}}$ and is part of the state space,

$$U^s := \{v^s \in H^1(D) \,|\, v^s = \overline{c}_l^s \text{ on } \Gamma_l \text{ and } v^s = \overline{c}_r^s \text{ on } \Gamma_r\}, \quad \text{for } 1 \le s \le S. \tag{11.29}$$

Next, we clarify what we mean by an optimal barycenter location/control $\mathbf{y}_b$.

**Optimal state**

**Definition 11.3.1.** A barycenter $\underline{\mathbf{y}}_b \in Y_{\text{ad}}$, together with its associated optimal states $\underline{\mathbf{j}}_r^s(\underline{\mathbf{y}}_b) \in U^s, 1 \le s \le S$, is called *optimal*, if it holds that,

$$C(\{\underline{\mathbf{j}}_r^s(\underline{\mathbf{y}}_b)\}_{s=1}^S, \underline{\mathbf{y}}_b) \le C\left(\{\underline{\mathbf{j}}_r^s(\underline{\mathbf{y}}_b)\}_{s=1}^S, \mathbf{y}_b\right), \tag{11.30}$$

for all $\mathbf{y}_b \in Y_{\text{ad}}$ and $c^s \in U^s, 1 \le s \le S$.

With this definition, we can state an elementary first-order requirement guaranteeing optimality.

**First order optimality condition**

**Lemma 11.3.2.** *Let* $Y_{\text{ad}} \subset \mathbb{R}^d$ *be the admissible set (11.28), let* $Y_c \subseteq Y_{\text{ad}}$ *be a convex set, and let* $\kappa(\mathbf{y}_b) := C(\{\mathbf{j}_r^s(\mathbf{y}_b)\}_{s=1}^S, \mathbf{y}_b) : Y_c \to \mathbb{R}$ *be a* $\mathbf{y}$*-differentiable for* $\mathbf{y} \in Y_{\text{ad}}$ *and real-valued functional in* $Y_c$*. Assume that* $\underline{\mathbf{y}}_b \in Y_c$ *is a solution of the problem,*

$$\mathbf{y}_b = \min_{\mathbf{y} \in Y_c} \kappa(\mathbf{y}). \tag{11.31}$$

*Then we have the following variational inequality,*

$$\nabla_{\mathbf{y}}^{\mathbf{y}-\mathbf{y}_b} \kappa(\underline{\mathbf{y}}_b) = \nabla_{\mathbf{y}} \kappa(\underline{\mathbf{y}}_b)(\mathbf{y} - \underline{\mathbf{y}}_b) \ge 0 \quad \textit{for all } \mathbf{y} \in Y_c. \tag{11.32}$$

*Proof.* Let $\mathbf{y} \in Y_c$ be arbitrary. Due to the convexity of $Y_c$, we have $\mathbf{y}(t) = \underline{\mathbf{y}}_b + t(\mathbf{y} - \underline{\mathbf{y}}_b) \in Y_c$ for arbitrary $t \in (0,1]$. Optimality (11.31) of $\underline{\mathbf{y}}_b$ in the sense of Definition (11.3.1) gives

$\kappa(\mathbf{y}(\mathbf{t})) \geq \kappa(\underline{\mathbf{y}}_b)$, and hence $(\kappa(\mathbf{y}(t)) - \kappa(\underline{\mathbf{y}}_b))/t \geq 0$. Finally, passing to the limit as $t \to 0$ implies $\nabla_{\mathbf{y}}\kappa(\underline{\mathbf{y}}_b)(\mathbf{y} - \underline{\mathbf{y}}_b) \geq 0$. □

Based on the Lagrangian (11.27), we can systematically determine the adjoint problem to (11.26). Taking the derivative of $L$ with respect to $\mathbf{y}_b$, for smooth enough $v \in C_0^\infty(D)$, we obtain (due to $\mathbf{j}_r^s(\mathbf{y}_b, \overline{c}_l^s, \overline{c}_r^s) = \mathrm{e}^{\psi_{\mathrm{HS}}^\theta(r(\mathbf{x},\mathbf{y}_b))}\nabla_n(\exp(-\psi_{\mathrm{HS}}^\theta)c^s)|_{\Gamma_r}$),

$$0 = \frac{\delta^v L(\{\mathbf{j}_r^s\}_{s=1}^S, \mathbf{y}_b, \{\lambda^s\}_{s=1}^S)}{\delta^v c^s} = \sum_{s=1}^S [(\overline{\mathbf{j}}_r^s - \mathbf{j}_r^s(\mathbf{y}_b, \overline{c}_l^s, \overline{c}_r^s), v)_{L^2(\Gamma_r)} + (\mathrm{div}(\mathrm{e}^{\psi_{\mathrm{HS}}^\theta(r(\mathbf{x},\mathbf{y}_b))}\nabla(\mathrm{e}^{-\psi_{\mathrm{HS}}^\theta}\lambda^s)), v)_D], \tag{11.33}$$

where the Dirichlet boundary conditions $\gamma(g^s)|_{\Gamma_l} = u_l^s$ and $\gamma(g^s)|_{\Gamma_r} = u_r^s$ have already been accounted for by transforming $u^s$ into the variable $\tilde{u}^s := u^s - g^s$ and by skipping the $\tilde{\cdot}$-notation. The operator $\gamma$ denotes the classical trace operator $\gamma : H^1(D) \to L^2(\partial D)$.

The variational equations (11.33) read for each $1 \leq s \leq S$ in classical form as follows,

$$\begin{cases} -\mathrm{div}(\mathrm{e}^{\psi_{\mathrm{HS}}^\theta(r(\mathbf{x},\mathbf{y}_b))}\nabla\Lambda^s(\mathbf{x})) = 0\,, & \text{in } D\,, \\ \Lambda^s = \overline{\Lambda}_l^s := \exp\big(\psi_{\mathrm{HS}}^\theta\big)\overline{c}_l^s\,, & \text{on } \Gamma_l\,, \\ \mathrm{e}^{\psi_{\mathrm{HS}}^\theta(r(\mathbf{x},\mathbf{y}_b))}\nabla_{\mathbf{n}}\Lambda^s = (\overline{\mathbf{j}}_r^s - \mathbf{j}_r^s(\mathbf{y}_b, \overline{c}_l^s, \overline{c}_r^s)) \cdot \mathbf{n}\,, & \text{on } \Gamma_r\,, \\ \nabla_{\mathbf{n}}\Lambda^s(\mathbf{x}) = 0\,, & \text{on } \partial D \setminus \{\Gamma_r \cup \Gamma_l\}\,, \end{cases} \tag{11.34}$$

where $\mathbf{n}$ is the outward normal to $\Gamma_r$. Equations (11.34) represent for each $1 \leq s \leq S$ the *adjoint equations* of (11.26), and the associated weak solution $\lambda^s := \exp(\psi_{\mathrm{HS}}^\theta)\Lambda^s \in H^1(D)$ is called the *adjoint state* to $c^s$ for $s = 1, 2, \ldots, S$. The existence and uniqueness of weak solutions $\Lambda^s \in H^1(D)$ is an immediate consequence of the Lax–Milgram theorem as applied in the forward problem (11.20), see also [137]. The adjoint solution of (11.34) allows us to immediately write the derivative of the cost function (11.25) in simplified form as detailed in the next lemma.

**Gradient representation**

**Lemma 11.3.3.** *The functional* $\kappa(\mathbf{y}_b) := C(\{\mathbf{j}_r^s(\mathbf{y}_b)\}_{s=1}^S, \mathbf{y}_b)$ *admits the following representation of its gradient,*

$$\nabla_{\mathbf{y}}^{\mathbf{z}}\kappa(\mathbf{y}) = \sum_{s=1}^S \big(\nabla\Lambda^s, \big(\mathrm{e}^{\psi_{\mathrm{HS}}^\theta}\mathbf{z} \cdot \nabla_{\mathbf{y}}\psi_{\mathrm{HS}}^\theta \nabla u^s\big)\big)_D + \alpha\mathbf{y} \cdot \mathbf{z}\,, \tag{11.35}$$

*where* $\Lambda^s \in H^1(D)$ *is the weak solution of* (11.34), *and* $u^s(\mathbf{y})$ *is the state belonging to the barycenter* $\mathbf{y} \in Y_{\mathrm{ad}}$.

*Proof.* Follows directly with the adjoint equation (11.34). □

Before we address the question about optimality in the sense of Definition 11.3.1 of the minimizers $(\underline{\mathbf{y}}_b, \{\underline{c}^s(\underline{\mathbf{y}}_b)\}_{s=1}^S)$ of the objective/cost function (11.25), we give a mathematical formulation of the formal Lagrangian (11.27).

**Lagrangian**

**Definition 11.3.4.** The inverse problem (11.25)–(11.26) admits a Lagrangian $L : [L^2(\Gamma_r)]^S \times Y_{ad} \times H^1(D) \to \mathbb{R}$ defined for $s = 1, 2, \ldots, S$ and $d = 1, 2, 3$ as follows,

$$L\left(\{\mathbf{j}_r^s(\mathbf{y}_b)\}_{s=1}^S, \mathbf{y}_b, \{\lambda^s\}_{s=1}^S\right) := C\left(\{\mathbf{j}_r^s(\mathbf{y}_b)\}_{s=1}^S, \mathbf{y}_b\right) - \sum_{s=1}^S \left(\left(e^{\psi_{HS}^\theta(r(\mathbf{x},\mathbf{y}_b))} \nabla\left(e^{-\psi_{HS}^\theta} c^s\right)\right), \nabla\left(e^{-\psi_{HS}^\theta} \lambda^s\right)\right)_D. \quad (11.36)$$

Based on Definition 11.3.4 and the well-posedness of the adjoint equation, we immediately obtain the following first-order optimality result.

**Optimality**

**Lemma 11.3.5.** *Let $\underline{\mathbf{y}}_b \in Y_{ad}$ be the optimal barycenter location, and let $\{\underline{c}^s\}_{s=1}^S$ be the optimal states associated with the data $(\bar{\mathbf{j}}_r^s, \bar{c}_l^s, \bar{c}_r^s) \in L^2(\Gamma_r) \times L^2(\Gamma_l) \times L^2(\Gamma_r)$ for $s = 1, 2, \ldots, S$ of the inverse problem (11.25)–(11.26). There exist unique weak solutions $\lambda^s$, $s = 1, 2, \ldots, S$, to the adjoint equations (11.34) such that due to the admissibility constraint, the following variational inequality holds,*

$$\nabla_{\mathbf{y}}^{\mathbf{y}-\mathbf{y}_b} L\left(c^s, \mathbf{y}_b, \lambda^s\right) = \nabla_{\mathbf{y}}^{\mathbf{y}-\mathbf{y}_b} K(\underline{\mathbf{y}}_b) = \alpha \underline{\mathbf{y}}_b \cdot (\mathbf{y} - \underline{\mathbf{y}}_b) \geq 0, \quad \text{for all } \mathbf{y} \in Y_{ad}. \quad (11.37)$$

*On the contrary, every barycenter $\underline{\mathbf{y}}_b \in Y_{ad}$ that, together with the states $c^s(\underline{\mathbf{y}}_b)$, $s = 1, 2, \ldots, S$, and the associated adjoint states $\{\lambda^s\}_{s=1}^S$, satisfies inequality (11.37), is optimal.*

## 11.4 Computational method: constrained gradient descent

To solve the inverse problem defined by (11.25)–(11.26), we employ a problem-specific gradient method. In abstract terms, this numerical approach reads as follows.

---

! **Steepest descent scheme**

*Assume that we have already computed the iterates $\mathbf{y}_b^0, \mathbf{y}_b^1, \ldots, \mathbf{y}_b^\ell$.*

1. *Compute $\{c_\ell^s(\mathbf{y}_b^\ell)\}_{s=1}^S$ from the state equations (11.26) and the adjoint state $\{\lambda_\ell^s\}_{s=1}^S$ from the adjoint equations (11.34).*
2. *Find $\mathbf{z}^\ell \in Y_{ad}$ that minimizes gradient (11.35), that is,*

$$\mathbf{z}^\ell = \min_{\mathbf{z} \in Y_{ad}} \sum_{s=1}^S \left(\nabla \Lambda_\ell^s, \left(e^{\psi_{HS}^\theta} \mathbf{z} \cdot \nabla_{\mathbf{y}} \psi_{HS}^\theta \nabla u_\ell^s\right)\right)_D + \alpha \mathbf{y} \cdot \mathbf{z}. \quad (11.38)$$

   *If $\nabla_{\mathbf{y}}^{\mathbf{z}^\ell - \mathbf{y}_b^\ell} K(\mathbf{y}) \geq 0$ for $(\mathbf{y}_b^\ell, \mathbf{z}^\ell)$, then $\mathbf{y}_b^\ell$ satisfies (11.32), and hence it is an optimal solution according to (11.32), and the algorithm stops.*

   *If, on the contrary, $\nabla_{\mathbf{y}}^{\mathbf{z}^\ell - \mathbf{y}_b^\ell} K(\mathbf{y}) < 0$, then $\mathbf{z}^\ell - \mathbf{y}_b^\ell$ defines our next direction of descent.*
3. *We determine the amount $\alpha_\ell \in (0, 1]$ of descent by,*

$$\alpha_\ell = \min_{\alpha \in (0,1]} K\left(\mathbf{y}_b^\ell + \alpha\left(\mathbf{z}^\ell - \mathbf{y}_b^\ell\right)\right), \quad (11.39)$$

*which allows us to define the next candidate for the optimal barycenter location by,*

$$\mathbf{y}_b^{\ell+1} := \mathbf{y}_b^{\ell} + a_\ell\left(\mathbf{z}^{\ell} - \mathbf{y}_b^{\ell}\right). \tag{11.40}$$

*Set* $\ell := \ell + 1$ *and go back to* 1.

# Bibliography


[1] H. Abels, H. Garcke, and G. Grün. Thermodynamically consistent, frame indifferent diffuse interface models for incompressible two-phase flows with different densities. *Mathematical Models and Methods in Applied Sciences*, 22(03):1150013, March 2012.

[2] A. Alexanderian, M. Rathinam, and R. Rostamian. Homogenization, symmetry, and periodization in diffusive random media. *Acta Mathematica Scientia*, 32(1):129–154, 2012.

[3] G. Allaire. Homogenization and two-scale convergence. *SIAM Journal on Mathematical Analysis*, 23(6):1482–1518, 1992.

[4] R. Arens. An axiomatic basis for classical thermodynamics. *Journal of Mathematical Analysis and Applications*, 6:207–229, 1963.

[5] D. Aurbach, B. Markovsky, M. D. Levi, E. Levi, A. Schechter, M. Moshkovich, and Y. Cohen. New insights into the interactions between electrode materials and electrolyte solutions for advanced nonaqueous batteries. *Journal of Power Sources*, 81-82:95–111, 1999.

[6] A.-L. Barabási and H. E. Stanley. *Fractal Concepts in Surface Growth*. Cambridge University Press, 1995.

[7] M. Z. Bazant, K. Thornton, and A. Ajdari. Diffuse-charge dynamics in electrochemical systems. *Physical Review E*, 70(2):1–24, August 2004.

[8] M. Z. Bazant. Theory of chemical kinetics and charge transfer based on nonequilibrium thermodynamics. *Accounts of Chemical Research*, 46(5):1144–1160, 2013. PMID: 23520980.

[9] M. Z. Bazant, M. S. Kilic, B. D. Storey, and A. Ajdari. Towards an understanding of induced-charge electrokinetics at large applied voltages in concentrated solutions. *Advances in Colloid and Interface Science*, 152:48–88, 2009.

[10] R. J. Bearman. The Onsager thermodynamics of galvanic cells with liquid-liquid junctions. *Journal of Chemical Physics*, 22(4):585–587, 1954.

[11] R. J. Bearman and J. G. Kirkwood. Statistical mechanics of transport processes. xi. equations of transport in multicomponent systems. *Journal of Chemical Physics*, 28(1):136–145, 1958.

[12] D. Bedeaux, S. Kjelstrup, and H. C. Öttinger. Nonlinear coupled equations for electrochemical cells as developed by the general equation for nonequilibrium reversible-irreversible coupling. *Journal of Chemical Physics*, 141(12), 2014.

[13] C. H. Bennett. Logical reversibility of computation. *IBM Journal of Research and Development*, 17(6), 1973.

[14] L. Berlyand and P. Mironescu. Two-parameter homogenization for a Ginzburg-Landau problem in a perforated domain. *Networks and Heterogeneous Media*, 3(3):461–487, 2008.

[15] D. Bernoulli. Exposition of a new theory on the measurement of risk. *Econometrica*, 22:23–36, 1738, 1954.

[16] A. L. Bertozzi, S. Esedoglu, and A. Gillette. Inpainting of binary images using the Cahn-Hilliard equation. *IEEE Transactions on Image Processing*, 16(1):285–291, 2007.

[17] J. R. Jr. Biden. Memorandum on Improving the Cybersecurity of National Security, Department of Defense, and Intelligence Community Systems. 2022.

[18] R. B. Bird, W. E. Stewart, and E. N. Lightfood. *Transport Phenomena*. Wiley International Edition, 1960.

[19] R. B. Bird, W. E. Stewart, E. N. Lightfood, and D. J. Klingenberg. *Introductory Transport Phenomena*. Wiley, 2014.

[20] A. Birolini. *Reliability Engineering - Theory and Practice*. 8 Ed., Springer, 2017.

[21] L. Boltzmann. *Vorlesung über Gastheorie - Erster Teil*. Outlook Verlag GmbH - 2022, 1896.

[22] L. Boltzmann. *Vorlesung über Gastheorie - Zweiter Teil*. Outlook Verlag GmbH - 2022, 1896.

[23] D. Bothe, A. Fischer, and J. Saal. Global well-posedness and stability of electrokinetic flows. *SIAM Journal on Mathematical Analysis*, 46(2):1263–1316, 2014.

[24] A. Bourgeat, A. Mikelic, and S. Wright. Stochastic two-scale convergence in the mean and applications. *Journal für die Reine und Angewandte Mathematik*, 456:19–52, 1994.


https://doi.org/10.1515/9783110579543-012

[25] J. B. Boyling. An axiomatic approach to classical thermodynamics. *Proceedings of the Royal Society of London. Series A*, 329:35–70, 1972.

[26] H. Brenner and L. Gaydos. The constrained brownian movement of spherical particles in cylindrical pores of comparable radius. *Journal of Colloid and Interface Science*, 58(2):312–356, 1977.

[27] H. Brezis. *Functional Analysis, Sobolev Spaces and Partial Differential Equations*. Springer, 2011.

[28] L. Brillouin. Maxwell's demon cannot operate: Information and entropy. i. *Journal of Applied Physics*, 22(334), 1951.

[29] L. Brillouin. Physical entropy and information. ii. *Journal of Applied Physics*, 22(338), 1951.

[30] G. W. Brown. Iterative solutions of games by fictitious play. In: *Activity Analysis of Production and Allocation*. John Wiley and Sons, Inc., pages 374–376, 1951.

[31] J. W. Cahn and J. E. Hilliard. Free energy of a nonuniform system. I. Interfacial free energy. *Journal of Chemical Physics*, 28(2):258, 1958.

[32] C. Carathéodory. Untersuchungen über die Grundlagen der Thermodynamik. *Mathematische Annalen*, 67:355–386, 1909.

[33] S. Carnot. *Réflexions sur la puissance motrice du feu et sur les machines propres à développer cette puissance*. Bachelier, à Paris, 1824.

[34] Y. A. Cengel. *Introduction to Thermodynamics and Heat Transfer*. McGraw-Hill, second edition, 2008.

[35] X. Chen and A. Jüngel. Analysis of an incompressible navier-stokes-maxwell-stefan system. *Communications in Mathematical Physics*, 340:471–497, 2015.

[36] Y.-C. Cheng. *Macroscopic and Statistical Thermodynamics*. World Scientific Publishing, 2006.

[37] D. Cioranescu and P. Donato. *An Introduction to Homogenization*. 2000.

[38] J. W. Cooley and J. W. Tukey. An algorithm for the machine calculation of complex Fourier series. *Mathematics of Computation*, 19:297–301, 1965.

[39] G. B. Dantzig and M. N. Thapa. *Linear Programming 1: Introduction*. Springer, 1997.

[40] S. R. de Groot and P. Mazur. *Non-equilibrium Thermodynamics*. North-Holland, 1969.

[41] D. Deutsch. Quantum theory, the church-turing principle and the universal quantum computer. *Proceedings of the Royal Society of London. Series A*, 400:97–117, 1985.

[42] D. Deutsch. *The Beginning of Infinity - Explanations that Transform the World*. Pinguin Books UK, 2012.

[43] H. Dickinson. *A Short History of the Steam Engine*. Cambridge University Press, Cambridge, 2011.

[44] M. Doyle, T. F. Fuller, and J. Newman. Modeling of galvanostatic charge and discharge of the lithium/polymer/insertion cell. *Journal of the Electrochemical Society*, 140(6):1526, 1993.

[45] W. Dreyer, C. Guhlke, and R. Muller. Overcoming the shortcomings of the Nernst-Planck model. *Physical Chemistry Chemical Physics*, 15:7075–7086, 2013.

[46] W. Dreyer, J. Jamnik, C. Guhlke, R. Huth, J. Moskon, and M. Gaberscek. The thermodynamic origin of hysteresis in insertion batteries. *Nature Materials*, 9:448–453, 2010.

[47] C. Eck, G. Grün, F. Klingbeil, and O. Vantzos. On a phase-field model for electrowetting. *Interfaces and Free Boundaries*, 11:259–290, 2009.

[48] P. Ehrenfest. Phasenumwandlungen im ueblichen und erweiterten sinn, classifiziert nach den entsprechenden des thermodynamischen potentials. *zu den Mitteilungen aus dem KAMERLINGH ONNES-Institut, Leiden*, Supplement(75b), 1933.

[49] M. H. Eikerling and P. Berg. Poroelectroelastic theory of water sorption and swelling in polymer electrolyte membranes. *Soft Matter*, 7(13):5976, 2011.

[50] B. Eisenberg, Y. Hyon, and C. Liu. Energy variational analysis of ions in water and channels: Field theory for primitive models of complex ionic fluids. *Journal of Chemical Physics*, 133(10):104104, September 2010.

[51] K. R. Elder and R. C. Desai. Role of nonlinearities in off-critical quenches as described by the Cahn-Hilliard model of phase separation. *Physical Review B*, 40(1), 1989.

[52] C. M. Elliott, B. Stinner, V. Styles, and R. Welford. Numerical computation of advection and diffusion on evolving diffuse interfaces. *IMA Journal of Numerical Analysis*, 31(3):786–812, May 2010.

[53] E. Estrada, M. Fox, D. J. Higham, and G.-L. Oppo (Editors). *Network Science - Complexity in Nature and Technology*. Springer, London, 2010.

[54] L. C. Evans. *Partial Differential Equations*, volume 19 of *Graduate Studies in Mathematics*. AMS, Providence, Rhode Island, 1998.

[55] K. Falconer. *Fractal Geometry - Mathematical Foundations and Applications*. John Wiley & Sons Ltd, 2003.

[56] J. M. Fraser. *Assoud Dimension and Fractal Geometry*. Cambridge University Press, 2021.

[57] P. Fratzl and J. L. Lebowitz. Universality of scaled structure functions in quenched systems undergoing phase-separation. *Acta Metallurgica*, 37:3245–3248, 1989.

[58] P. Fratzl, J. L. Lebowitz, O. Penrose, and J. Amar. Scaling functions, self-similarity, and the morphology of phase separating systems. *Physical Review B*, 44:4794–4811, 1991.

[59] T. Gagliardoni. Quantum attack resource estimate: using shor's algorithm to break rsa vs dh/dsa vs ecc. *Kudelski Security Research*, 2021.

[60] R. E. García, Y. M. Chiang, W. C. Carter, P. Limthongkul, and C. M. Bishop. Microstructural modeling and design of rechargeable lithium-ion batteries. *Journal of the Electrochemical Society*, 152(1):A255–A263, 2005.

[61] Gaudin, O. - CEO & Co-Founder of SonarSource. Sonarqube - code quality and security.

[62] M. Gell-Mann and S. Lloyd. Information measures, effective complexity, and total information. *Complexity*, 2:44–52, 1996.

[63] F. Golfier, M. Quintard, and B Wood. Comparison of theory and experiment for solute transport in weakly heterogeneous bimodal porous media. *Advances in Water Resources*, 34:899–914, 2011.

[64] W. Gray and P. C. Lee. On the theorems for local volume averaging of multiphase systems. *International Journal of Multiphase Flow*, 3:333–340, 1977.

[65] M. Grmela and H. C. Öttinger. Dynamics and thermodynamics of complex fluids. i. development of a general formalism. *Physical Review E*, 56:6620–6632, Dec 1997.

[66] E. A. Guggenheim. Mixtures. 1952.

[67] E. A. Guggenheim. *Thermodynamics*. North Holland Publishing Co., 5th edition, 1967.

[68] B. C. Han, A. Van der Ven, D. Morgan, and G. Ceder. Electrochemical modeling of intercalation processes with phase field models. *Electrochimica Acta*, 49(26):4691–4699, October 2004.

[69] H. Harbrecht and C. Schwab. Sparse tensor finite elements for elliptic multiscale problems. *Computer Methods in Applied Mechanics and Engineering*, 200(45-46):3100–3110, 2011.

[70] A. W. Harrow, A. Hassidim, and S. Lloyd. Quantum algorithm for linear systems of equations. *Physical Review Letters*, 103(150502), 2009.

[71] Z. Hashin and S. Shtrikman. A variational approach to the theory of the elastic behaviour of multiphase materials. *Journal of the Mechanics and Physics of Solids*, 11(2):127–140, 1963.

[72] J. D. Hidary. *Quantum Computing: An Applied Approach*. Springer, 2nd edition, 2021.

[73] U. Hornung. *Homogenization and Porous Media*. Interdisciplinary Applied Mathematics. Springer, 1997.

[74] R. J. Hunter. *Foundations of Colloid Science*. Oxford University Press, 2001.

[75] Y. Hyon, D. Y. Kwak, and C. Liu. Energetic variational approach in complex fluids: Maximum dissipation principle. *Discrete and Continuous Dynamical Systems*, 26(4):1291–1304, 2010.

[76] Wilton E. Scott Institute. Solid oxide fuel cells. *Carnegie Mellon University, Policy Brief*, 2020.

[77] A. Jelic. Bridging scales in complex fluids out of equilibrium. PhD thesis, ETH Zurich, 2009.

[78] J. W. Jerome. Analytical approaches to charge transport in a moving medium. *Transport Theory and Statistical Physics*, 31(4-6):333–366, January 2002.

[79] A. M. Johnson and J. Newman. Desalting by means of porous carbon electrodes. *Journal of the Electrochemical Society*, 118(3), 1971.

[80] R. Jordan, D. Kinderlehrer, and F. Otto. The variational formulation of the Fokker-Planck equation. *SIAM Journal on Mathematical Analysis*, 29(1):1, 1998.

[81] D. Kahneman and A. Tversky. Prospect theory: An analysis of decision under risk. *Econometrica*, 47:263–291, 1979.

[82] R. E. Kalman. A new approach to linear filtering and prediction problems. *Transactions of the ASME - Journal of Basic Engineering*, 1960.

[83] P. M. Kekenes-Huskey, A. K. Gillette, and J. A. McCammon. Predicting the influence of long-range molecular interactions on macroscopic-scale diffusion by homogenization of the smoluchowski equation. *Journal of Chemical Physics*, 140(17), 2014.

[84] S. Kjelstrup, D. Bedeaux, E. Johannessen, and J. Gross. *Non-Equilibrium Thermodynamics for Engineers*. World Scientific Publishing Co. Pte. Ltd., Singapore, 2010.

[85] V. Klika and C. Öttinger. On the compatibility of sharp and diffuse interfaces out of equilibrium. *SIAM MMS*, accepted, 2023.

[86] V. R. Kohn and F. Otto. Upper bounds on coarsening rates. *Communications in Mathematical Physics*, 229(3):375–395, 2002.

[87] R. Krishna and J. A. Wesselingh. The Maxwell-Stefan approach to mass transfer. *Chemical Engineering Science*, 52(6):861–911, 1997.

[88] H. Kronmüller. Fundamentals of parameter estimation. *Periodica Polytechnica Electrical Engineering*, 28(2-3):95–111, 1984.

[89] J. H. Kruhl. Fractal-geometry techniques in the quantification of complex rock structures: A special view on scaling regimes, inhomogeneity and anisotropy. *Journal of Structural Geology*, 2013.

[90] R. Kubo. The fluctuation-dissipation theorem. *Reports on Progress in Physics*, 29(255), 1966.

[91] R. Kümmel. *The Second Law of Economics*. Springer, 2011.

[92] F. Kuypers, H. Hummel, J. Kempf, and E. Wild. *Physik für Ingenieure - Band 2: Elektrizität und Magnetismus, Wellen, Atom- und Kernphysik*. VCH Weinheim, 1997.

[93] R. Landauer. Irreversibility and heat generation in the computing process. *IBM Journal of Research and Development*, 5(31), 1961.

[94] B. Li. Continuum electrostatics for ionic solutions with non-uniform ionic sizes. *Nonlinearity*, 22:811–833, 2009.

[95] G. Li, J. Woo, and S. B. Lim. Hpc cloud architecture to reduce hpc workflow complexity in containerized environments. *Applied Sciences*, 11(923), 2021.

[96] E. H. Lieb and J. Yngvason. The physics and mathematics of the second law of thermodynamics. *Physics Reports*, 310:1–96, 1999.

[97] I. M. Lifshitz and V. V. Slyozo. The kinetics of precipitation from supersaturated solid solutions. *Journal of Physics and Chemistry of Solids*, 19:35–50, 1961.

[98] C. Liu and J. Shen. A phase field model for the mixture of two incompressible fluids and its approximation by a Fourier-spectral method. *Physica D: Nonlinear Phenomena*, 179(3-4):211–228, May 2003.

[99] M. López-Suárez, I. Neri, and L. Gammaitoni. Sub-$k_B T$ micro-electromechanical irreversible logic gate. *Nature Communications*, 7(12068), 2016.

[100] G. J. Lord, C. E. Powell, and T. Shardlow. *An Introduction to Computational Stochastic PDEs*. CUP, 2014.

[101] J. Lowengrub and L. Truskinovsky. Quasi-incompressible Cahn-Hilliard fluids and topological transitions. *Proceedings of the Royal Society A*, 454(1978):2617–2654, 1998.

[102] H.-W. Lu, K. Glasner, A. L. Bertozzi, and C.-J. Kim. A diffuse-interface model for electrowetting drops in a Hele-Shaw cell. *Journal of Fluid Mechanics*, 590:411–435, 2007.

[103] M. Malanowski. *Signal Processing for Passive Bistatic Radar*. Artech, 2019.

[104] B. B. Mandelbrot. *The Fractal Geometry of nature*. W.H. Freeman and Company, New York, 1982.

[105] B. B. Mandelbrot and R. L. Hudson. *The MisBehavior of Markets*. Basic Books, 2004.

[106] J. C. Maxwell. *Theory of Heat*. Appleton, London, 1871.

[107] T. McCabe. A complexity measure. *IEEE Transactions on Software Engineering*, 2(4):308–320, 1976.

[108] C. C. Mei, J. L. Auriault, and C.-O. NG. Some applications of the homogenization theory. *Advances in Applied Mechanics*, 32:277–348, 1996.

[109] C. C. Mei and B. Vernescu. *Homogenization Methods for Multiscale Mechanics*. World Scientific, 2010.

[110] L. Modica and S. Mortola. Un esempio di γ-convergenza. *Bollettino dell'Unione Matematica Italiana*, 14:285–299, 1977.
[111] J. Molla and M. Schmuck. Basic and extendable framework for effective charge transport in electrochemical systems. *Applied Mathematics Letters*, 95:85–91, 2019.
[112] NASA. Fast facts about nasa fuel cells. 2010.
[113] G. Nguetseng. A general convergence result for a functional related to the theory of homogenization. *SIAM Journal on Mathematical Analysis*, 20(3):608–623, 1989.
[114] J. Nolen, G. A. Pavliotis, and A. M. Stuart. Multiscale modelling and inverse problems. In: Graham, I., Hou, T., Lakkis, O., Scheichl, R. (eds). *Numerical Analysis of Multiscale Problems*. Lecture Notes in Computational Science and Engineering, 83, 2012.
[115] A. Novick-Cohen. The Cahn-Hilliard equation. In: *Handbook of Differential Equations - Evolutionary Equations*, pages 201–228, 2008.
[116] The Ministry of Defence (MOD). *Wargaming Handbook*. Development, Concepts and Doctrine Centre, 2017.
[117] M. J. Osborne and A. Rubinstein. *A Course in Game Theory*. The MIT Press, 1994.
[118] M. Oskin, F. T. Chong, and I. L. Chuang. A practical architecture for reliable quantum computers. *Computer*, 35(1):79–87, 2002.
[119] H. C. Öttinger. *Beyond Equilibrium Thermodynamics*. Wiley, 2004.
[120] H. C. Öttinger. Constraints in nonequilibrium thermodynamics: general framework and application to multicomponent diffusion. *Journal of Chemical Physics*, 130:114904, 2009.
[121] H. C. Öttinger and M. Grmela. Dynamics and thermodynamics of complex fluids. ii. illustrations of a general formalism. *Physical Review E*, 56:6633–6655, Dec 1997.
[122] R. L. Pego. Front migration in the nonlinear Cahn-Hilliard equation. *Proceedings of the Royal Society of London. Series A*, 422(1863):261–278, 1989.
[123] O. Penrose. *Foundations of Statistical Mechanics - a Deductive Treatment*. Dover, 2005.
[124] M. Planck. *Einführung in die Theorie der Wärme*. Hirzel, Leipzig, 1930. pages 188–189.
[125] V. N. Pokrovskii. *Econodynamics*. Springer, 2012.
[126] K. Promislow and B. Wetton. Pem fuel cells: a mathematical overview. *SIAM Journal on Applied Mathematics*, 70:369–409, 2009.
[127] S. Puri, A. J. Bray, and J. L. Lebowitz. Phase-separation kinetics in a model with order-parameter-dependent mobility. *Physical Review E*, 56:758–765, 1997.
[128] J. Robinson. An iterative method of solving a game. *Annals of Mathematics*, 54(2):296–301, 1951.
[129] T. M. Rogers and R. C. Desai. Numerical study of late-stage coarsening for off-critical quenches in the Cahn-Hilliard equation of phase separation. *Physical Review B*, 39(0):11956–11964, 1989.
[130] R. Rose and C. Lanier. Final Report: Emerging Technologies Subcommittee Quantum Information Science. *Home Land Security Advisory Council*, 2020.
[131] E. Rosenberg. *Fractal Dimensions of Networks*. Springer, Switzerland, 2020.
[132] I. Rubinstein. *Electro-Diffusion of Ions*. SIAM Studies in Applied Mathematics, SIAM, 1990.
[133] T. Sagawa. Thermodynamics of information processing in small systems. *Progress of Theoretical Physics*, 1(127), 2012.
[134] T. Sagawa. Second law, entropy production, and reversibility in thermodynamics of information. *arXiv:1712.06858v1*, 2017.
[135] M. Schmuck. Analysis of the Navier-Stokes-Nernst-Planck-Poisson System. *Mathematical Models and Methods in Applied Sciences*, 19(06):993, 2009.
[136] M. Schmuck. New porous medium Poisson-Nernst-Planck equations for strongly oscillating electric potentials. *Journal of Mathematical Physics*, 54(2):021504, 2013.
[137] M. Schmuck. Heterogeneous hard-sphere interactions for equilibrium transport processes beyond perforated domain formulations. *Applied Mathematics Letters*, 49:78–83, 2015.
[138] M. Schmuck. Upscaling of solid-electrolyte composite intercalation cathodes for energy storage systems. *Applied Mathematics Research eXpress*, pages 1–29, 2017.

[139] M. Schmuck and M. Z. Bazant. Homogenization of the Poisson-Nernst-Planck equations for ion transport in charged porous media. *SIAM Journal on Applied Mathematics*, 75(3):1369–1401, 2015.

[140] M. Schmuck and P. Berg. Homogenization of a catalyst layer model for periodically distributed pore geometries in PEM fuel cells. *Applied Mathematics Research eXpress*, (Cl):1–22, July 2012.

[141] M. Schmuck and P. Berg. Effective macroscopic equations for species transport and reactions in porous catalyst layers. *Journal of the Electrochemical Society*, 161(8):E3323–E3327, 2014.

[142] M. Schmuck and S. Kalliadasis. General framework for adsorption processes on dynamic interfaces. *Journal of Physics A: Mathematical and Theoretical*, 125502, 2016.

[143] M. Schmuck and S. Kalliadasis. Rate of convergence of general phase field equations in strongly heterogeneous media toward their homogenized limit. *SIAM Journal on Applied Mathematics*, 77(4):1471–1492, 2017.

[144] M. Schmuck, G. A. Pavliotis, and S. Kalliadasis. Recent advances in the evolution of interfaces: thermodynamics, upscaling, and universality. *Computational Materials Science*, 156:441–451, 2019.

[145] M. Schmuck, M. Pradas, S. Kalliadasis, and G. A. Pavliotis. New stochastic mode reduction strategy for dissipative systems. *Physical Review Letters*, 110:244101, Jun 2013.

[146] M. Schmuck, M. Pradas, G. A. Pavliotis, and S. Kalliadasis. Upscaled phase-field models for interfacial dynamics in strongly heterogeneous domains. *Proceedings of the Royal Society A*, 468(2147):3705–3724, 2012.

[147] M. Schmuck, M. Pradas, G. A. Pavliotis, and S. Kalliadasis. Derivation of effective macroscopic Stokes–Cahn–Hilliard equations for periodic immiscible flows in porous media. *Nonlinearity*, 26(12):3259, 2013.

[148] N. R. Schneidewind and J. Heineman. MCTSSA Software Reliability Handbook Volume I - Software Reliability Engineering, Process and Modelling for a Single Process, 1996.

[149] N. R. Schneidewind and J. Heineman. MCTSSA Software Reliability Handbook Volume II - Data Collection Demonstration and Software Reliability Modeling for a Multi-Function Distributed System, 1996.

[150] M. Schroeder. *Fractals, Chaos, Power Laws: Minutes From an Infinite Paradise*. W.H. Freeman, New York, 1991.

[151] Z. Schuss, B. Nadler, and R. Eisenberg. Derivation of Poisson and Nernst-Planck equations in a bath and channel from a molecular model. *Physical Review E*, 64(3):1–14, August 2001.

[152] J. P. Sethna. *Statistical Mechanics: Entropy, Order Parameters, and Complexity*. Oxford University Press, 2006.

[153] E. Shannon. A mathematical theory of communication. *The Bell System Technical Journal*, 27:379–423, 1948.

[154] M. L. Shooman. Software reliability: A historical perspective. *IEEE Transactions on Reliability*, vol R-33:48–55, 1984.

[155] D. N. Sibley, A. Nold, N. Savva, and S. Kalliadasis. The contact line behaviour of solid-liquid-gas diffuse-interface models. *Physics of Fluids*, 25(092111), 2013.

[156] D. N. Sibley, A. Nold, N. Savva, and S. Kalliadasis. On the moving contact line singularity: Asymptotics of a diffuse-interface model. *The European Physical Journal E*, 36(26), 2013.

[157] L. E. (Laurence E.) Sigler. *Fibonacci's Liber Abaci: A Translation into Modern English of Leonardo Pisano's Book of Calculation*. Sources and Studies in the History of Mathematics and Physical Sciences, 2002.

[158] J. C. Slattery. Flow of viscoelastic fluids through porous media. *AIChE Journal*, 13:1066–1071, 1967.

[159] A. Smith. *An Inquiry into the Nature and Causes of the Wealth of Nations - Vol. 1*, W. Strahan, London, 1 edition, 1776.

[160] W. H. Smyrl and J. Newman. Potentials of cells with liquid junctions. *The Journal of Physical Chemistry*, 72(13):4660–4671, 1968.

[161] M. Strand and J. H. Wiik. Security for autonomous and unmanned devices: Cryptography and its limits. NATO, STO-MP-AVT-337, 2022.

[162] J. Telega and W. Bielski. Stochastic homogenization and macroscopic modelling of composites and flow through porous media. *Theoretical and Applied Mechanics*, 28(28-29):337–378, 2002.

[163] P. R. Thie and G. E. Keough. *An Introduction To Linear Programming and Game Theory*. John Wiley & Sons, New Jersey, 2008.

[164] K. E. Thomas-Alyea and J. Newman. *Electrochemical Systems*. John Wiley & Sons, Incorporated, Third edition, 2010.

[165] A. N. Tikhonov. Regularization of incorrectly posed problems. *Soviet Mathematics. Doklady*, 4:1624–1627, 1963.

[166] C. Truesdell and S. Bharatha. *The Concepts and Logic of Classical Thermodynamics as a Theory of Heat Engines*. Springer, 1977.

[167] A. M. Turing. On computable numbers, with an application to the entscheidungsproblem. *Proceedings of the London Mathematical Society*, 42(2):230–265, 1937.

[168] J. D. Van Der Waals. The thermodynamic theory of capillarity under the hypothesis of a continuous variation of density. *Verhandel Konink. Akad. Weten. Amsterdam (Sec. 1)*, 1:1–56, 1979. Translation by J. S. Rowlingson, *J. Stat. Phys.* 20,197–233, 1892.

[169] A. Ververis and M. Schmuck. Computational investigation of porous media phase field formulations: Microscopic, effective macroscopic, and langevin equations. *Journal of Computational Physics*, 344(Supplement C):485–498, 2017.

[170] J. von Neumann and O. Morgenstern. *Theory of Games and Economic Behavior*. Interbooks, 1953.

[171] C. Wagner. Theorie der Alterung von Niederschlägen durch Umlösen. *Zeitschrift für Elektrochemie*, 65:581–594, 1961.

[172] W. Wang and P. Alexandridis. Composite polymer electrolytes: Nanoparticles affect structure and properties. *Polymers*, 8(387), 2016.

[173] Y. Wang, Z. Hu, B. C. Sanders, and S. Kais. Qudits and high-dimensional quantum computing. *Frontiers in Physics*, 8(589504), 2020.

[174] S. Whitaker. Flow in porous media i: A theoretical derivation of darcy's law. *Transport in Porous Media*, 1:3–25, 1986.

[175] J. D. Williams. *The Compleat Strategyst*. Dover Publications, New York, 1954.

[176] S. Wolfram. *Cellular Automata and Complexity: Collected Papers*. Addison-Wesley, 1994.

[177] V. V. Zhikov, S. M. Kozlov, and O. A. Oleĭnik. *Homogenization of Differential Operators and Integral Functionals*. Springer, 1994.

[178] J. Zhu, L.-Q. Chen, J. Shen, and V. Tikare. Coarsening kinetics from a variable-mobility Cahn-Hilliard equation: Application of a semi-implicit fourier spectral method. *Physical Review E*, 60:3564–3572, Oct 1999.

[179] B. Zygelman. *A First Introduction to Quantum Computing and Information*. Springer, 2018.

# Index



https://doi.org/10.1515/9783110579543-013

www.ingramcontent.com/pod-product-compliance
Lightning Source LLC
Chambersburg PA
CBHW061340210326
41598CB00035B/5843